Prentice Hall Advanced Reference Series

Polymer Science and Engineering

PRENTICE HALL
Polymer Science and Engineering

James E. Mark, Series Editor

Computer Simulation of Polymers

R. J. Roe, Editor

PRENTICE HALL, Englewood Cliffs, New Jersey 07632

Library of Congress Cataloging-in-Publication Data

Computer simulation of polymers / R.J. Roe, editor.
      p.   cm. -- (Prentice Hall advanced reference series. Physical
   and life sciences) (Polymer science and engineering series)
   Includes index.
   ISBN 0131614800
   1. Polymers--Computer simulation--Congresses.   I. Roe, R. J.
   (Ryong-Joon) II. Series.   III. Series: Polymer science and
   engineering series.
   QD381.9.E4C67   1991
   547.7'01'13--dc20   90-40720
                                                            CIP

Editorial/production supervision
  and interior design: *Carol Atkins*
Cover design: *Wanda Lubelska Design*
Manufacturing buyers: *Kelly Behr and Patrice Friaccio*

Prentice Hall Advanced Reference Series

Prentice Hall Polymer Science and Engineering Series

 © 1991 by Prentice-Hall, Inc.
A Division of Simon & Schuster
Englewood Cliffs, New Jersey 07632

Printed in the United States of America
10 9 8 7 6 5 4 3 2 1

ISBN 0-13-161480-0

Prentice-Hall International (UK) Limited, *London*
Prentice-Hall of Australia Pty. Limited, *Sydney*
Prentice-Hall Canada Inc., *Toronto*
Prentice-Hall Hispanoamericana, S.A., *Mexico*
Prentice-Hall of India Private Limited, *New Delhi*
Prentice-Hall of Japan, Inc., *Tokyo*
Simon & Schuster Asia Pte. Ltd., *Singapore*
Editors Prentice-Hall do Brasil, Ltda., *Rio de Janeiro*

CONTENTS

Foreword
Polymer Science and Engineering Series

J. E. Mark, University of Cincinnati, Series Editor

One of the most exciting areas in chemistry, chemical engineering, and materials science is the preparation, characterization, and utilization of polymers. The growing importance of polymers has been truly astounding, to the point that it is difficult to imagine our lives without them. They are under development in virtually every industrialized country in the world with activities accelerating rather than abating.

Not surprisingly, the amount of information relevant to polymer science and engineering is increasing correspondingly, making it more and more difficult to enter new fields in this area or even to remain abreast of developments in one's current field. There is thus a real need for authoritatively written, easily accessible books on polymer science and engineering, both for the relatively uninitiated and for the better-informed professional.

The present series of books was inaugurated to help meet this need. It will cover the organic chemistry of polymers, the relevant physical chemistry and chemical physics, polymer processing and other engineering aspects, and the applications of polymers as materials. The level will range from highly introductory treatments that are tutorial and therefore particularly useful for self-study, to rather advanced treatments of more specialized subjects. Many of these books will be exceedingly useful as textbooks in formal courses at colleges and universities.

Considerable attention will be paid to polymers as "high-tech" materials. This is in response to the fact that the most exciting applications of polymers no longer involve huge tonnage amounts of materials. Rather, they involve situations in which polymers generally are not present in large amounts, but are absolutely critical for the functioning of the system. Examples are polymer matrices and encapsulants for the controlled release of drugs and agricultural chemicals, biopolymers and synthetic polymers for biomedical applications, conducting polymers for batteries and other electrical devices, polymers having non-linear optical properties for optoelectronic applications, high-temperature polymers for use in outer space and in other hostile environments, ultra-oriented polymers for high strength materials, new types of polymer-based composites, photosensitive polymers for microlithography, and inorganic and organometallic polymers for use as ceramic precursors. These are all rapidly developing fields, and there is particularly a great need for authoritative, comprehensive treatments of these subject areas.

It is hoped that this series of volumes will meet these needs, and be of lasting value to the polymer community.

Preface

Computer simulation, or molecular modeling, is fast emerging as an important technique of research in the study of polymers—for investigating not only single chain behavior but also bulk properties of amorphous, crystalline, and liquid crystalline polymers. In other fields of science such as biochemistry and pharmaceutical chemistry, molecular modeling has already established itself as an extremely valuable tool for research. Although conformational properties of single polymer molecules has long been studied by computers, the importance of simulations in the polymer field is perhaps only now beginning to be recognized widely. The properties of interest with polymers more often arise from interaction of segments and molecules with their surroundings in the bulk state. Requirements of computational resources for such investigations are much more demanding. The wider availability of supercomputers and the generally increased capabilities of computers which has been achieved in recent years have, however, made it possible for a much wider range of problems in polymers to be investigated by computer simulation techniques.

The papers collected in this book originally were presented as part of the five-day symposium entitled Computer Simulation of Polymers, September 11–15, 1989, sponsored by the Division of Polymer Chemistry at the American Chemical Society Meeting in Miami Beach, Florida. The symposium was preceded by a one-day workshop, Technique of Computer Simulation of Polymers, in which four speakers presented introductions to basic techniques of simulations such as the atomic and molecular force field representation, crystalline polymer modeling, and molecular dynamics and Monte Carlo methods. The symposium itself dealt with a wide range of topics, as the chapter titles in this book indicate. Among others the topics covered include: application of Monte Carlo, molecular dynamics, and brownian dynamics method to simulations of polymer liquids and glasses to study their rheological and conformational properties, transitions, and free volume distributions; diffusion of polymer molecules, and diffusion of small molecules through polymers; conformation of rigid polymers investigated by molecular mechanics or quantum mechanical calculations; transitions of liquid crystalline polymers; conformational properties of polymers at interfaces; and phase separation of polymer blends. Topics *not* covered in this book (and in the symposium) are applications to biological macromolecules and simulations of polymer processing and polymerization reactions.

Finally I would like to acknowledge the generous financial support provided by the following institutions, which helped make organization of the symposium and publication of this book possible: Petroleum Research Fund administered by the American Chemical Society, Allied-Signal Corporation, IBM Corporation, E.I. du Pont de Nemours & Co., Tripos Associates, Inc., BioSym Technologies, Inc., and BioDesign, Inc.

Ryong-Joon Roe
University of Cincinnati

Calculation of *Ab Initio* Intrasegmental Potential-Energy Functions for Use in Modeling Polymer Properties

RICHARD L. JAFFE
NASA Ames Research Center
Moffett Field, California 94035

DO Y. YOON and A. D. McLEAN
IBM Almaden Research Center
650 Harry Road
San Jose, California 95120

ABSTRACT

Ab initio quantum chemical methods offer a reliable and nonempirical approach for the calculation of molecular geometries and conformational energies as well as for torsional and vibrational potentials. We have studied a series of mono- and diphenyl molecules containing up to about 35 atoms which serve as model compounds for a variety of thermoplastic polymers, including polycarbonate, poly(ether sulfone), polyimides, aromatic polyesters, and aromatic polyamides. We have studied basis set effects and explored the role played by electron correlation. The resulting molecular geometries and torsional potentials are often significantly different from those predicted from minimal basis set, semiempirical, and molecular mechanics calculations. Examples are given illustrating the use of these potential-energy functions for the modeling of low-temperature relaxation phenomena and the radius of gyration in polycarbonate and the chain persistence length in poly(*p*-hydroxybenzoic acid).

INTRODUCTION

The utility of *ab initio* quantum chemical methods has clearly been demonstrated for obtaining geometric parameters, fundamental vibrational frequencies, and relative energies for small molecules [1]. These calculations are open-ended in that more accurate results are generally obtained if larger atomic orbital basis sets or more extensive treatments of electron correlation are utilized. For most small molecules, using medium-sized calculations, one can determine bond lengths to within 0.03 Å and bond angles to within 0.5°, energy differences between conformers can be determined to within 1 kcal/mol and harmonic vibrational frequencies can be determined to within ±10% for stretches and bends. Relative errors in weaker modes such as torsions are generally larger. Until recently, such calculations have not generally been applied to model compounds representing polymeric systems. In specific cases quantitative insights into macroscopic polymer properties have been obtained from detailed studies of model compounds representing small chain segments.

At the same time, interest in the computational modeling of polymeric systems has increased greatly. It is becoming routine to study the structure and properties of polymers through simulations of ensembles comprised of 5000 to 10,000 atoms [2,3]. In these simulations, geometry optimizations are performed by minimizing the total energy under static or dynamic conditions [4]. Most of this work is carried out using empirical force fields based on molecular mechanics (MM) potential-energy functions [5]. The force field is built up as a combination of segmental vibrational interactions (bond stretching, angle bending, torsion, and inversion), van der Waals nonbonded interactions, and electrostatic interactions based on the partial charges of the atoms. This approach has been demonstrated to be successful for many molecular and polymeric systems. However, there are also many examples for which the existing MM potential energy functions give qualitatively incorrect results or for which the necessary vibrational force field and atomic partial charge data are not available. Fortunately, the results of *ab initio* quantum chemistry calculations can provide a reliable source for some of these data.

We have carried out *ab initio* quantum chemistry studies for a series of mono- and diphenyl molecules which were selected to represent segments of various thermoplastic organic polymers with *p*-phenylene groups embedded in the chain. For example, diphenyl carbonate (DPC) and 2,2-diphenyl propane (DPP) were chosen as models for Bisphenol-A polycarbonate (PC), and phenyl benzoate (PB) was chosen as a model for poly(*p*-hydroxybenzoic acid) (PHBA). One advantage of these calculations over semiempirical MO treatments [6,7] such as MNDO is that they do not contain any empiricism and thus can be used for predicting the structures and properties of novel systems that have not yet been created in the laboratory, as well as for assisting in the interpretation of experiments and for the development of structure–property relationships on a molecular level. In the next section we provide a brief description of the computational approach. Following that, results for polycarbonate and PHBA are presented.

DESCRIPTION OF THE CALCULATIONS

The *ab initio* quantum chemistry calculations described in this work represent approximate solutions of the Schrödinger equation describing the electronic structure of a molecule with fixed atomic positions. Most of the calculations are carried out at the Hartree Fock or self-consistent-field (SCF) level, where the electronic energy is determined under the approximation that each electron feels the average field of all the other electrons. At times, further calculations are carried out to account for the correlated motion of the electrons. For these calculations the electronic wavefunction is expanded in an atomic orbital basis set. In general, the more basis functions included for each atom in the molecule or the more extensive the treatment of electron correlation, the more accurate are the calculations. *Ab initio* quantum chemistry calculations do not contain any empiricism.

For the present study we used the computer program system called GRADSCF at NASA Ames, which was developed by Komornicki [8], and GAUSSIAN86 as implemented at IBM [9]. With these codes, the SCF energy of the molecule under study is calculated along with the analytic first and second derivatives of the energy with respect to displacements of the atomic centers. These gradient calculations are used both for the efficient location of equilibrium geometries (i.e., energy minimization) of stable conformers and energy barriers and for the determination of harmonic vibrational force fields and fundamental infrared frequencies. At a stationary-point geometry the first derivatives are all zero and the second derivatives are the harmonic force constants. Efficient algorithms have been implemented in GRADSCF and GAUSSIAN86 to iteratively locate stationary-point geometries starting from approximate atomic coordinates. The approximate contribution of the electron correlation energy is calculated from second-order Møller Plesset perturbation theory (MP2). It is important to note that *all our calculations have included complete geometry optimization.* Previous quantum chemical calculations [10] have led to inaccurate conclusions because geometries were not fully optimized or inadequate basis sets were used.

Calculations carried out with these codes utilize atomic orbital basis sets comprised of linear combinations of gaussian-type functions. The smallest possible basis set (denoted "minimal") contains one cartesian gaussian function for each core and valence atomic orbital of the molecule ($1s$ on H, $1s$, $2s$, $2p_x$, $2p_y$, and $2p_z$ on C, etc.). Larger basis sets include "split valence" ($1s$ and $2s$ on H, $1s$, $2s$, $2p$, $2s'$, and $2p'$ on C, etc.) and "split valence plus polarization" (split valence with the addition of $2p$ for H, $3d$ for C, etc.). Generally, calculations performed using minimal basis sets only provide a qualitative description of the molecular geometry and relative energies. For most closed-shell molecules comprised of first- and second-row atoms, calculations performed using split valence plus polarization basis sets provide accurate molecular geometries and energetics. In the present study we used the following basis sets: STO-3G (minimal), 4-31G (split valence), 6-31G* (split valence plus $3d$ polarization functions on C and O atoms), and 6-31G* (d_O) (split valence with $3d$ functions only for the O atoms). The numbers in the basis set labels refer to the number of gaussian terms in each basis function. For each C, O, and N atom in a molecule there are 5 basis functions in a STO-3G basis set, 9 in a 4-31G set, and 15 in a 6-31G* set. For H atoms the numbers of basis functions are 1, 2, and 5, respectively. Currently, calculations for a molecule with 175 to 200 basis functions are tractable on a Cray X-MP or Y-MP computer. Thus for the diphenyl compounds considered in this study, calculations are possible with either 4-31G or 6-31G* (d_O) basis sets.

A summary of some of the phenylene linkages studied is given in Table 1. It can be seen that most phenyl groups prefer skewed conformations but exhibit widely different energy barriers for rotation of one or both phenyl groups. Our primary focus has been to determine geometric parameters, relative energies, atomic charges, and harmonic force fields for stationary points on the potential energy surface for torsional motion. These include all

TABLE 1 Library of Linkages Between Phenyl Groups[a]

$\mathcal{R}$	Preferred Conformation	Skew Angle (deg)	Single Rotation Barrier[b]	Double Rotation Barrier[b]
—O—	Skew	43	2.5	0.5
—S—	Skew	58	1.8	0.5
>C(CH$_3$)$_2$	Skew	51	≈ 7.0	1.9
>C=O	Skew	29		2.9
—O(C=O)O—	Skew	57	0.7	1.4
—(C=O)O—	Planar/Skew	0/59	4.8/1.8	6.6
>SO$_2$	(Perp.)[c]	90	3.6	20.3

[a] Based on SCF calculations carried out in collaboration with B. C. Laskowski, A. Komornicki, and A. C. Scheiner, with the largest basis set reported.

[b] Energies in kcal/mol.

[c] Investigation of skewed conformations not completed.

stable conformers (i.e., energy minima) and rotation barriers. We have also mapped out selected torsional potentials in greater detail by freezing individual torsional angles at values away from equilibrium and then minimizing the energy with respect to all the other degrees of freedom. Finally, in a similar manner we also determined the deformation potentials for selected valence angle bending modes.

These calculations can affect polymer simulations in a number of ways and can lead to more realistic simulations that better represent the behavior of actual polymers. The geometries and conformational energies for model compounds can be used for direct validation of MM potential energy functions, and the *ab initio* harmonic force constants and atomic partial charges can be used to improve or extend the parameterization of the MM models. While partial charges can be determined from a Mulliken population analysis [11], other schemes can be used that more accurately partition the electron density residing between the atoms [12]. The results of the *ab initio* quantum chemical calculations can also provide insights into the physical and mechanical properties of polymers. Various measures of chain stiffness, such as radius of gyration [13], modulus [14], and persistence length [15,16] can be determined for isolated individual chains based on the calculated torsional and deformation potentials. Rotational isomeric state models (and other statistical models) for single chains and atomistic Monte Carlo and molecular dynamics simulations for ensembles of chains can be set up based on the calculated chain geometries and energetics. In the multiple-chain case, empirical van der Waals potentials such as a Lennard-Jones (12-6) or exponential-6 function [3] can be used to describe interchain interactions. The results of these studies should be more reliable polymer simulations.

RESULTS FOR BISPHENOL-A POLYCARBONATE

The glassy polymer Bisphenol-A polycarbonate, which is shown in Fig. 1, has been the subject of many experimental studies (see, e.g., Refs. 17–25), owing to its high impact strength at temperatures far below the glass transition. The γ relaxation occurs near 173 K and is characterized by large-amplitude molecular motions [17]. Solid-state nuclear magnetic resonance [20] has also been used to identify the occurrence of 180° phenylene ring flips in the low-temperature glass. The polymer is thought to exist in a predominately all-trans conformation of the carbonate groups, but a trans-cis conformer is also possible (see Fig. 1). The experimental studies have not succeeded in elucidating the nature of these chain motions, but several hypotheses have been presented. These include (1) interchain motions in the glass that serve to open cavities in which the ring flips can occur [20]; (2) propagation of a trans-cis "defect" along the chain by a crankshaft motion [21]; and (3) torsional oscillations or rotations of the carbonate groups [25]. Various studies have indicated that the activation energies for the phenylene flips and the γ relaxation are approximately 10 to 12 kcal/mol [18,19].

In our study, the polycarbonate chain has been represented as an alternating series of DPC and DPP molecules [13]. The latter can accurately be described using STO-3G basis set calculations. However, O-atoms (with their lone-pair electrons) are considerably more difficult to treat computationally, making the results for DPC somewhat dependent on the basis set and level of theory used. Therefore, we have carried out a detailed study of DPC and related molecules. Previous computational studies have used semiempirical molecular orbital methods, such as MNDO [7,10,26] or minimal basis set calculations without complete geometry optimization [7,10]. The latter include PRDDO calculations [7] which use an approximate molecular orbital method designed to reproduce *ab initio* minimal basis set calculations [27].

To illustrate the computational approach and the effect of systematically improving the basis set, results are presented here for the phenyl formate (PF) and monophenyl carbonate (MPC) molecules. Energy minimizations for these molecules using the STO-3G basis set lead to structures having the O—C^{ar}

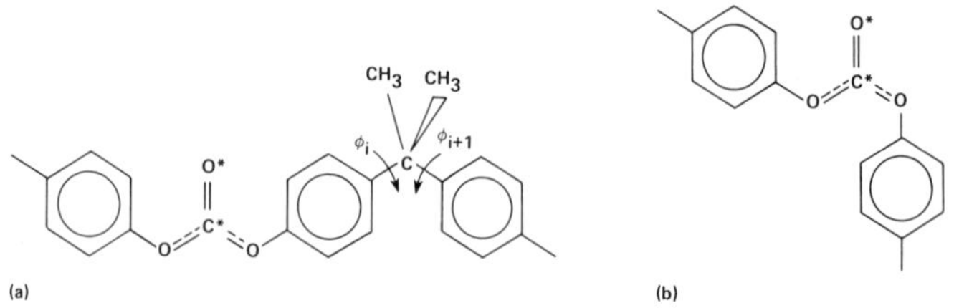

(a)   (b)

Figure 1   Schematic representation of the PC chain showing the trans-trans (a) and trans-cis (b) conformations of the carbonate group. The low-energy phenylene rotation pathway is denoted by the arrows.

bond cis to the carbonyl and the plane of the phenyl ring skewed with respect to the O—C*=O plane. If the cis PF or MPC structure were replicated to make an oligomer chain, the resulting molecule would have a completely trans conformation. However, ψ, the preferred O—C^{ar} torsional angle, is very sensitive to basis set. Using a 4-31G basis set, the phenyl ring is found to be coplanar with the carbonate moiety. Adding polarization functions to the O atoms causes the minimum energy structure to shift to perpendicular. Finally, adding polarization functions to the C atoms returns a skewed structure, in agreement with experimental x-ray diffraction data for diphenyl carbonate (DPC) [28,29]. The phenyl torsional potential for skewed geometries is extremely flat (using the 6-31G* basis set, the perpendicular structure is less than 0.05 kcal/mol higher in energy than the skewed structure), making it difficult to assign an equilibrium value to the torsional angle. However, it seems likely that the preferred orientation of the phenyl group is skewed and that it should be subject to large-amplitude vibrations at thermal energies.

The C^{ar}—O—C* bond angle and O—C* bond length are also sensitive to basis set, as can be seen in Table 2. Similar observations have been made for ether linkages in other molecules. In general, the 6-31G* basis sets give accurate bond lengths and valence bending angles. Interestingly, the average of the STO-3G and 4-31G values for C—O—C bond angles is close to both the 6-31G* and experimental values [13]. These aspects of the phenyl formate geometry are similar to those found for monophenyl carbonate (MPC) and DPC. They also match the experimentally determined values for DPC and the results of MNDO calculations [7].

The energy barriers for phenyl rotation and for rotation about the O—C* bond are also strongly dependent on the basis set used in the calculation. The computed energetics of PF, MPC, and DPC are shown in Table 3. For phenyl formate the calculated phenyl rotation barrier is 1.42 kcal/mol using the STO-3G basis set. However, it is lowered to 0.72 kcal/mol in our best calculation (6-31G* basis set). For this case the SCF and MP2 energy barriers are identical. The best estimate of $\Delta E_{cis \rightarrow trans}$ is 2.2 kcal/mol (6-31G* basis set

TABLE 2 Calculated Geometric Parameters for Phenyl Formate

Basis Set	Torsion Angle, ψ	C^{ar}—O—C* Angle	R(C^{ar}—O) (Å)	R(O—C*) (Å)
STO-3G	58.	114.9	1.41	1.39
4-31G	0.0	129.3	1.40	1.34
6-31G* (d_O)	90.0	118.4		
6-31G*	64.	120.5	1.39	1.32
MPC (STO-3G)	59.6	114.3	1.42	1.39
DPC (STO-3G)	56.7	114.6	1.42	1.32
DPC (expt., Refs. 28, 29)	45.	120.0	1.41	1.33
DPC (MNDO, Ref. 7)	≈0			
DPC (PRDDO, Ref. 7)	44.	120.0		

TABLE 3   Energetics[a] of Phenyl Formate and Phenyl Carbonate Molecules

Molecule/Basis Set	ψ	$E_{\text{rot}}^{\text{b}}$	$\Delta E_{\text{cis} \rightarrow \text{trans}}$	$E_{\text{barr}}^{\text{c}}$
MPC/STO-3G	59.6	1.13	1.43	4.54
MPC/4-31G	0.0	1.09	1.69	6.23
			(1.33)	(6.09)
MPC/6-31G*	91.7		2.74	7.79
			(2.21)	(7.45)
DPC/STO-3G	56.7	1.23	1.68	4.52
DPC (MNDO, Ref.7)	≈0		≈1	
DPC (PRDDO, Ref. 7)	44.		1.1	5.4
PF/STO-3G	58.0	1.42		
PF/4-31G	0.0	1.34		
PF/6-31G* (d_O)	90.0	0.96		
		(0.92)		
PF/6-31G*	64.0	0.72		
		(0.72)		

[a] SCF energies in kcal/mol. MP2 results given in parentheses.
[b] Energy barrier for phenyl group rotation.
[c] Energy barrier for interconversion between cis and trans conformers.

MP2 calculation), which is 50% larger than the STO-3G result and approximately double the value obtained by semiempirical MO calculations [7]. For this case the inclusion of electron correlation in the calculations lowers the energy difference by 0.5 kcal/mol. Results for mono- and diphenyl carbonate are similar. The differences in vibrational zero point energy between the trans-trans, trans-cis, and trans-to-cis barrier and between various phenyl orientations are negligible (<0.05 kcal/mol). The fraction of the higher-energy trans-cis conformers (or "defects") in fully annealed thermal samples of PC is given by the Boltzmann factor $2 \exp(-\Delta E_{\text{cis} \rightarrow \text{trans}}/RT)/ [1 + 2 \exp(-\Delta E_{\text{cis} \rightarrow \text{trans}}/RT)]$. The factor of 2 represents the statistical weight of the trans-cis conformer. If $\Delta E_{\text{cis} \rightarrow \text{trans}} \approx 1$ kcal/mol (the MNDO value) there would be 27% population of defects at 300 K (10% at the γ relaxation temperature). However, for our best estimate of $\Delta E_{\text{cis} \rightarrow \text{trans}}$ (2.2 kcal/mol) the defect population is only 5% at 300 K.

The barrier for interconversion of the trans-cis and trans-trans forms is found to be 7.5 kcal/mol from MP2 calculations of MPC using a 6-31G* basis set. The STO-3G values for MPC and DPC are nearly identical and 60% smaller. The larger basis set calculations have been carried out only for MPC. There has not been any experimental confirmation of these results. However, they seem to cast doubt on the hypothesis that a tctt→ tttc crankshaft-type motion is responsible for the low-temperature relaxation process in polycarbonate.

Ab initio quantum chemistry calculations have shown that the torsional motion of the two phenyl groups in DPP is strongly coupled [13]. It takes less energy (1.9 kcal/mol; see Table 1) to rotate both groups simultaneously than it does to rotate one while holding the other fixed (≈ 7 kcal/mol). Similar results have been found (see Table 1) for diphenyl ether [30] and diphenyl thioether. On the other hand, in DPC the motions of the phenyl groups are found to be completely uncoupled (it takes twice the energy to rotate both groups simultaneously as it does to rotate only one). Assuming that the barrier for 1,4-phenylene (ϕ) rotation in $\mathcal{R} - \phi - \mathcal{R}'$ is given by the sum of $\phi - \mathcal{R}$ and $\phi - \mathcal{R}'$ rotation barriers, the torsional energy required in polycarbonate to rotate the two phenyl groups simultaneously at a DPP moeity would be 3.3 kcal/mol. Interestingly, it would require ≈ 7.7 kcal to rotate a single phenyl group. These values, along with the torsional energy for the crankshaft motion, constitute the intrinsic barriers for the torsional motions considered. As the low-temperature PC glass is a dense system, it is likely that there is also a significant interchain contribution to these energies. Thus the lowest-energy pathway for the double phenylene rotation is consistent with the observed 10-kcal/mol activation energy for the ring flip process.

The calculated geometric parameters and $\Delta E_{\text{cis} \to \text{trans}}$ were used as input for a rotational isomeric state (RIS) calculation of the unperturbed chain dimension of polycarbonate ($\langle r^2 \rangle_0 / M$) and a value of 1.11 was obtained at 300 K [13]. This result compares favorably with recent small-angle neutron scattering data (1.28 and 1.25 ± 0.5) [31,32] from bulk amorphous samples. The theoretical value is most sensitive to changes in the C^{ar}—O—C^* angle and $\Delta E_{\text{cis} \to \text{trans}}$. For example, decreasing the angle and energy difference from the 6-31G* to the STO-3G values results in about a 25% reduction in the chain dimension.

RESULTS FOR POLY(p-HYDROXYBENZOIC ACID)

Poly(p-hydroxybenzoic acid) is a member of the class of polymers that exhibits liquid-crystal properties. Its nematic phase is stable even at temperatures exceeding 700 K [33]. Similar high-temperature properties have been observed in the commercial polyesters Xydar and Bektra, which are random copolymers of hydroxybenzoic acid and hydroxynaphthoic acid [34], in poly(phenylene terephthalate), and in the corresponding polyamides poly(p-benzamide) and Kevlar. All of these macromolecules are known to exist as stiff extended chains that require high temperatures for processing. The chain persistence length provides a measure of the importance of these extended structures in the liquid-crystalline state. For example, experimental determinations of this quantity for poly(p-benzamide) yield values between 325 and 1050 Å [35]. While the reported values are sensitive to the measurement technique used, all the data indicate long persistence lengths. Hummel and Flory [36] attempted to compute the persistence length of aromatic polyesters and polyamides using a RIS model. They assumed that no rotation was possible about the ester or amide bond and obtained values greater than 1500 to 2000 Å. It has been argued that small torsional oscillations about these bonds would greatly reduce the calculated persistence length.

To further investigate the degree of chain stiffness in these polymers, we have carried out *ab initio* quantum chemical calculations for phenyl benzoate (PB) as a model compound for PHBA. Similar calculations on the corresponding amide have been carried out by Scheiner et al. [37]. The equilibrium geometry of phenyl benzoate is shown in Fig. 2. The ester phenyl group is found to be cis to the carbonyl bond which corresponds to an all-trans or zigzag conformation for PHBA. The computed geometric parameters and the ester phenyl rotation barrier are similar to those determined for MPC and DPC at the same level of calculation. At the STO-3G level, this energy barrier is 1.8 kcal/mol, compared to 1.7 kcal/mol for DPC. The plane of the phenyl ring is skewed with respect to the O—C*=O plane (59.5° for STO-3G and 58.1° for 6-31G* (d_O) compared to an x-ray diffraction value of 65.1° [38]). The alpha phenyl group can experience a favorable conjugation interaction with the carbonyl and prefers a coplanar orientation with respect to the O—C*=O group. It has a rotation barrier of 4.8 kcal/mol at the STO-3G level of calculation. Previous semiempirical calculations [7,39,40] have determined similar equilibrium bond angles and bond lengths. Studies using the NDDO [39], AM1 [7,40], and PRDDO [7] methods obtained the same torsional conformation as the *ab initio* quantum chemical calculations. However, calculations using the MNDO method [7] predicted that both phenyl rings would prefer perpendicular orientations with respect to the plane of the ester group. These calculations completely missed the effect of conjugation in stabilizing the benzoate group. The MNDO and AM1 phenyl rotational barriers were comparable for the two rings, yielding values of ≈2 kcal/mol in all cases.

Unlike DPC, there does not exist a stable conformer with a trans orientation. Instead, that structure corresponds to the saddle point (i.e., maximum in the minimum energy path) for rotation about the C*—O bond. At the STO-3G level, this barrier is 9.6 kcal/mol. Improving the basis set to 6-31G* (d_O) yields a barrier of 13.9 kcal/mol (10.1 at the MP2 level). The difference in energetics between cis and trans conformers for PB and DPC is due in part to the conjugation interaction that is possible only for the cis form of PB. Steric repulsion interactions are also greater for trans PB than for trans-cis DPC. Similar behavior was observed in calculations using the MNDO

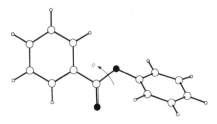

Figure 2 Equilibrium geometry conformation of phenyl benzoate. Carbon atoms are denoted by the larger unfilled circles, hydrogen atoms by the small circles, and oxygen atoms by the filled circles. The ester phenyl group is skewed by 58.1° with respect to the plane of the carboxy group. Rotation about the ester bond is shown schematically by the arrow.

method [7], but the torsional barrier (18 kcal/mol) is much greater. In contrast, the AM1 method [40] yielded a local minimum structure with a trans geometry at an energy only 5 kcal/mol above the cis conformer. These results illustrate the difficulty in using semiempirical molecular orbital methods for conformational studies of aromatic molecules. The orientations and energetics of the phenyl rings arise from a delicate balance between opposing steric and conjugation effects and are extremely sensitive to deficiencies in the computational method being used.

The recent calculations carried out by Scheiner et al. [37] for phenyl benzamide reveal major differences in the intrinsic conformational energy surfaces between these polyesters and polyamides as shown in Table 4. In phenyl benzamide the rotational barriers are somewhat smaller than in phenyl benzoate, and both phenyl rings are skewed with respect to the plane of the amide group. In addition, the torsional potential for rotation about the amide bond shows a barrier of 13.6 kcal/mol and a local minimum at the trans orientation that is 4.0 kcal/mol above the more stable cis conformer. These differences in relative energies between the ester and amide are in part due to the stronger conjugation effects in phenyl benzoate.

It is thought that the stiffness of the PHBA chain is controlled by the amplitude of the torsional oscillations at the C*—O bond under thermal conditions. Random small displacements of the torsional angles along an otherwise fully extended chain will result in a reduction of the persistence length and could even lead to a random-coil conformation for the polymer. To analyze this possibility further, we computed the torsional deformation potential for ester-bond rotation in phenyl benzoate. In our calculations the molecular geometry was reoptimized subject to the constraint that the ester torsional angle was fixed at various values over a range of $\pm 30°$ from the equilibrium value. The results are shown in Fig. 3. Analysis of the deformation potential based on MP2 calculations using the 6-31G* (d_O) basis set indicates that the

TABLE 4 Comparison of Results for Phenyl Benzoate and Phenyl Benzamide[a]

Parameter	Phenyl Benzoate	Phenyl Benzamide
Skew angle (α-phenyl) (deg)	0	30
E_{rot}^b (α-phenyl)	4.8	3.4
Skew angle (ester/amide phenyl) (deg)	58	10
E_{rot}^b (ester/amide phenyl)	1.8	0.6
$\Delta E_{cis \rightarrow trans}^c$	10.1	4.0
E_{barr}^d	$(10.1)^e$	13.6

[a] MP2 energies in kcal/mol from calculations using the largest basis sets reported.

[b] Energy barrier for phenyl group rotation.

[c] Energy difference between cis and trans conformers.

[d] Energy barrier for interconversion between cis and trans conformers.

[e] Energy barrier occurred at trans geometry.

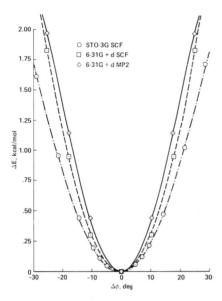

Figure 3 Torsional deformation potential for the ester bond in phenyl benzoate. Calculated points shown by symbols. Curves represent fourth-order polynomial fits. $\Delta\phi = \phi - \phi_e$ where ϕ is the $\overline{C\text{—}C^*\text{—}O\text{—}C}$ dihedral angle. ϕ_e is the value of ϕ at the equilibrium geometry ($\phi_e = 180°$).

phenyl benzoate molecule can undergo torsional displacement of $\pm 12°$ with an energy equal to RT (at 300 K) and $\pm 18°$ with an energy equal to $2RT$. Jung and Schürmann [16] have carried out parametric RIS calculations of the chain persistence length for PHBA. They permitted several different amplitude torsional fluctuations about the ester bond. For the totally stiff case they reported a computed persistence length of 706 Å, but for fluctuations of 8.8° and 15.1° this value was reduced to 256 and 142 Å, respectively. Based on our computed deformation potential, which allows a $\pm 12°$ fluctuation at room temperature, we estimate a chain persistence length of approximately 200 Å. However, the thermal fluctuations would be ± 18 to $20°$ at temperatures between 600 and 700 K, where important phase transitions of aromatic polyesters normally occur, and as a consequence, the persistence length would become much shorter (probably less than 100 Å).

Comparison of the torsional potentials for phenyl benzoate and phenyl benzamide indicates that the latter molecule should be somewhat stiffer because its torsional potential rises to a maximum of 13.9 kcal/mol over a span of $\approx 90°$ as compared to 10.1 kcal/mol over a span of 180° for the ester. This finding is consistent with the longer persistence lengths observed for aromatic polyamides than for the corresponding polyesters.

SUMMARY

The examples given above describe the basic approach used for the determination of segmental properties of polymers based on *ab initio* quantum chemistry calculations. In this work we conclude (1) that the γ relaxation in polycarbonate probably involves the coupled rotation of neighboring phenylene groups with a common propane linkage, and (2) that the chain stiffness, as measured by the persistence length, for PHBA is controlled by thermal oscillations of ± 12 to $15°$ about the ester bond. Additional calculations have been carried out for a variety of polyethers, polysulfones, polyamides, and polyimides in addition to the two case studies presented here. In general, calculations using a 6-31G* basis set accurately reproduce experimentally determined x-ray diffraction data for geometric structures. However, one must be cautious if minimal basis set (STO-3G or PRDDO) calculations are used. In particular, ether bond angles and bond lengths are very sensitive to the level of calculation. Relative energies between conformers and torsional barrier heights are also basis set dependent. Although the larger basis set values seem more reasonable, direct experimental determination of these quantities is generally not possible. Semiempirical molecular orbital methods such as MNDO appear to be less reliable for the determination of the conformational geometries and relative energies of these aromatic polymers.

Statistical mechanical and dynamical modeling of polymeric systems requires quantitative information about the intrasegmental, intersegmental, and possibly interchain interactions of the system. The calculations described above provide an effective method for determining the first of these. In addition, the results of these quantum chemistry calculations can be used to parameterize and test molecular mechanics potential-energy functions before large-scale simulations are attempted.

ACKNOWLEDGMENTS

The authors thank A. Komornicki (Polyatomics Research Inst.), B. C. Laskowski (Analatom Inc.), and A. C. Scheiner (IBM) for their assistance during a long and fruitful series of collaborations.

REFERENCES

1. W. A. Lathan, L. A. Curtiss, W. G. Hehre, J. B. Lisle, and J. A. Pople, *Prog. Phys. Org. Chem.* **11,** 175 (1974).
2. R. A. Sorensen, W. B. Liau, and R. H. Boyd, *Macromolecules* **21,** 194 (1988).
3. N. Karasawa and W. A. Goddard III, *J. Phys. Chem.* **92,** 5828 (1988).
4. See, for example, R. A. Sorensen, W. B. Liau, L. Kesner, and R. H. Boyd, *Macromolecules* **21,** 200 (1988).
5. U. Burket and N. L. Allinger, *Molecular Mechanics,* American Chemical Society, Washington, D.C., 1982.
6. M. J. S. Dewar and W. Thiel, *J. Am. Chem. Soc.* **99,** 4899 (1977).
7. J. Bicerano and H. Clark, *Macromolecules* **21,** 585 (1988).
8. GRADSCF is an *ab initio* gradient program system designed and written by A. Komornicki at Polyatomics Research Institute, Inc. and supported on grants through NASA.

9. GAUSSIAN86 is an *ab initio* electronic structure program system written by M. J. Frisch, J. S. Binkley, H. B. Schlegel, K. Raghavachari, C. F. Melius, R. L. Martin, J. J. P. Stewart, F. W. Bobrowicz, C. M. Rohlfing, L. R. Kahn, D. J. DeFrees, R. Seeger, R. A. Whiteside, D. J. Fox, E. M. Fluder, and J. A. Pople, Carnegie–Mellon Quantum Chemistry Publishing Unit, Pittsburgh, 1984.
10. J. T. Bendler, *Ann. N.Y. Acad. Sci.* **371,** 299 (1981).
11. R. S. Mulliken, *J. Chem. Phys.* **23,** 1833 (1955).
12. U. Dinur and A. T. Hagler, *J. Am. Chem. Soc.* **III,** 5149 (1989).
13. B. C. Laskowski, D. Y. Yoon, A. D. McLean, and R. L. Jaffe, *Macromolecules* **21,** 1629 (1988).
14. R. L. Jaffe, B. C. Laskowski, D. Y. Yoon, and A. D. McLean (to be published).
15. R. L. Jaffe, D. Y. Yoon, and A. D. McLean (to be published).
16. B. Jung and B. L. Schürmann, *Macromolecules* **22,** 477 (1989).
17. J. Heijboer, *J. Polym. Sci.* **C16,** 3755 (1968).
18. A. A. Jones, F. F. O'Gara, P. T. Inglefield, J. T. Bendler, A. F. Yee, and K. L. Ngai, *Macromolecules* **16,** 658 (1983).
19. J. Schaefer, E. O. Stejskal, R. A. McKay, and W. T. Dixon, *Macromolecules* **17,** 1479 (1984).
20. J. Schaefer, E. O. Stejskal, D. Perchak, J. Skolnick, and R. Yaris, *Macromolecules* **18,** 368 (1985).
21. A. A. Jones, *Macromolecules* **18,** 902 (1985).
22. A. K. Roy, A. A. Jones, and P. T. Inglefield, *Macromolecules* **19,** 1356 (1986).
23. P. Tekely, *Macromolecules* **19,** 2544 (1986).
24. J. J. Connolly, P. T. Inglefield, and A. A. Jones, *J. Chem. Phys.* **86,** 6602 (1987).
25. P. M. Henrichs, H. R. Luss, and R. P. Scaringe, *Macromolecules* **22,** 2731 (1989).
26. M. A. Mora, M. Rubio, and C. A. Cruz-Ramos, *J. Polym. Sci., Polym. Phys. Ed.* **24,** 239 (1986).
27. T. A. Halgren, D. A. Kleier, J. H. Hall, Jr., L. D. Brown, and W. Lipscomb, *J. Am. Chem. Soc.* **100,** 6595 (1978).
28. D. Y. Yoon and P. J. Flory (in preparation).
29. S. Perez and R. P. Scaringe, *Macromolecules* **20,** 68 (1987).
30. B. C. Laskowski, R. L. Jaffe, and A. Komornicki, *Int. J. Quantum Chem.* **29,** 563 (1986).
31. W. Gawrisch, M. G. Brereton, and E. W. Fischer, *Polym. Bull.* **4,** 687 (1981).
32. D. G. H. Ballard, A. N. Burgess, P. Cheshire, E. W. Janke, and A. Nevin, *Polymer* **22,** 1353 (1981).
33. J. Economy, W. Volksen, C. Viney, R. Geiss, R. Seimens, and T. Karis, *Macromolecules* **21,** 2777 (1988).
34. J. Blackwell, G. A. Gutierrez, and R. A. Chivers, *Macromolecules* **17,** 1219 (1984).
35. K. Zero and S. M. Aharoni, *Macromolecules* **20,** 1957 (1987).
36. J. P. Hummel and P. J. Flory, *Macromolecules* **13,** 996 (1980).
37. A. C. Scheiner, R. L. Jaffe, A. D. McLean, and D. Y. Yoon (in preparation).
38. J. M. Adams and S. E. Morsi, *Acta Crystalogr.* **B32,** 1345 (1976).
39. P. Birner, S. Kugler, K. Simon, and G. Náray-Szabó, *Mol. Cryst. Liq. Cryst.* **80,** 11 (1982).
40. P. Coulter and A. H. Windle, *Macromolecules* **22,** 1129 (1989).

Conformational Energy and Molecular Dynamics Studies of Models for the Conducting Polymer Poly(phenylenevinylene)

CORAL GETINO and JESUS SANTAMARIA
Dpto de Quimica Fisica, Facultad de CC Quimicas,
Universidad Complutense de Madrid
28040-Madrid
Spain

JERRY A. DARSEY
Department of Physical Sciences
Tarleton State University
Stephenville, Texas 76402

BOBBY G. SUMPTER
Chemistry Division
Oak Ridge National Laboratory
Oak Ridge, Tennessee 37831
and Department of Chemistry
The University of Tennessee
Knoxville, Tennessee 37996

ABSTRACT

Ab initio calculations are used to study the geometric properties of the two isomeric states of stilbene. A reaction pathway for isomerization is determined giving a barrier height of $E_0 = 104$ kcal/mol and $\Delta H = 4.43$ kcal/mol. Analytical fits to the *ab initio* data are then used to carry out molecular dynamics calculations of the isomerization process. Intramolecular energy transfer pathways for both nonrotating and rotating (300 K) *trans*- and *cis*-stilbene are discussed. The calculated lifetimes for isomerization are in good agreement with the RRK theory.

INTRODUCTION

Conducting polymers have attracted considerable attention in the past two decades [1,2]. Much of the interest comes from the unique properties of polymers, such as flexibility and low mass, compared to those of other conductive materials [1,2]. Since their discovery [1], a number of polymers have been shown to be conductors [2], although their conductivity is usually low. A considerable amount of research has been devoted to finding new methods or new polymers that can conduct electricity in a more efficient way. Poly(*p*-phenylenevinylene) (PPV) appears to be a promising conducting polymer with the desirable property of conjugation [3]. Its structure can be considered as an alternating copolymer of acetylene and *p*-phenylene (see Fig. 1). PPV (or at least oligomers of the polymer) have been synthesized since 1969 [4] and Wudl and co-workers [5] have developed methods to process PPV into desirable shapes.

The electrical properties of PPV have been shown to converge rapidly toward a limiting value for three monomers [6]. In this light, stilbene (Fig. 2) is a reasonable model for PPV. The structure and isomerization of stilbene have been the subject of numerous experimental [7–18] and theoretical [19–24] studies. The experimental determination of the geometry of an isolated molecule seems to be difficult because of the floppiness of the phenyl rings in stilbene about the ethylenic carbon–phenyl bonds. In a gas-phase electron diffraction experiment [8] it was found that the two phenyl rings are rotated about 30° in *trans*-stilbene and 43° in *cis*-stilbene. However, the most recent measurements [10a] suggest that the equilibrium structure of *trans*-stilbene is planar. On the theoretical side, *ab initio* and semiempirical calculations [19–24] yield a variety of nonplanar structures for *trans*- and *cis*-stilbene (see Table 1, for example).

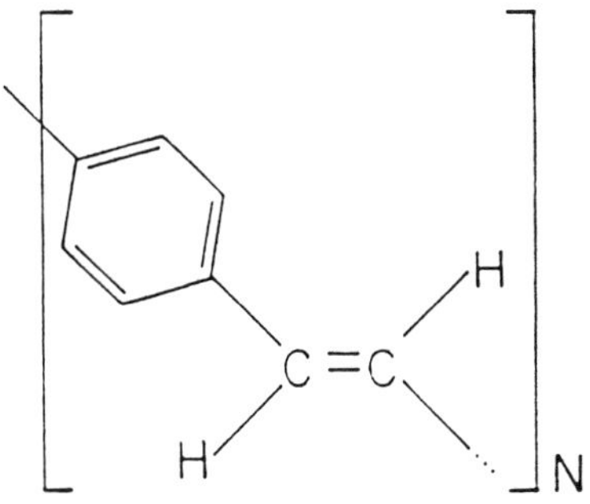

Figure 1 Structure of poly (phenylenevinylene).

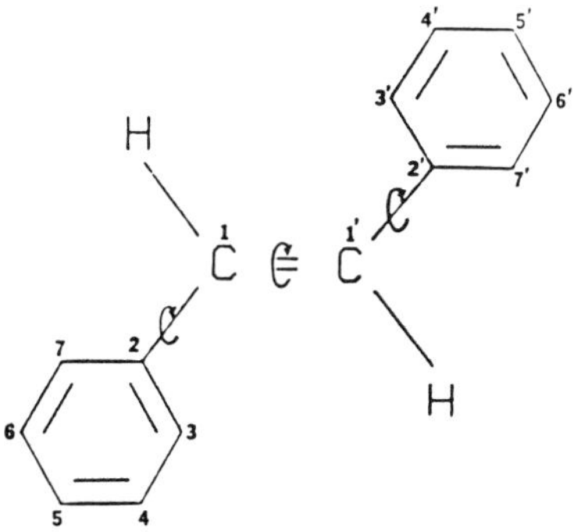

Figure 2 Structure of *trans*-stilbene.

In this chapter we are interested in determining a potential energy function for the isomerization of stilbene (*trans*/*cis* or *cis*/*trans*). This involves the determination of the energy and the geometry as a function of the dihedral angle about the ethylenic C=C bond. Molecular dynamics simulations will then be used to elucidate intramolecular energy transfer processes and isomerization rates.

In the next section the methods that we have employed to carry out the calculations are described, and in the following section a discussion of the results is given.

METHODS

Ab Initio Calculations

Ab initio molecular orbital calculations were carried out at the Hartree–Fock level of theory using both the minimal STO-3G basis set and the split-valence 4-31G basis set, employing the GAUSSIAN suite of programs [25]. In the initial step, the geometry of *trans*-stilbene was calculated using the 4-31G basis set with full optimization of all the geometric parameters (see the table in Ref. 24). This geometry was then used to carry out more restricted calculations. In a first series of calculations the phenyl rings were rotated about the ethylenic C=C bond and the energy at each point was calculated [24] optimizing only the ethylenic C=C bond length. The results of this calculation revealed that steric repulsion of the two phenyl rings (see Fig. 3 of Ref. 24) made the planar cis conformation impossible. This has also been discussed in previous studies of stilbene (cf. Refs. 20–23).

In a second series of calculations, the phenyl rings were rotated in opposite directions 45° out of the plane (90° with respect to each other), and the potential for the torsion about the ethylenic C=C bond for optimized

TABLE 1 Equilibrium Geometry for Stilbene[a]

Coordinate	STO-3G[b]		4-31G[c]		ab initio[d]		MNDO[e]		CNDO[e]		Experiment[f]	
	trans	cis	trans	cis	trans	cis	trans	cis	trans	cis	trans	cis
$r(H-C)$	1.0823	1.0854	1.074		1.08		1.096	1.098	1.08	1.08	1.095	1.095
$r(C_2=C_1')$	1.3208	1.3187	1.3262		1.32		1.347	1.347	1.34	1.35	1.330	1.340
$r(C_1-C_2)$	1.4990	1.5048	1.4777		1.47		1.482	1.481	1.54	1.54	1.480	1.490
$\theta(H-C_1=C_1')$	113.47	114.16	113.26				120.8	118.3	121	113		
$\theta(H-C_1'=C_1)$	113.46	114.16	113.26									
$\theta(C_2-C_1=C_1')$	127.00	128.40	127.76		127.4	126	125.5	128.5	128	138	128	130
$\theta(C_1'-C_1'-C_1)$	126.97	128.40	127.76									
$\omega(C_2-C_1=C_1'-C_2')$	180.0	3.42	180.0		180.0	0	180.0	0	180	0	180	0
$\phi(Ph-C_1-C_1'-Ph)$	3.54	88.53	0	90	19	52	70.6	75.1	40	40	33	43
$\chi(H-C_1'-C_2-C_7)$	-2.53	44.26	0									
$\chi(H-C_1'-C_2-C_7)$	-0.49	44.27	0'									

[a] Bond lengths are given in Å and angles in degrees. Figure 2 shows the definition of the geometric parameters.
[b] This work.
[c] Ref. 24.
[d] Ref. 20.
[e] Ref. 23.
[f] Ref. 8.

C=C bond lengths was then recalculated. The shape of the potential joining the trans and cis conformations of twisted stilbene (see Fig. 4 of Ref. 24) was found to be significantly different from that calculated for a planar molecule. The results of these preliminary studies also indicated a strong coupling of the ethylenic C=C stretch to the torsional motion.

In a second approach, more detailed calculations were performed to develop a better description of the pathway between the trans and cis isomers of stilbene. All of the geometrical parameters, except for the bond lengths and bend angles in the phenyl rings, were optimized in the determination of the torsional potential about the ethylenic C=C bond. Specifically, the parameters (see Fig. 2): $r(H-C_1)$, $r(H-C_1')$, $r(C_1=C_1')$, $r(C_1-C_2)$, $r(C_1'-C_2')$, $\theta(C_2-C_1=C_1')$, $\theta(C_2'-C_1'=C_1)$, $\theta(H-C_1=C_1')$, $\theta(H-C_1'=C_1)$, $\omega(C_2-C_1=C_1'-C_2')$, $\phi(C_3-C_2-C_1=C_1')$, $\phi(C_7'-C_2'-C_1'=C_1)$, $\chi(H-C_1-C_2-C_7)$, and $\chi(H-C_1'-C_2'-C_7')$ are optimized, while the parameters for the phenyl rings are set to their equilibrium values obtained for the fully optimized trans geometry [24]. A fundamental understanding of how various modes of the molecule are coupled to the isomerization reaction coordinate (torsional motion about the ethylenic C=C bond) is elucidated by this procedure [26].

The potential obtained in this approach represents a very reasonable approximation to the "absolute" minimum energy pathway (MEP) for the isomerization of stilbene. The absolute MEP for the isomerization would require full optimization of *all* the geometric parameters for each value of the dihedral angle about the ethylenic C=C bond. While this type of calculation is feasible, it requires a remarkably large amount of computer time and does not actually improve overall results sufficiently enough to justify the computational cost.

The barrier to isomerization (trans to cis) is calculated as the difference in energy between the trans conformation and the transition state. Similarly, ΔH is taken as the energy of the cis conformation relative to the trans. In future calculations we plan to study in more detail the effects of higher levels of theory and larger basis sets with diffuse and polarization functions.

Molecular Dynamics

To perform molecular dynamics calculations, a potential-energy surface must be formulated. In the present study we have coupled the *ab initio* calculations reported here, with previous experimental results [7–19], to determine an approximate semiempirical global potential-energy surface for a four-atom model of stilbene (see below). In particular, the MEP obtained from the *ab initio* energies for the rotation of the phenyl groups about the ethylenic C=C bond is used as the potential describing the internal rotation which leads to isomerization. This is accomplished by a nonlinear least-squares fit of the energies to a six-term Fourier cosine series:

$$V_{tor} = \sum a_i \cos i\omega \tag{1}$$

where ω is the torsional dihedral angle. The Fourier coefficients are given in Table 2, along with the other potential parameters and the normal-mode frequencies for the trans and cis isomers obtained using the potential given below. The remaining $3N - 7$ vibrational degrees of freedom are described using the following functions:

$$V_{str} = \sum D_e\{1 - \exp[-\beta(r_i - r_i^0)]\}^2 \tag{2a}$$

$$V_{bend} = \tfrac{1}{2}\sum S(r_i)S(r_j)K_\theta(\theta - \theta^0)^2 \tag{2b}$$

The parameters for the potential functions [Eq. (2)] were taken from a harmonic force field fit to experiment [16–18]. The Morse parameters are obtained from $\beta = (K/2D_e)^{1/2}$. The bending potential [Eq. (2b)] is a harmonic oscillator with the force constant attenuated by switching functions $S(r)$. The switching functions describe the nonlinear coupling of the bending motion to the stretching of the bonds defining the angle θ. We have found in previous molecular dynamics studies [27] that this type of stretch–bend coupling is

TABLE 2 Potential-Energy Parameters and Calculated Normal-Mode Frequencies for the Tetratomic Model of Stilbene

Stretch	
C—Ph	C=C
$\beta = 1.961$ Å^{-1}	$\beta = 2.132$ Å^{-1}
$D_e = 88$ kcal/mol	$D_e = 146.9$ kcal/mol
$r_e = 1.48$ Å	$r_e = 1.33$ Å

Bends
$K_\theta = 1.1507$ mdyn-Å/rad
$\theta = 128°$

Torsion (kcal/mol)
$a_0 = 25.0207$
$a_1 = -21.2462$
$a_2 = -2.2553$
$a_3 = 8.2467$
$a_4 = -12.3614$
$a_5 = 13.4799$

Normal-Mode Frequencies (cm^{-1})		
	trans-Stilbene	*cis*-Stilbene
sym. Ph—C stretch	983	681
asym. Ph—C stretch	847	1011
sym. Ph—C=C bend	240	546
asym. Ph—C=C bend	283	126
C=C stretch	1646	1651
Ph—C=C—Ph torsion	964	1190

necessary for correctly describing the overall intramolecular and reaction dynamics of polyatomic molecules. The switching function used here is [27,28]

$$S(r) = 1 - \tanh[1.531368 \times 10^{-7} \Delta r (r + 4.669625)^8] \qquad (3)$$

The same switching function is used to attenuate the isomerization barrier to zero as the C=C bond breaks. The molecular Hamiltonian is then

$$H = T + V_{str} + V_{bend} + V_{tor} \qquad (4)$$

where the first term represents the kinetic energy.

All calculations were carried out in Cartesian coordinates, which give an exact definition of the kinetic coupling. The phenyl rings were collapsed into single point masses of 77 amu and, similarly, the C—H's on the ethylenic C=C were represented by masses of 13 amu. Thus the stilbene molecule is treated in this study as a four-atom model (Ph—C=C—Ph). In future calculations we will consider the full dimensionality of the problem.

The molecular dynamics method gives the time evolution of the coordinates and momenta for a given set of initial conditions. To simulate a process such as photoisomerization, it is necessary to carry out ensemble averages of many trajectories. Single trajectories from an ensemble provide information on the details of the intramolecular dynamics, while the ensemble averages are necessary to calculate the rates associated with the isomerization process. In the present study we examine the details of the isomerization of stilbene induced by local excitations of particular modes of the molecule (with or without molecular rotation). Initial conditions are chosen by placing zero point energy into all of the normal modes of the molecule plus the additional energy for the excitation of a specific bond mode. Overall molecular rotation is studied by including angular momenta to give a rotational energy of the $\frac{3}{2}RT$ ($T = 300$ K). Further details of the various methods to select initial conditions can be found in other references [29].

Analysis of single trajectories out of the ensembles consisted primarily of examining internal mode energies as a function of time. This analysis is useful in determining the nature of energy redistribution (either specific energy-flow pathways or statistical randomization of the energy) leading to isomerization. Instantaneous G^{-1} matrix elements [30] associated with the kinetic coupling of the internal vibrational modes were also examined to help clarify the type of dynamic coupling responsible for intramolecular energy redistribution prior to reaction.

For the present model of stilbene, there are four different possible reaction pathways (two Ph—C bond fissions, one C=C bond fission, one isomerization). The unimolecular decay constant K_d for each of these reactions was determined by least-squares fitting the lifetime distributions to

$$-K_d t = \ln \frac{N_t}{N_0} \qquad (5)$$

where N_t is the number of nonreactive molecules for each pathway at time t and N_0 is the total number of trajectories in the ensemble (200 for each case). Hamilton's equations of motion were integrated using a fixed step (1×10^{-16} s) sixth-order hybrid Gear predictor-corrector routine. Trajectories were calculated for a total time of 5 ps or until a reaction had occurred. Conservation of at least six digits for all the constants of motion (total energy, linear, and angular momenta) was required. Reactions were considered to have occurred when either the bond lengths had exceeded 6 Å or the dihedral angle about the ethylenic C=C bond had passed through the transition state in the direction of the products.

Qualitative comparison to experimental results for the photoisomerization of stilbene was made using the RRK (Rice–Ramsberger–Kassel) [31] classical formula,

$$K_{nr}(E_v) = A \left(\frac{E_v - E_T}{E_v} \right)^{s-1}$$

(6)

where E_T is the barrier height, E_v the total energy, and s the effective number of oscillators. The values determined from the supersonic jet studies of Amirav and Jortner [14] are $A = 1 \times 10^{11}$ s^{-1}, $E_T = 900$ cm^{-1}, and $s = 7.3$. These values were used to extend the experimental measurements to the energies appropriate in our theoretical calculations.

RESULTS AND DISCUSSION

Ab initio Calculations

Figure 3a is a plot of the approximate minimum energy pathway (MEP) for the isomerization of stilbene at the HF/STO-3G level. The transition state is shown to lie at 47° with a barrier height of 104 kcal/mol separating the trans from the cis isomer. The position of the transition state is significantly closer to the cis configuration than to the trans (this type of transition state is not uncommon; for example, the barrier to the isomerization of methyl cyanide is 10.8° closer to the less stable product ($\Delta H = 23.7$ kcal/mol) methyl isocyanide [32]). The calculated enthalpy of isomerization for stilbene is $\Delta H = 4.43$ kcal/mol. This is in excellent agreement with the most recent experimental measurements [10b], $\Delta H = 4.53$ kcal/mol. In the present calculation, the excessive barrier height, $E_0 = 104$ kcal/mol compared to experimental estimates of $E_0 \approx 50$, could be due to both the neglect of electron correlation and/or an insufficiently large basis set [26]. The applicability of the minimal basis set (STO-3G) to calculations of equilibrium geometries and torsional potentials has been discussed in Refs. 33 and 34.

The dependence of the geometric parameters on the dihedral angle about the ethylenic C=C bond is also shown in Fig. 3b–h. All of the parameters show some coupling to the rotation about the ethylenic C=C bond. In particular, the C=C bond length (Fig. 3b) the H—C=C (Fig. 3d), Ph—C=C (Fig. 3e) bends, and the dihedral angle between the two phenyl rings (Fig. 3g). In Fig. 3g it is shown that the phenyl rings rotate relative to each other as stilbene undergoes isomerization. This rotation is due to the steric repulsion between the two phenyl rings. The optimized angle found for the trans conformation is 3.5° and for the cis configuration is 89°. These results are in contrast with previous *ab initio* calculations (see Table 1), where more restricted geometry optimizations were carried out but support the most recent experimental measurements [10a] on the planarity of *trans*-stilbene.

Molecular Dynamics

Classical trajectories calculations are carried out for the molecular Hamiltonian given in Eq. (4). The objectives of the calculations are to determine the lifetimes for isomerization of the model of stilbene for different initial excitations (with or without overall molecular rotation) and develop qualitative descriptions for the pathways of energy flow.

Two different types of initial conditions are studied for both the cis and trans conformations and for the vibrational excitation of a Ph—C bond and a C=C bond. The energy of excitation is chosen so that the total vibrational energy (bond excitation + zero point energy) is 125 kcal/mol. Lifetimes associated with the addition of overall molecular rotation of $\frac{3}{2}RT$ ($T = 300$ K) are also calculated.

Unimolecular reaction lifetimes.
A summary of the calculations for the model of stilbene is given in Table 3. Lifetimes were computed from microcanonical ensembles of 200 quasiclassical trajectories for each case. The room-temperature isomerization lifetime for the model of *trans*-stilbene with an initial excitation of a Ph—C bond is 10.67 ps. A qualitative comparison to experiment can be accomplished by using an RRK fit to experimental data [see Eq. (6)] and the vibrational energy associated with our trajectories. From this type of calculation, a lifetime of 11.4 ps was determined. There is excellent agreement with the value that was obtained from the trajectory results. This agreement does not necessarily mean that the model for stilbene is sufficiently accurate to quantitatively describe the reaction dynamics, but it provides evidence which indicates that qualitatively accurate results can be obtained. In particular, the trends in the reaction dynamics, such as the effects of overall molecular rotation, different vibrational excitations, or initial conformation, should be correct.

For the initial excitation of the Ph—C bond, the unimolecular reactions appear to be statistical in nature. That is, the energy is rapidly redistributed on a time scale compared to the reaction. This can be seen by examining the lifetimes of the trans conformation for the fission of the initially excited bond

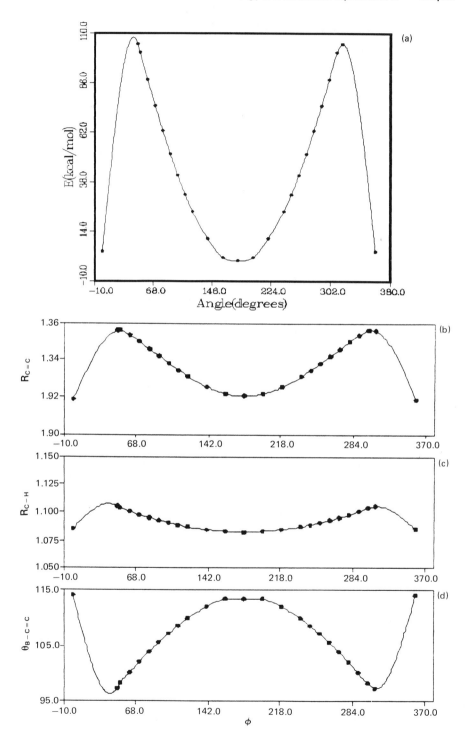

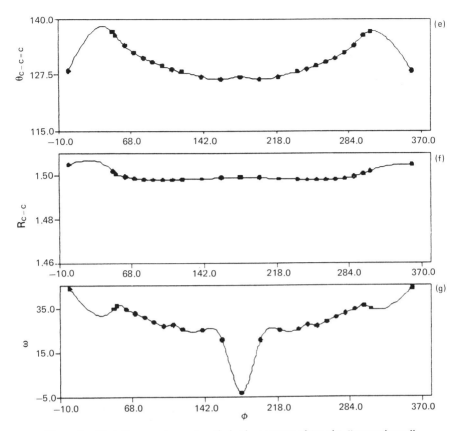

Figure 3 *Ab initio* energies and optimized geometry along the "approximate" minimum energy pathway: (a) energy versus the dihedral angle about the C=C bond; (b) $C_1=C_1'$ bond length versus the dihedral angle about the C=C bond; (c) C—H bond length versus the dihedral angle about the C=C bond; (d) H—$C_1=C_1'$ bend angle versus the dihedral angle about the C=C bond; (e) C_2—C_1=C_1' bend angle versus the dihedral angle about the C=C bond; (f) C_2—C_1 bond length versus the dihedral angle about the C=C bond; (g) dihedral angle between the two phenyl rings versus the dihedral angle about the C=C bond.

versus that for the unexcited C—Ph′ bond (see also Fig. 4). For both rotating and nonrotating cases, the lifetimes for the two Ph—C bond fissions are approximately the same. Similar results are obtained for C=C excitation.

While overall molecular rotation causes the lifetime for bond fission to increase, the lifetime for isomerization is decreased by a factor of 2 for Ph—C excitation and by a factor of 3 for C=C excitation! These results indicate that *overall molecular rotation plays an important role in the reaction dynamics of trans-stilbene.*

The mechanisms for the interaction of internal rotations with molecular rotation has been discussed for a model of HOOH [35]. The internal rotation about the O—O in HOOH is similar in many ways to the internal rotation

TABLE 3 Lifetimes for Unimolecular Reactions of *cis* and *trans*-Stilbene for Different Types of Excitation

Mode	Excitation Energy	τ_{isom}	τ_{Ph-C}	$\tau_{C-Ph'}$	T_{rot}
		Trans			
Ph—C	80	20.80	10.20	10.60	0
Ph—C	80	10.67	16.57	16.51	300
C=C	120	42.32	13.67	12.87	0
C=C	120	13.80	15.70	13.24	300
		Cis			
Ph—C	80	13.74	8.86	16.43	0
Ph—C	80	16.26	7.63	18.11	300
C=C	120	7.75	14.89	15.81	0
C=C	120	9.10	18.07	14.49	300

[a] Total energy = 125 kcal/mol. Energies are given in kcal/mol, lifetimes in picoseconds, and rotational temperature in kelvin.

about the ethylenic C=C bond in stilbene. In particular, both internal rotations are about a bond containing two identical atoms. For HOOH it was shown that there can be significant interactions through low-order nonlinear resonances which lead to dramatic changes in the internal dynamics. The vibrational phase space was found to become increasingly chaotic as the total angular momentum was increased, leading to a significant enhancement in the rate of trans to cis torsional motion. For *trans*-stilbene it is apparent that a similar mechanism could be responsible for the observed rotational effects; however, since the rotational energies considered here are very small, it is more reasonable to believe that the effects are due to rotationally enhanced vibrational coupling (Coriolis).

For the cis conformation, the lifetimes of isomerization are noticeably shorter than for the trans conformation. This suggests that the mechanism for energy flow from the initial excitation(s) into the reaction coordinate has changed, leading to a more facile pathway for energy flow. Kinetic or resonant couplings are most commonly associated to such changes [27]. Since there are no low-order nonlinear resonances at the energies studied, kinetic coupling of the modes is a probable cause for the difference of the two conformational isomers. Examination of the elements of the G^{-1} matrix shows that initially the Ph—C bond is strongly coupled to the Ph—C=C bends for the cis conformation, while it is strongly coupled to the unexcited C—Ph' bond for the trans. Thus the initial flow of energy out of an excited Ph—C bond should be into the Ph—C=C bends for the cis conformation, which in turn must lead to a more direct pathway for energy flow into the reaction coordinate.

The lifetimes for isomerization of the cis conformation are not as strongly influenced by overall molecular rotation. This is probably due to a significant

TABLE 4 Elements of the Inverse G-matrix (G_{ij}^{-1}) for the Equilibrium Geometry of the Models for *cis* and *trans*-Stilbene

	r(Ph—C)	r(C=C)	r(C—Ph′)	θ(Ph—C=C)	θ(C=C—Ph′)	ω
			Trans			
r(Ph—C)	41.36361	29.70718	30.24139	14.39451	14.39451	0.00000
r(C=C)	29.70717	31.63931	29.70718	12.86367	12.86367	0.00000
r(C—Ph′)	30.24139	29.70718	41.36361	14.39451	14.39451	0.00000
θ(Ph—C=C)	14.39451	12.86367	14.39451	19.69554	−4.66576	0.00000
θ(C=C—Ph′)	14.39451	12.86367	14.39451	−4.66576	19.69554	0.00000
ω	0.00000	0.00000	0.00000	0.00000	0.00000	7.56338
			Cis			
r(Ph—C)	25.95674	23.70401	10.13793	25.96701	21.33429	0.00000
r(C=C)	23.70401	45.00000	23.70401	44.89870	44.89870	0.00000
r(C—Ph′)	10.13793	23.70401	25.95674	21.33429	25.96701	0.00000
θ(Ph—C=C)	25.96701	44.89870	21.33429	59.26303	54.69332	0.00000
θ(C=C—Ph′)	21.33429	44.89870	25.96701	54.69332	59.26303	0.00000
ω	0.00000	0.00000	0.00000	0.00000	0.00000	1.52764

reduction in the centrifigual and/or Coriolis coupling in the cis conformation as compared to the trans conformation. A second and more noticeable difference for the cis conformation is that the lifetime for isomerization for Ph—C excitation is longer than for C=C excitation. This is opposite to that observed for trans. There is also an apparent nonstatistical redistribution of energy between the two C—Ph bonds. The initially excited bond has a lifetime that is a factor of 2 shorter than the unexcited bond. From examining the instantaneous G^{-1} matrix for the trans and cis conformations (see Table 4) it can be seen that the kinetic coupling between the two Ph—C bonds is much larger for the trans conformation. Thus energy redistribution is promoted through the kinetic coupling in the trans conformation and is delayed in the cis conformation. This delayed Ph—C energy flow is also seen in the energy transfer plots discussed below (see Figs. 4 and 5).

The lifetimes for the various reactions of stilbene given in Table 3 demonstrate a number of interesting points. First, the lifetimes for non-rotating stilbene are shorter for the cis conformation. This is attributed to a more facile pathway for energy flow into the reaction coordinate. Second, the effects of overall molecular rotation are very large for the trans conformation but not for the cis conformation. The lifetime for isomerization for the trans conformation is decreased while the lifetime for bond fissions is increased. The opposite is observed for the cis conformation (this is probably due to a centrifugal barrier, which increases the threshold energy to isomerization). Third, there is an apparent nonstatistical energy distribution for Ph—C excitation in the cis conformation. Fourth, the lifetime of isomerization for the trans conformation without overall molecular rotation is a factor of 2 longer for C=C excitation than C—Ph excitation. Fourth, C=C excitation leads to a faster isomerization rate for the cis conformation (this is, of course, related to the nonstatistical redistribution of the Ph—C excitation).

Intramolecular energy flow. Figures 4 and 5 are plots of the six internal mode energies for the trans (Fig. 4) and cis (Fig. 5) conformations, respectively. These plots are for single trajectories which lead to isomerization. Each frame of an individual plot is for a particular vibrational mode and is appropriately labeled on the y-axis: Ph_1—C stretch, Ph_1—C=C bend, C=C stretch, Ph_2—C=C bend, Ph_2—C stretch, and Ph_1—C=C—Ph_2 torsion, where Ph_1 and Ph_2 are the two phenyl groups. Figure 4a is a plot of the energy in the six internal modes as a function of time for the model of *trans*-stilbene with an initial excitation of the Ph—C bond and no overall molecular rotation. The energy flow out of the excited bond is very rapid (<0.1 ps) and flows predominately into the unexcited C—Ph bond (labeled Ph_2—C). Initially there is some reversibility to the energy flow. After approximately 0.2 ps, the energy accumulates in the C=C bond and remains localized in this bond for approximately 1 ps. Following this period of time, the energy in the C=C appears to be redistributed among all the vibrational modes. Prior to isomerization, there is a "pulse" of energy from the two Ph—C=C bending modes into the torsion. The energy pulse causes an isomerization to occur.

Addition of overall molecular rotation (Fig. 4b) changes the qualitative nature of the energy flow. The initial excitation energy is partially transferred into the C=C bond, followed by significant flow into the unexcited Ph—C bond. The energy beats back and forth between the two Ph—C bonds, with little participation of the C=C bond. The two Ph—C=C bending modes slowly become more active, apparently receiving the energy that was initially transferred to the C=C bond. Similar to the nonrotating case, the energy responsible for causing isomerization comes from the bending modes.

For initial excitation of the C=C bond (Fig. 4c), the energy appears to flow into all modes. The rate at which the energy flows out of the C=C bond is slower than that for the Ph—C bond (cf. Fig. 4a and c). This is also clearly shown in Table 3, where the lifetime for C=C excitation is a factor of 2 longer than for Ph—C excitation. The flow of energy out of the C=C bond exhibits some reversibility for the model of nonrotating *trans*-stilbene (Fig. 4c), but there is essentially no reversibility in the rotating case (Fig. 4d). Molecular rotation significantly reduces the reversibility of the energy flow out of the C=C bond. This increases the rate at which the energy accumulates in the bending modes, which are strongly coupled to the torsion and the primary source for isomerization. Thus, as was shown in Table 3, the lifetime of isomerization of *trans*-stilbene is much faster for the rotating case.

The energy flow out of an initially excited Ph—C bond for the cis conformation is significantly different from that for the trans conformation. This can be seen by comparing Figs. 4a and 5a. For the cis conformation (Fig. 5a), the energy in the excited bond remains for at least three times as

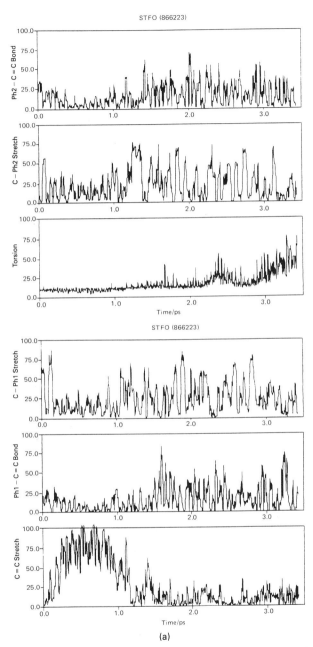

Figure 4 Time evolution of mode energies for the modal of *trans*-stilbene for single isomerizing trajectories. Initial conditions correspond to: (a) excitation of the Ph—C stretch, no overall molecular rotation; (b) same as part (a) except for an overall molecular rotational energy of $\frac{3}{2}RT$ ($T = 300$ K); (c) excitation of the C=C stretch, no overall molecular rotation; (d) same as part (c) except for an overall molecular rotational energy of $\frac{3}{2}RT$ ($T = 300$ K).

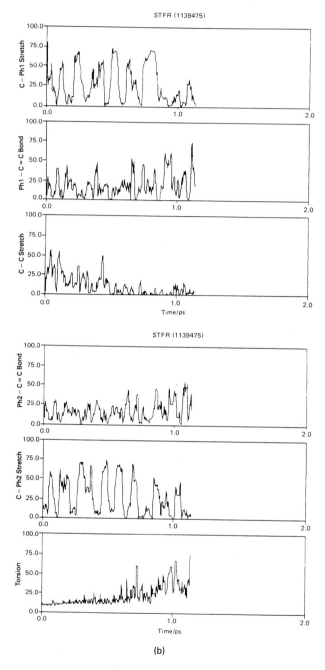

(b)

Figure 4 (continued)

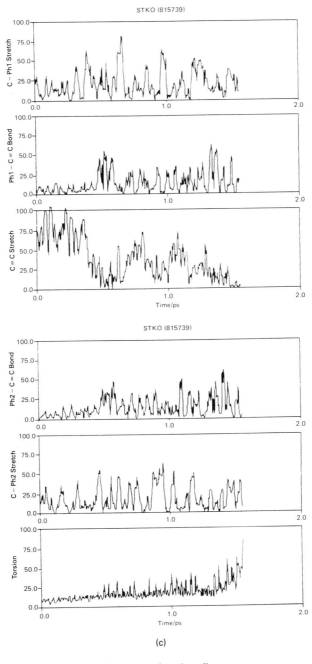

(c)

Figure 4 (continued)

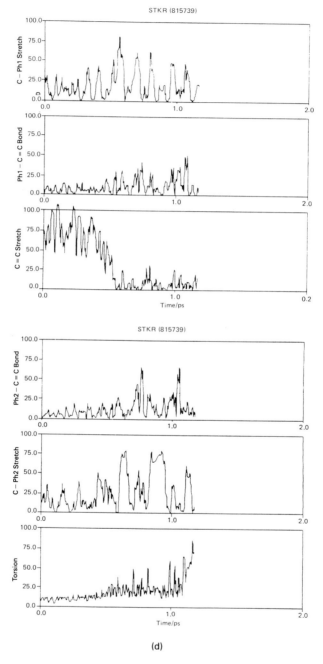

(d)

Figure 4   (continued)

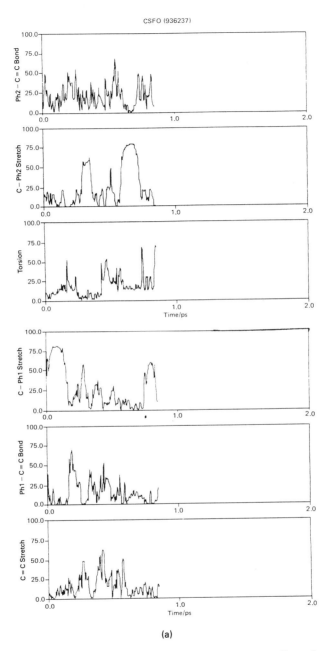

CSFO (936237)

(a)

Figure 5 Time evolution of mode energies for the model of *cis*-stilbene for single isomerizing trajectories. Initial conditions correspond to: (a) excitation of the Ph—C stretch, no overall molecular rotation; (b) same as part (a) except for an overall molecular rotational energy of $\frac{3}{2}RT$ ($T = 300$ K); (c) excitation of the C=C stretch, no overall molecular rotation; (d) same as part (c) except for an overall molecular rotational energy of $\frac{3}{2}RT$ ($T = 300$ K).

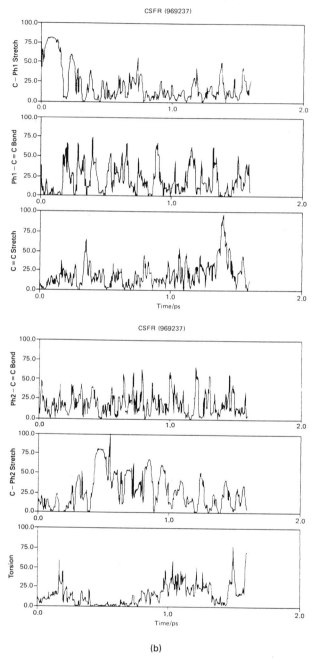

(b)

Figure 5   (continued)

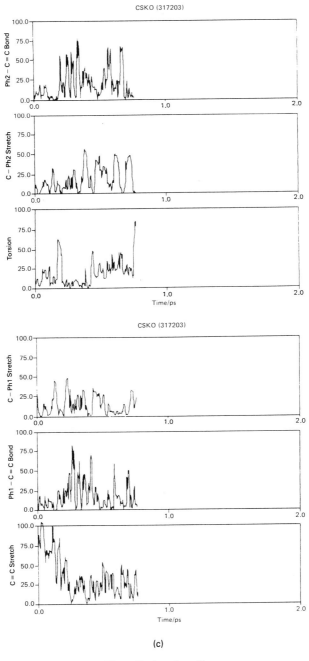

(c)

Figure 5 (continued)

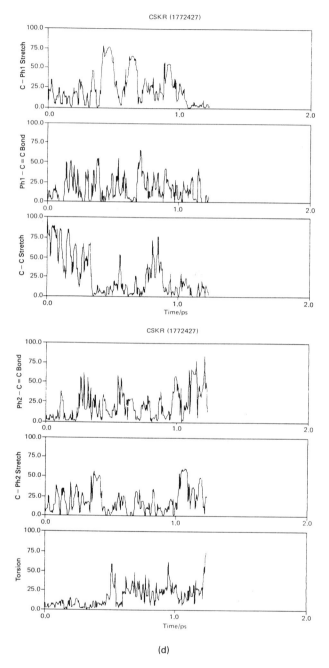

(d)

Figure 5   (continued)

long as for trans conformation. This is due to the reduction of the kinetic coupling between the two Ph—C bonds in the cis conformation (see Table 4). For cis, the initial energy flows predominately into the adjacent Ph—C=C bending mode. In contrast to the trans conformation, the bending modes do not appear to be responsible for the promotion of isomerization. The energy comes predominantly from the two Ph—C bonds. Addition of molecular rotational energy (Fig. 5b) does not appear to change the energy transfer pathways significantly; however, it causes an increase in the lifetime for isomerization (see Table 3).

The energy flow out of the excited C=C bond for the model of *cis*-stilbene is shown in Fig. 5c and d. For the nonrotating case, the majority of energy flows from the C=C bond into the Ph—C=C bending modes. Some of the energy in the bending modes flows into the Ph—C bonds. It is the energy in the Ph—C bonds that causes isomerization to occur. Addition of molecular rotation (Fig. 5d) leads to subtle changes in the energy-flow pathways.

The results shown in Table 3 for the unimolecular reaction lifetimes of the models for *trans*- and *cis*-stilbene are consistent with the qualitative intramolecular energy pathways obtained from the internal-mode energy plots. Many unimolecular isomerization reactions appear to have similar dynamics prior to reaction [36]. The energy flow into the reaction coordinate shown in Figs. 4 and 5 is similar to that recently described as "impulsive energy flow" [36].

In this chapter we have reported the details of a preliminary study of the intramolecular dynamics for a model of stilbene. Although our model does not treat all of the vibrational degrees of freedom of stilbene, it is a reasonable first step to developing a fundamental understanding of the isomerization reaction and bond fissions.

ACKNOWLEDGMENTS

This work was supported in part by the Education Ministry of Spain and by the Petroleum Research Fund of the American Chemical Society (JAD), Robert A. Welch Foundation of Houston, Texas (JAD), and the Organized Research Grant of Tarleton State University (JAD). Computations reported here were performed on the Cornell National Supercomputer Facility, which is supported by the NSF and IBM Corporation, the IBM 3090 at the University of Tennessee (UTCC), and the Cray X-MP/48 supported by the National Center for Supercomputing Applications (NCSA) Grant TRA890046N, projects YPO and XCK, at the University of Illinois at Urbana–Champaign. The majority of this work was completed while Coral Getino was a Visiting Scientist at Oak Ridge National Laboratory.

REFERENCES

1. Walatka, Labes, and Perlstein (1973); C. M. Mikulski, P. J. Russo, M. S. Saran, A. F. Garito, and A. J. Heeger, *J. Am. Chem. Soc.* **97**, 6358 (1975); A. G. MacDiarmid, C. M. Mikulski, M. S. Saran, P. J. Russo, M. J. Cohen, A. A. Bright, A. F. Garito, and A. J. Heeger, in *Inorganic Compounds with Unusual Properties,* ed. R. B. King, Advances in Chemistry Series, No. 150, American Chemical Society, Washington, D.C., 1976; R. L. Greene, G. B. Street, and L. J. Suter, *Phys. Rev. Lett.* **34**, 577 (1975); H. Shirakawa, E. J. Louis, A. J. Mac-Diarmid, C. K. Chiang, and A. J. Heeger, *J. Chem. Soc., Chem. Commun.* 578 (1977); C. K. Chiang, C. R. Fincher, Jr., Y. W. Park, A. J. Heeger, H. Shirakawa, E. J. Louis, S. C. Gau, and A. G. MacDiarmid, *Phys. Rev. Lett.,* **39**, 1098 (1977).

2. For example, see the *Handbook of Conducting Polymers,* Vols. 1 and 2, ed. T. A. Skotheim, Marcel Dekker, New York, 1986, and references cited therein.

3. S. Tokito, T. Tsutsui, S. Saito, and R. Tanaka, *Polym. Commun.* **27**, 333, (1986).

4. H. H. Horhold and J. Opfermann, *J. Makromol. Chem.* **130**, 105 (1970).

5. S. Stinson, *Chem. Eng. News,* Oct. 17, 1988, p. 37.

6. G. Drefahl, R. Kuhmstedt, H. Oswald, and H. Horhold, *J. Macromol. Chem.* **131**, 89 (1970).

7. J. M. Robertson and I. Woodward, *Proc. R. Soc. London* **A162**, 568 (1937); C. J. Finder, M. G. Newton, and N. L. Allinger, *Acta Crystallogr.* **B30**, 411 (1974); J. Bernstein, *Acta Crystallogr.* **B31**, 1268 (1975); A. Bylima and Z. R. Garbowski, *Trans. Faraday Soc.* **65**, 458 (1969); H. Stegemeyer, *J. Phys. Chem.* **66**, 458 (1969); D. F. Evans, *J. Chem. Soc.* 1351 (1957); R. H. Dyck and D. S. McClure, *J. Chem. Phys.* **36**, 2326 (1962); S. Malkin and E. Fischer, *J. Phys. Chem.* **68**, 1153 (1964); G. S. Hammond, J. Saltiel, A. A. Lamola, J. J. Turro, J. S. Bradshaw, D. O. Cowan, R. C. Counsell, V. Vogt, and C. Dalton, *J. Am. Chem. Soc.* **86**, 3197 (1964); W. G. Herckstroeter and G. S. Hammond, *J. Am. Chem. Soc.* **88**, 4769 (1966); G. Heinrich, H. Blume, and D. Schulte-Frohline, *Tetrahedron Lett.* 4693 (1967); W. G. Herckstroeter and D. S. McClure, *J. Am. Chem. Soc.* **90**, 4522 (1968); J. Saltiel, *J. Am. Chem. Soc.* **90**, 6394 (1968); J. Saltiel, J. T. D'Agostino, E. D. Megarity, L. Metts, K. R. Neuberger, M. Wrighton, and O. C. Zafiriou, *Org. Photochem.* **3**, 1 (1971); F. Dainton, E. A. Robinson, and G. A. Saimon, *J. Phys. Chem.* **76**, 3897 (1972); G. Heinrich, H. Gusten, F. Mark, G. Olbrich, and D. Schulte-Frohlinde, *Ber. Bunsenges. Phys. Chem.* **77**, 103 (1973); J. Saltiel, D. N. L. Chang, E. D. Megarity, A. D. Rousseau, P. T. Shannon, B. Thomas, and A. K. Uriarte, *Pure Appl. Chem.* **41**, 559 (1975); M. Sumitani, K. Yoshihara, and S. Nagakura, *Bull. Chem. Soc. Jpn.* **51**, 2503 (1978); T. M. Stachelek, T. A. Pazoha, W. M. McClain, and R. P. Drucker, *J. Chem. Phys.* **66**, 4540 (1977); R. Benson and D. F. Willams, *J. Phys. Chem.* **81**, 215 (1977); J. B. Birks, *Chem. Phys. Lett.* **38**, 437 (1976); D. J. S. Birch and J. B. Birks, *Chem. Phys. Lett.* **38**, 432 (1976); J. B. Birks, *Chem. Phys. Lett.* **54**, 430 (1978); M. Edelson and A. Bree, *Chem. Phys. Lett.* **41**, 562 (1976); E. Heumann, W. Triebel, and B. Wilheimi, *Chem. Phys. Lett.* **32**, 589 (1975); E. Heumann, W. Triebei, R. Uhimann, and B. Wilheimi, *Chem. Phys. Lett.* **45**, 425 (1977).

8. M. Traetteberg and E. B. Frantsen, *J. Mol. Struct.* **26**, 69 (1975); M. Traetteberg and E. B. Frantsen, *J. Mol. Struct.* **26**, 169 (1975).

9. G. Fischer, G. Seger, K. A. Muszkat, and E. Fischer, *J. Chem. Soc., Perkin Trans.*

2, 1569 (1975); V. Santoro, E. J. Barrett, and H. H. Hoger, *J. Am. Chem. Soc.* **89,** 4545 (1967).

10. (a) L. H. Spangler, *J. Phys. Chem.* **91,** 6077 (1987); (b) J. Saltiel, S. Ganapathy, and C. Werking, *J. Phys. Chem.* **91,** 2755 (1987); (c) S. K. Kim and Graham R. Fleming, *J. Phys. Chem.* **92,** 2168 (1988); E. L. Eliel and E. Brunet, *J. Org. Chem.* **51,** 1902 (1986); M. Sumitani and K. Yoshihara, *J. Chem. Phys.* **76,** 738 (1982); B. I. Green, R. M. Hochstraser, and R. B. Weisman, *Chem. Phys.* **48,** 289 (1980); T. Urano, M. Maegawa, K. Yamanouchi, and S. Tsuchiya, *J. Phys. Chem.* **93,** 3459 (1989); J. M. Hicks, M. T. Vandersall, E. V. Sitzmann, and K. B. Eisenthal, *Chem. Phys. Lett.* **135,** 413 (1987); S. K. Kim, S. H. Courtney, and G. R. Fleming, *Chem. Phys. Lett.* **159,** 543 (1989); A. Kruppa, O. I. Mikhailovskaya, and T. V. Leshina, *Chem. Phys. Lett.* **147,** 65 (1988); N. S. Park, N. Sivakumar, E. A. Hoburg, and D. H. Waldeck, *Ultrafast Phenomena VI,* Springer Series in Chem. Phys., Vol. 48, (1988), p. 551; M. Lee and R. M. Hochstrasser, *Ultrafast Phenomena VI,* Springer Series in Chem. Phys., Vol. 48, 1988, p. 344; B. I. Greene, R. M. Hochstrasser, and R. B. Weisman, *J. Chem. Phys.* **71,** 544 (1979); *Chem. Phys.* **48,** 289 (1980); B. I. Greene and R. C. Farrow, *J. Chem. Phys.* **78,** 3336 (1983).

11. J. W. Perry, N. F. Scherer, and A. H. Zewail, *Chem. Phys. Lett.* **103,** 1 (1983); N. F. Scherer, J. P. Shepanski, and A. H. Zewail, *Chem. Phys. Lett.* **81,** 2181 (1984); J. A. Syage, P. M. Felker, and A. H. Zewail, *J. Chem. Phys.* **81,** 4685 (1984); J. A. Syage, P. M. Felker, and A. H. Zewail, *J. Chem. Phys.* **81,** 4685 (1984); J. A. Syage, P. M. Felker, and A. H. Zewail, *J. Chem. Phys.* **81,** 4706 (1984); P. M. Felker and A. H. Zewail, *J. Phys. Chem.* **89,** 5402 (1985); L. R. Khundkar, R. A. Marcus, and A. H. Zewail, *J. Phys. Chem.* **87,** 2473 (1983).

12. D. Bahatt, U. Even, and J. Jortner, *Chem. Phys. Lett.* **117,** 527 (1985).

13. A. Amirav and J. Jortner, *Chem. Phys. Lett.* **95,** 295 (1983); T. J. Majors, U. Even, and J. Jortner, *J. Chem. Phys.* **81,** 2330 (1984); T. S. Zwier, B. Carrasquillo, and D. H. Levy, *J. Chem. Phys.* **78,** 5493 (1983).

14. A. Amirav and J. Jortner, *Chem. Phys. Lett.* **95,** 295 (1983).

15. J. Troe, *Chem. Phys. Lett.* **114,** 241 (1985); J. Schroeder and J. Troe, *J. Phys. Chem.* **90,** 4215 (1986).

16. F. Negri, G. Orlandi, and F. Zerbetto, *J. Phys. Chem.* **93,** 5124 (1989).

17. K. Palmo, *Spectrochim. Acta* **A44,** 341 (1988).

18. J. Syage, W. R. Lambert, P. M. Felker, A. H. Zewail, and R. Hochstrasser, *Chem. Phys. Lett.* **88,** 268 (1982).

19. F. J.Adrian, *J. Chem. Phys.* **28,** 608 (1958); D. L. Beveridge and H. H. Jaffé, *J. Am. Chem. Soc.* **87,** 5340 (1965); P. Borrell and H. H. Greenwood, *Proc. R. Soc. London,* **A298,** 453 (1967); D. H. Lo and M. A. Whitehead, *Can. J. Chem.* **46,** 2041 (1968); L. Pedersen, D. G. Whitten, and M. T. McCall, *Chem. Phys. Lett.* **3,** 569 (1969); S. Ljunggren and G. Wettermark, *Theor. Chim. Acta* **19,** 326 (1968); J. V. Knop and L. Knop, *Z. Phys. Chem.* **71,** 9 (1970); C.-H. Ting and D. S. McClure, *J. Chin. Chem. Soc. (Peking)* **18,** 95 (1971); F. F. Momicchioli, M. C. Brun, I. Baraldi, and G. R. Corradini, *J. Chem. Soc., Faraday Trans. 2,* **70,** 1325 (1974); F. F. Momicchioli, G. R. Corradini, M. C. Bruni, and I. Baraldi, *J. Chem. Soc., Faraday Trans. 2,* **71,** 215 (1975); A. Warshel, *J. Chem. Phys.* **62,** 214 (1975); T. Bally, E. Haselbach, S. Lanyiova, F. Marchner, and M. Rossi, *Helv. Chim. Acta* **59,** 486 (1976); H.-J. Hofmann and P. Birner, *J. Mol. Struct.* **39,** 145 (1977); P. Tavan and K. Schulten, *Chem. Phys. Lett.* **58,** 200 (1978); A. Wolf, H.-H.

Schmidtke, and J. V. Knop, *Theor. Chim. Acta* **48**, 37 (1978); K. L. Yip, N. O. Lipare, C. B. Duke, B. S. Hudson, and J. Diamond, *J. Chem. Phys.* **64**, 4020 (1976); M. Hochstrasser, *Chem. Phys. Lett.* **88**, 266 (1982); G. Olbrich, *Ber. Bunsenges. Phys. Chem.* **86**, 209 (1982); G. Olbrich and W. Siebrand, *Chem. Phys. Lett.* **30**, 352 (1975).

20. G. Orlandi, P. Palmieri, and G. Poggi, *J. Am. Chem. Soc.* **101**, 3492 (1979).

21. G. Orlandi and G. C. Marconi, *Chem. Phys.* **8**, 458 (1975); G. Orlandi, P. Palmiere, and G. Poggi, *J. Am. Chem. Soc.* **101**, 3492 (1979); G. Orlandi and F. Zerbetto, *J. Mol. Struct.* **138**, 185 (1986).

22. A. R. Gregory and D. F. Willams, *J. Phys. Chem.* **85**, 2652 (1979), and references cited therein.

23. J. Troe and K.-M. Weitzel, *J. Chem. Phys.* **88**, 7030 (1988).

24. J. A. Darsey and B. G. Sumpter, *Polym. Prepr., Am. Chem. Soc., Div. Polym. Chem.* **30**, 5 (1989).

25. M. J. Frisch, J. S. Binkley, H. B. Schlegel, K. Raghavachari, C. F. Melius, R. L. Martin, J. J. P. Stewart, F. W. Bobrowicz, C. M. Rohlfing, L. R. Kahn, D. J. Defrees, R. Seegar, R. A. Whiteside, D. J. Fox, E. M. Fluder, S. Topeol, and J. A. Pople. This work (GAUSSIAN 86 system of programs) is based on the GAUSSIAN 82 system, which is copyright (©) 1983 by Carnegie–Mellon University.

26. C. Getino, B. G. Sumpter, J. Santamaria, and G. S. Ezra, *J. Phys. Chem.* (submitted).

27. C. Getino, B. G. Sumpter, J. Santamaria, and G. S. Ezra, *J. Phys. Chem.* **93**, 3877 (1989).

28. R. Duchovic, W. L. Hase, and H. B. Schlegel, *J. Chem. Phys.* **88**, 1339 (1984).

29. C. S. Sloane and W. L. Hase, *J. Chem. Phys.* **66**, 1523 (1977); G. C. Schatz, in *Molecular Collision Dynamics*, ed. J. M. Bowman, Springer-Verlag, Berlin, 1983; B. G. Sumpter and G. S. Ezra, *Chem. Phys. Lett.* **142**, 142 (1989); B. G. Sumpter and G. S. Ezra, *J. Chem. Phys.* (submitted).

30. E. B. Wilson Jr., J. C. Decius, and P. C. Cross, *Molecular Vibrations*, Dover, New York, 1979.

31. P. J. Robinson and K. A. Holbrook, *Unimolecular Reactions*, Wiley, New York, 1972; W. Forst, *Theory of Unimolecular Reactions*, Academic Press, New York, 1973.

32. B. G. Sumpter and D. L. Thompson, *J. Chem. Phys.* **87**, 5809 (27)

33. B. C. Laskowski, D. Y. Yoon, D. McLean, and R. L. Jaffe, *Macromolecules* **21**, 1629 (1988); R. L. Jaffe, *Poly. Prepr., Am. Chem. Soc., Div. Polym. Chem.* **30**, 1 (1989).

34. J. A. Darsey, J. F. Kuehler, N. R. Kestner, and B. K. Rao, *J. Macromol. Sci. Chem.* **A25**, 159 (1988); J. A. Darsey, J. F. Kuehler, and B. K. Rao, *J. Macromol. Sci. Chem.* (in press); B. K. Rao and J. A. Darsey, *Phys. Rev.* **B31**, 1187 (1985).

35. B. G. Sumpter, C. C. Martens, and G. S. Ezra, *J. Phys. Chem.* **92**, 7193 (1988).

36. R. P. Muller, J. S. Hutchinson, and T. A. Holme, *J. Chem. Phys.* **90**, 4582 (1989).

3

Conformational Analysis of Diphenyl Linkages

A. ANWER, R. LOVELL, and A. H. WINDLE
Department of Materials Science and Metallurgy
University of Cambridge
Pembroke Street
Cambridge CB2 3QZ
United Kingdom

ABSTRACT

A comparative study, using molecular mechanics, is made for seven molecules of the form Ph—X—Ph, where the link X = —S—, —O—, —CO—, —SO₂—, —CH₂—, —C(CH₃)₂— or —C(CF₃)₂—. They are shown to have two main types of minimum energy conformation, albeit with a wide range of flexibilities dependent on the level of conjugation and steric interaction between linking groups and phenyls. The persistence lengths of chains of these linkages are determined and shown to depend mainly on the bond angle of the link.

INTRODUCTION

High-performance polymers are often based on *para*-phenylene groups linked via small groups. For example, polyetheretherketone (PEEK) contains —O— and —CO— linkages. Other common linking groups are —S— and —SO₂—. Diphenyls are also common in components of epoxy systems such as diglycidyl ether of bisphenol-A (DGEBA), diaminodiphenylmethane (DDM), and diaminodiphenylsulfone (DDS), which contain linking groups —C(CH₃)₂—, —CH₂—, and —SO₂—, respectively. Analogs of DGEBA containing —CH₂—, —SO₂—, and —C(CF₃)₂— have also been prepared [1]. In this chapter we report a comparative study using molecular mechanics on the conformational characteristics of seven such linkages. The main emphasis of the work is on the minimum energy conformations and on the chain flexibility.

As discussed by de Gennes [2], there are two components of flexibility: static and dynamic. Static (or equilibrium) flexibility depends on the number, position, and extent of the low-energy regions in conformational space. For polymers in noncrystalline states (amorphous solid, melt, or θ-solution), a high static flexibility leads to increased conformational disorder and reduced persistence length and radius of gyration. In contrast, dynamic (or kinetic) flexibility increases as the height of the lowest barriers between minima falls. Dynamic flexibility determines the ease with which conformational transitions such as the glass transition and deformation can take place.

RELEVANT CONFORMATIONAL ANALYSES

The majority of published work has been on the diphenyl propane linkage which occurs, for example, in polycarbonate and DGEBA, although other linkages have also been investigated. The principal variables (Fig. 1) are the

PLANAR

BUTTERFLY

TWIST

MORINO

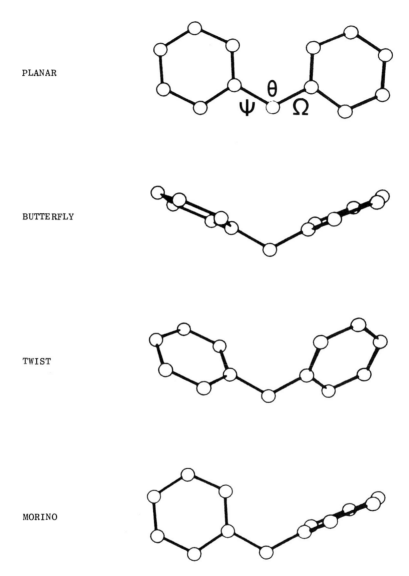

Figure 1   Conformations of diphenyl linkages.

bond angle at the linking group (θ) and the rotation angles of the phenyls (Ψ, Ω). Minimum energy conformations are generally near one of the symmetrical conformations ("butterfly" or "twist") [3] shown in Fig. 1. The "planar" conformation ($0°, 0°$) is not accessible for any of the groups considered here. The Morino [4] conformation ($0°, 90°$) is usually at a saddle point between two equivalent minima.

2,2-Diphenyl Propane (DPP) (Ph—C(CH$_3$)$_2$—Ph)

There have been numerous reported studies of the conformation of DPP, which are summarized in Table 1. Although some of these show the Morino conformation as having the lowest energy, the most reliable of recent reports favor a twist conformation ($\Psi = \Omega \approx 50°$) or near twist (with $\Psi \approx \Omega$). This is in agreement with conformations found in crystalline materials [17].

In general, conformational energy maps of DPP (e.g., Fig. 8) show a broad minimum energy region which lies roughly along the line $\Psi + \Omega = 90°$, with a barrier (at $0°, 90°$) between minima of about 1.5 kcal/mol. Thus there are low-energy "corridors" along which conformational changes can occur by coordinated rotation of the phenyls [7,9,18].

Other Linkages

Studies on other linkages are summarized in Table 2. Most have been on ether or sulfide linkages, which again adopt twist conformations ($\Psi = \Omega \approx 40°$). The most comprehensive study was by Welsh et al. [20], who made calculations for all the linkages in Table 2 together with —C(CF$_3$)$_2$—. The only maps shown in this work were for the —S— and —O— linkages, although it is claimed that all the other links [including —C(CF$_3$)$_2$—] have flexibility equal to or greater than —S— and —O—.

TABLE 1   Earlier Conformational Analysis of 2,2-Diphenyl Propane

	Ref.	θ	Ψ	Ω
Morino				
Tonelli	5	112	10	90
Tekely and Turska	6	—	0	90
Jones	7	111	0	90
Schaefer et al.	8	—	0	90
Sundararajan	9	112	10	80
Symmetric twist				
Williams and Flory	10	112	30	30
Banks and Ellis	11	109.5	40	40
Erman et al.	12	109.5	46	46
Jones	7	120	27	27
Bicerano and Clark	13	109.8	48	48
Laskowski et al.	14	109.4 to 109.9	51	51
Asymmetric twist				
Sundararajan	9	109.5	25	75
			45	75
Sundararajan	15	109.5	40	70
Perez and Scaringe	16	109.5	20	60
Present work		108.0	52	52
Average for crystals	17	108.9	50	60

TABLE 2   Earlier Conformational Analysis for Other Diphenyls

Authors	Ref.	—S— θ	Ψ	Ω	—O— θ	Ψ	Ω	—CO— θ	Ψ	Ω	—SO₂— θ	Ψ	Ω	—CH₂— θ	Ψ	Ω
Galasso et al.	4	109.5	38	38	123	52	52									
Zubkov et al.	19		35	35		45	45	126	30	30				111	90	90
Welsh et al.	20	109	43	43	123	43	43	126.5			100			111		
Jones et al.	7, 21	110	35	35	121	40	40				104.3	45	45			
												90	90			
Barnes et al.	22													110.9	90	90
Boon and Magré	23				124	40	40									
Hay and Kemmish	24				120	65	65	117.3	90	90						
Sundararajan	25														90	90
Jaffe	31		58	58		43	43					90	90			
Present work		105.5	40	40	120.3	38	38	120.9	33	33	104.8	90	90	109.9	90	90
Average for crystals	17	105.4	12	69	120.3	24	55	120.9	30	30	104.8	78	86	114.4	58	58

MODELING METHODS

The symmetry of the phenyl groups means that the linkage is unaltered if either of the phenyls is turned through ±180°. Thus the energy calculations need only be performed for Ψ and Ω in the range 0 to 180°. Energy maps were produced using the MM2 molecular mechanics routines (version 6.0 from Molecular Design Ltd, San Leandro, California) to calculate energy as a function of the two torsional angles Ψ and Ω (in 10° steps). For each point on the map, the rest of the structure was minimized. The parameters used in the calculations were optimized by comparison with crystal structures [17]. Electrostatic interactions were not included.

The symmetry of the phenyls and of the bridging groups leads to diagonal mirror lines on the maps. Conformations lying on the SW-NE diagonal (Ψ = Ω) have twofold (C₂) symmetry axes (twist conformations) and energy minima occur in pairs at (Ψ, Ψ) and (180° − Ψ, 180° − Ψ), corresponding to mirror-image conformations, whereas conformations lying on the NW-SE diagonal (Ψ = 180° − Ω) have mirror symmetry. The intersection of the two diagonals (90°, 90°) has both and is the butterfly conformation.

MODELING RESULTS

—S— and —O— Linkages

The conformational energy maps for these linkages are shown in Figs. 2 and 3. They are very similar, with distinct minima at (40°, 40°) and (38°, 38°), respectively. These are twist conformations, which agree with previous work [4,7,19–21,23,24] and with conformations found in polymer crystals [23,26].

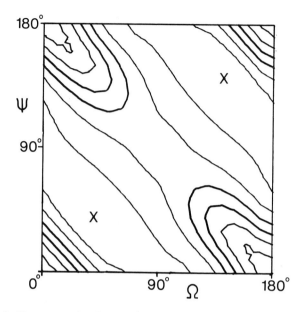

Figure 2   Energy map for the —S— linkage, showing minima near (40°, 40°) and (140°, 140°). Contours are at 1-kcal/mol intervals above the minima.

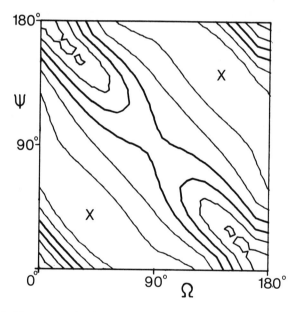

Figure 3   Energy map for the —O— linkage, showing minima near (38°, 38°) and 142°, 142°).

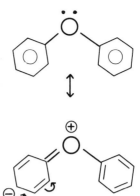

Figure 4 Conjugation in diphenyl ether is enhanced for the coplanar conformation, which gives the maximum overlap between the π lone pair of the oxygen and the π-electrons of the phenyls.

However, crystals of small-molecule analogs tend to show asymmetric conformations. For example, Garbarczyk [27] found that a model compound for poly(p-phenylene sulfide) had a conformation with torsional angles (4°, 62°).

The prediction of a twist conformation for diphenyl sulfide and diphenyl ether can be interpreted as being due to electron delocalization where the lone electron pairs of the sulfur or oxygen tend to stabilize the planar conformation (0°, 0°) by a conjugative effect [4,21], as shown in Fig. 4. The planar conformation is excluded because of overlap of hydrogens, so that the minimum near (40°, 40°) represents a balance between opposing forces. Although neither of the low-energy regions (<3 kcal/mol) of the two linkages are extensive in conformational space, that for —O— is slightly less localized, indicating a larger degree of conformational freedom for the diphenyl ether than for the diphenyl sulfide linkage.

—CO— Linkage

This linkage has a minimum energy conformation near (33°, 33°) (Fig. 5). The low-energy regions are smaller than for —S— or —O—. Crystal data [17] and earlier work [19] also favor a minimum near (30°, 30°). Again the shape of the map and the positions of minima are determined by a balance between conjugation, which is maximized in the planar conformation, and steric effects, which prohibit the planar conformation.

—SO₂— and —CH₂— Linkages

Figure 6 shows the energy map for the —SO₂— linkage. Although this has only a single minimum at (90°, 90°) corresponding to the butterfly conformation, there is a large region of conformational space within the 3-kcal/mol contour, indicating that rotation of the phenyls is effectively unhindered from

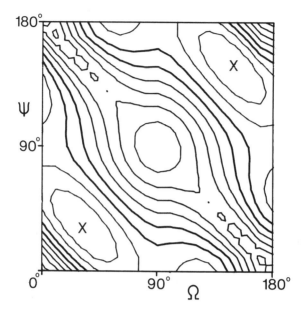

Figure 5 Energy map for the —CO— linkage, showing minima near (33°, 33°) and (147°, 147°).

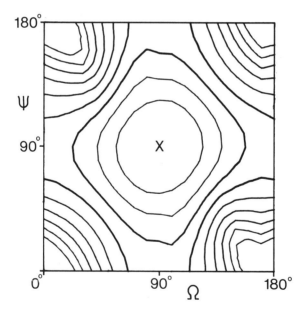

Figure 6 Energy map for the —SO$_2$— linkage, showing a minimum at (90°, 90°).

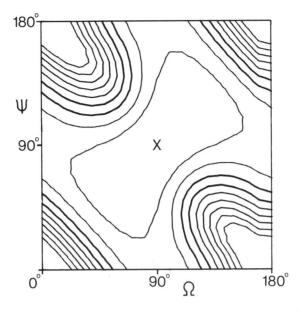

Figure 7   Energy map for the —CH₂— linkage, showing a minimum at (90°, 90°).

45 to 135°. Most crystal data [17] show both torsional angles between 80 and 90°. In some cases, one angle is less than this (although never less than 65°).

The energy map for —CH₂— is similar (Fig. 7), although with an even more extensive low-energy region. Crystal data [17] show that one angle can take values as low as 20° (although the other lies in the range 50 to 90°), thus confirming the prediction of the present work that the —CH₂— linkage imparts greater conformational freedom than does the —SO₂— linkage.

For —SO₂—, conjugation in the phenyl–sulfur bonds and steric effects both favor the butterfly conformation [28]. Since there are no lone pairs of electrons in —CH₂—, whatever conjugation there is between the phenyls will be very weak indeed, so the form of the map for —CH₂— is determined essentially by steric interactions that prohibit conformations near planar (0°, 0°).

—C(CH₃)₂— and —C(CF₃)₂— Linkages

Figure 8 shows the energy map for the —C(CH₃)₂— linkage. In the optimum geometry the methyl groups were slightly (about 5°) off the staggered positions. The minimum at (52°, 52°) corresponds to a twist conformation, which is similar to those found by some other workers [12–14]. The position of the minima and the shape of the accessible regions of the map (<3 kcal/mol) are very similar to those found by Laskowski et al. [14] using *ab*

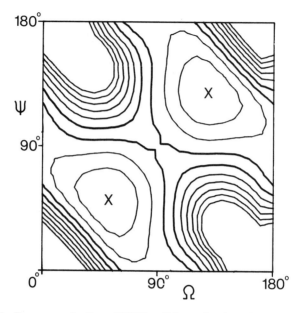

Figure 8   Energy map for the —C(CH₃)₂— linkage, showing minima near (52°, 52°) and (128°, 128°).

initio methods. However, crystal data [17] favor asymmetric structures, typical conformations being near (50°, 60°).

The map for the —C(CF₃)₂— linkage (Fig. 9) shows much smaller low-energy regions than for —C(CH₃)₂—, although the minimum is in a similar position (46°, 46°). In the optimum geometry the CF₃ groups are about 9° off the staggered positions and the bond angle (θ) is 108.9°. No map had previously been published for this linkage, although Welsh et al. [20] have reported some energy calculations. Moreover, so far only one crystal structure containing this linkage has been analyzed [29]. This shows a conformation with torsional angles (42.3°, 42.3°) in satisfying agreement with the present work. The form of the map is determined by the strong steric interactions between all four substituents at the quaternary carbon atom. The only low-energy conformation is one in which the two phenyls and the six fluorines of the CF₃'s interlock, leading to a very rigid conformation (Fig. 10).

PERSISTENCE LENGTHS

A good measure of the static flexibility of polymer chains is the persistence length, which is defined as the average projection of all subsequent bonds onto the direction of the first bond [32]. We have determined the persistence lengths of chains consisting of *para*-phenylene units alternating with linkages

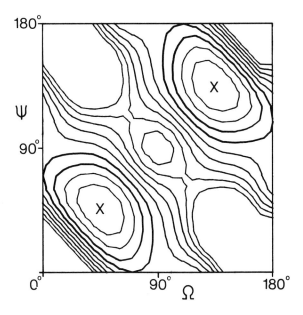

Figure 9 Energy map for the —C(CF₃)₂— linkage, showing minima near (46°, 46°) and (134°, 134°).

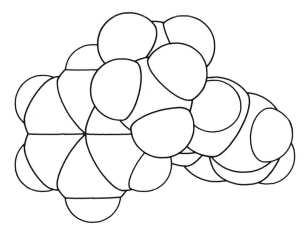

Figure 10 Minimum energy conformation for Ph—C(CF₃)₂—Ph, showing interlocking phenyls and CF₃'s.

of the various forms considered earlier. For example, the —S— linkage forms chains of poly(*para*-phenylene sulfide), —Ph—S—Ph—S—Ph—S—. The method used has been described by Coulter and Windle [33]. It consists of generating many random chains built using the energy maps derived here and then calculating the mean persistence length. All the modeling assumed that

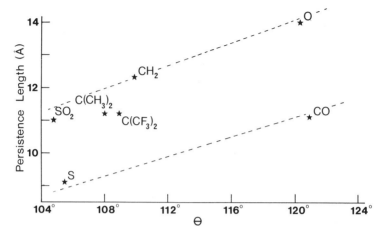

Figure 11   Persistence lengths plotted against bond angle (θ) for all the linkages.

the temperature was 20°C. The values of persistence length for the various linkages are shown in Fig. 11, plotted against the bond angle in the linking group (θ). There is a broad correlation between the persistence length and θ, with persistence length increasing as θ increases.

DISCUSSION

The energy maps calculated for the different linkages are of two basic forms: (1) for the $-SO_2-$ and $-CH_2-$ linkages, there are very broad low energy regions centered on a single minimum at (90°, 90°) (the butterfly conformation); and (2) for the other linkages, the twist conformations have the minimum energy with torsion angles in the range 30 to 50°. The low-energy region is an elongated valley lying along the NW-SE direction ($\Psi + \Omega$ = constant = 60 to 90°). The freedom of rotation of the phenyls depends on the form of the map. For group (1) the phenyls can rotate independently, whereas for group (2) only coordinated rotation of both phenyls (maintaining $\Psi + \Omega \approx$ constant) can occur [9,18,21,30].

CONCLUSIONS

1. For the link ages considered here, the minimum energy conformations of diphenyls are of two types:

         Butterfly (90°, 90°): $-SO_2-$, $-CH_2-$

         Twist ($\Psi = \Omega$ = 30 to 50°): $-S-$, $-O-$, $-CO-$,
                                         $-C(CH_3)_2-$, $-C(CF_3)_2-$

2. The form of the energy maps and positions of minima are determined by the occurrence of conjugation and/or steric interactions between linking groups and phenyls.

3. The persistence lengths for chains of linkages all lie in the range 9 to 14 Å. The persistence length increases with increasing bond angle (θ).

ACKNOWLEDGMENTS

We are grateful to Patrick Coulter for advice and assistance with the persistence length calculations, and to the Air Force Office of Scientific Research, Air Force Systems Command, USAF, who sponsored part of the research under Grant AFOSR 87-0220.

REFERENCES

1. S. Kumar and W. W. Adams, *Polymer* **28,** 1497 (1987).
2. P.-G. de Gennes, *Scaling Concepts in Polymer Physics,* Cornell University Press, Ithaca, N.Y., 1979, p. 21.
3. S. P. N. van der Heijden, E. A. H. Griffith, W. D. Chandler, and B. E. Robertson, *Can. J. Chem.* **53,** 2084 (1975).
4. V. Galasso, G. de Alti, and A. Bigotto, *Tetrahedron* **27,** 6151 (1971).
5. A. E. Tonelli, *Macromolecules* **5,** 558 (1972).
6. P. Tekely and E. Turska, *Polymer* **24,** 667 (1983).
7. T. P. H. Jones, Ph.D. thesis, University of Cambridge, 1983.
8. J. Schaefer, E. O. Stejskal, D. Perchak, J. Skolnick, and R. Yaris, *Macromolecules* **18,** 368 (1985).
9. P. R. Sundararajan, *Can. J. Chem.* **63,** 103 (1985).
10. A. D. Williams and P. J. Flory, *J. Polym. Sci.,* Part A-2, **6,** 1945 (1968).
11. L. Banks and B. Ellis, *J. Polym. Sci. Phys.* **20,** 1055 (1982).
12. B. Erman, D. C. Marvin, P. A. Irvine, and P. J. Flory, *Macromolecules* **15,** 664, 670 (1982).
13. J. Bicerano and H. A. Clark, *Macromolecules* **21,** 585, 597 (1988).
14. B. C. Laskowski, D. Y. Yoon, D. McLean, and R. L. Jaffe, *Macromolecules* **21,** 1629 (1988).
15. P. R. Sundararajan, *Macromolecules* **20,** 1534 (1987).
16. S. Perez and R. P. Scaringe, *Macromolecules* **20,** 68 (1987).
17. A. Anwer, P. D. Coulter, and A. H. Windle, *Macromolecules* (to be submitted).
18. J. T. Bendler, *Ann. N.Y. Acad. Sci.* **371,** 299 (1981).
19. V. A. Zubkov, T. M. Birshtein, and I. S. Milevskaya, *Polym. Sci. USSR* **17,** 2252 (1975).
20. W. J. Welsh, D. B. Bhaumik, and J. E. Mark, *J. Macromol. Sci.* **B20,** 59 (1981).
21. T. P. H. Jones, G. R. Mitchell, and A. H. Windle, *Colloid Polym. Sci.* **261,** 110 (1983).
22. J. C. Barnes, J. D. Paton, J. R. Damewood, and K. Mislow, *J. Org. Chem.* **46,** 4975 (1981).

23. J. Boon and E. P. Magré, *Makromol. Chem.* **126,** 130 (1969).
24. J. N. Hay and D. J. Kemmish, *Polym. Commun.* **30,** 77 (1989).
25. P. R. Sundararajan, *Macromolecules* **22,** 2149 (1989).
26. B. J. Tabor, E. P. Magré, and J. Boon, *Eur. Polym. J.* **7,** 1127 (1971).
27. J. Garbarczyk, *Makromol. Chemie* **187,** 2489 (1986).
28. H. P. Koch and W. E. Moffitt, *Trans. Faraday Soc.* **47,** 7 (1951).
29. V. E. Shklover, Yu. T. Struchkov, M. M. Dzhanashvili, and V. A. Vasaev, *J. Gen. Chem. USSR* **50,** 2075 (1980).
30. R. Hoffman and J. R. Swenson, *J. Phys. Chem.* **74,** 415 (1970).
31. R. L. Jaffe, *Polym. Prepr. Am. Chem. Soc., Div. Polym. Chem.* **30**(2), 1 (1989).
32. P. J. Flory, *Statistical Mechanics of Chain Molecules,* Wiley-Interscience, New York, 1969, p. 111.
33. P. D. Coulter and A. H. Windle, *Polym. Prepr. Am. Chem. Soc., Div. Polym. Chem.* **30**(2), 67 (1989).

4

Characterization of the Molecular Mechanisms Involved in $\beta_0 \rightarrow \alpha_h$ Phase Transition in C_{21} n-Alkane Crystals

D. C. DOHERTY and A. J. HOPFINGER
Department of Chemistry
The University of Illinois at Chicago
P.O. Box 6998, m/c 111
Chicago, Illinois 60680

ABSTRACT

The $\beta_0 \rightarrow \alpha_h$ order–order phase transition was investigated using molecular dynamics for C_{21} n-alkane crystals. The $(1, 3)$ intrachain torsion angle pairs are highly correlated, leading to cooperative motions that give rise to "kink" deformations. The cooperative torsion angle motions persist over the lengths of the chains, resulting in oscillating-wave motion behavior for the chains. The velocity of these waves is about 4700 m/s. In the α_h phase, the intramolecular backbone cooperativity leads to an overall twisting motion as the chains rotate.

INTRODUCTION

A most interesting time is upon us in the chemical sciences. Methods in theoretical and computational chemistry have evolved to a point where doing chemical modeling on a computer can be quite accurate and reliable. In addition, advances in computer software and hardware technologies make it possible to treat effectively chemical systems of considerable size and complexity. One important consequence of all this progress is that some chemical experiments, not possible to perform in the laboratory, can be done on the computer.

A class of chemical phenomena very difficult to investigate in the laboratory comprises the time-dependent mechanisms at play in phase transitions. Computer simulation modeling, using molecular dynamics (MD) [1], appears to be a useful tool for probing the molecular behavior of phase transitions.

Many chemical systems exhibit phase transitions. We have begun our MD-phase transitions simulation studies on lower-molecular-weight n-alkanes which exhibit crystal–crystal and crystal–melt transitions. The reasons for selecting these chemical systems are:

1. The size, in terms of the number of atoms, of the simulation samples is well within practical computation limits.
2. MD force fields are well developed and most reliable for saturated hydrocarbons [2].
3. Crystal–melt transitions in the n-alkanes are first order [3], and many have been investigated extensively, including analyses of premelt solid–solid phase transitions [4].
4. The n-alkanes can serve as prototypes for understanding the transition properties of linear polymers, most notably polyethylene.

The phase behavior of the n-alkanes is complex. All odd alkanes between C_9 and C_{45} undergo at least one premelting phase transition, and the longer members show as many as four high-temperature solid phases in addition to the low-temperature crystalline structure [5]. Müller [6] first observed that many n-alkanes undergo solid–solid phase transitions near their melt temperatures. Numerous experimental investigations of these high-temperature phases have led to the "rotater" model to describe interconversion between phases [6–13]. The rotator model envisions the chain molecules executing coupled rotations about their long axes while maintaining a fully extended (all-trans) conformation. In the case of the first premelt phase transition of the odd n-alkanes ($>C_9$) this corresponds to the chains initially packed in an orthorhombic lattice [14] rotating and laterally realigning into a hexagonal framework.

Various studies now suggest that the rotator model is not entirely adequate to describe completely the premelt phase behavior in n-alkanes [5,15–21]. Blasenberry and Pechhold [15] first demonstrated that a number of observed properties of the rotator phase could be explained by specific torsion angle conformational defects termed "kinks." Kink defects have received considerable attention in modeling polymers and lipids but not n-alkanes. The Schatzki and Boyer crankshaft models are examples of the application of kink defects to polymer phase transition theory [22,23].

Ewen and co-workers [17], using small-angle x-ray scattering data, suggest that the rotator phase may contain a large fraction of molecules in nonplanar conformations. Zerbi et al. [20] have presented evidence for the presence of gauche bonds near the ends of C_{19} in its high-temperature ordered phase.

Snyder and co-workers have carried out several elegant structural investigations of the solid phases of n-alkanes [5,21], including the high polymer, polymethylene (PM) [24–26]. Much of their work has involved applications of infrared spectroscopy. Overall, they conclude that the n-alkanes adopt phases containing an increasing number of nonplanar conformations as the melting point is approached. In addition, they have postulated specific classes of conformational defects.

Most of the computational efforts on n-alkanes have been directed toward the high-polymer state. A considerable number of simulation studies of polymethylene-like chains have been carried out. Conformational energy minimization (EM) [27], Monte Carlo (MC) modeling [28], and MD [1,29] methods have been used. Many simulation studies have been done using lattice models [30], although nonlattice calculations have also been performed [31]. Crystal structure refinements have been carried out using EM [32]. Pant and co-workers [33] have investigated orientational diffusion motions in chain-molecule models using MD. Theodorou and Suter [34] have done simulation modeling of highly dense amorphous polymeric systems, including PM. Helfand and co-workers [35] carried out MD simulation studies of PM-like chains at medium densities. Rigby and Roe [36] have used MD to simulate the behavior of linear chains, resembling PM in structure parameters, as polymer liquids and gases. Karasawa and Goddard [37] used MC simulation modeling to investigate phase transitions of an isolated PM chain in free space.

Ryckaert and Bellemans [38] employed semirigid MD to simulate liquid systems of n-alkanes. Ryckaert has continued to investigate solid phases of n-alkanes near the melt temperature. He and Ciccotti [39] observed rotator motions in α_h monolayers of n-alkanes using MD. MC simulation studies by Yamamoto [40] first suggested a four-site orientational distribution of chains in the rotator phases of n-alkane monolayers. Recently, Ryckaert et al. [41] have observed, from MD simulations, torsional defects and kinks within n-alkane chains that are distributed among the four orientational sites in the rotator phase of α_h bilayers of C_{23}.

METHODS

Molecular Dynamics

Full Cartesian MD simulation requires iteratively solving Newton's equations of motion over time. If there are k atoms in the system, and the initial conditions (geometry, temperature, pressure, volume) have been defined, then

$$\frac{-\partial P(r_1, \ldots, r_k)}{\partial \mathbf{r}_i} = \mathbf{F}_i(r_1, \ldots, r_k) = m_i \frac{d^2 r_i}{dt^2} \qquad i = 1, \ldots, K \qquad (1)$$

must be time integrated (numerically). P is the potential energy, $\mathbf{F}_i$ is the force acting on atom i having mass m_i, and t is the time. The $\mathbf{r}$ are the displacement vectors for the atoms. The integration of the set of equations defined in (1) allows a small step in time, Δt, to be taken, which, in turn, permits definition of a new set of velocities and atomic displacements (schematically):

$$\mathbf{v}_i(t + \Delta t) = \mathbf{v}_i(t) + \frac{d^2 \mathbf{r}_i}{dt^2} \Delta t \qquad (2)$$

and

$$\mathbf{r}_i(t + \Delta t) = \mathbf{r}_i(t) + \frac{d \mathbf{r}_i}{dt} \Delta t \qquad (3)$$

The dynamic trajectory of the system is achieved by cycling through equations (1), (2), and (3) over some time period. Equations (2) and (3) are general algebraic representations and do not reflect any specific algorithm. We have implemented the "leapfrog" numerical integration algorithm [49].

Central to what reliable information can be gleaned from a simulation study is the quality of the energy calculation. Application of EM, MC, and/or MD to n-alkanes chain assemblies necessitates the use of molecular mechanics (MM) [42]. This is the only practical means of computing the energetics of chemical systems composed of large numbers of atoms. Simulation studies involving molecules only containing saturated carbons and hydrogens should be the most reliable of any molecular system. Different force fields have been proposed from experimental fitting analyses [43] for $C(S_p^3)$ and H. In addition to this strong coupling to experimental data, the energy of an n-alkane system is predominantly due to atom-pair dispersive attractions and steric repulsions. Thus, unresolved complications of potential function representation for electrostatic, hydrogen bonding, resonance, and strain interactions are minimized for n-alkanes. We have used both the MM2 force field [43], and a modified MM2 force field using the nonbonded potentials of Flory and co-workers [44] in place of the MM2 potentials in the work reported here.

Model *n*-Alkane System

We have focused our attention on the beta-orthorhombic (β_0) to alpha-hexagonal (α_h) transition in the C_{21} n-alkane crystals. Twelve 21-carbon chains were placed in each of two layers of the β_0 paraffin bilayer crystal structure. A schematic illustration of the model crystal is shown in Fig. 1. The bilayer structure allowed us to study the effects due to the interface between adjacent layers. The model includes explicit hydrogen atoms. Following a period of equilibration of the β_0 structure, MD simulations, using a loose coupling ($\tau = 0.1$ ps) to a temperature bath [49] at 305 K, were performed on a consecutive series of six unit-cell microstates for 5 ps each. Changes in the unit cell cross-sectional area were varied linearly over the microstates to simulate a quasi-static volume expansion from the β_0 to the α_h unit cell geometry. The chain axis dimension remained constant as in a previous alkane MD simulation [41]. Full Cartesian MD simulation, as opposed to Brownian dynamics (BD), was performed. The last point in each trajectory was used as the starting point for the next microstate in the β_0 to α_h pathway. Minimum image periodic boundary conditions were used to simulate the bulk crystal.

RESULTS AND DISCUSSION

Visual analysis of a "movie" representing the MD trajectories of the C_{21} microstates indicated periodic, wavelike motions within the chains. This periodic motion became more pronounced as the α_h unit cell geometry was approached. The MD trajectories were analyzed for evidence of cooperativity between individual torsional motions in the alkane chains.

Figure 2 displays the autocorrelation function of all torsion angles averaged over all 24 chains, and Fig. 3, the (1, 3)-torsion angle pair cross-correlation function for all chains, over the entire 30-ps trajectory. The auto-correlation function shows that the individual torsional motions are periodic, while the cross-correlation function indicates strong reverse coupling between

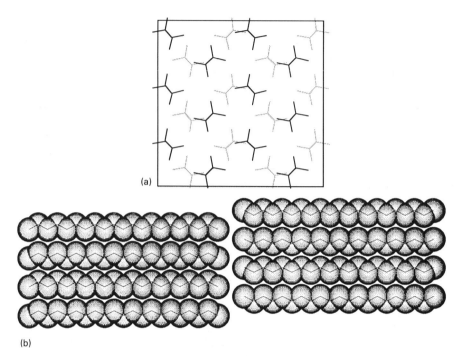

Figure 1   (a) C_{21} bilayer crystal model employed in the calculation, as viewed down the crystallographic c-axis. (b) Space-filled representation of the model with hydrogens removed for clarity, as viewed along the crystallographic a-axis.

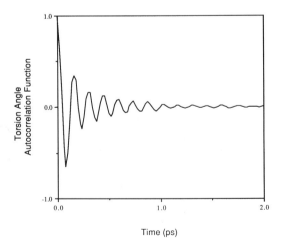

Figure 2   Autocorrelation function of all torsion angles, averaged over the entire simulation.

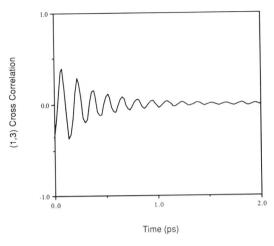

Time (ps)

Figure 3 Cross-correlation function of all torsion angle pairs separated by one (1, 3) bond, averaged over the entire simulation.

(1, 3)-torsion angle pairs. Both the autocorrelation and (1, 3) cross-correlation functions persist for about 1 ps (within three standard errors of the data). According to Fig. 3, when one torsion angle rotates in one direction, the other torsion angle of the (1, 3) pair shows a tendency to rotate in the opposite direction. A probable consequence of such behavior is "kink" [45] formation of the type $(g^{+/-} tg^{-/+})$. This cooperative behavior is consistent with the findings of Helfand [46], who inferred (1, 3)-torsion angle pair cooperativity from statistical measures of torsion angle motions over "long" time scales (nanoseconds) using BD.

Very little coupling is apparent in the (1, 2)-torsion angle pair crosscorrelation function, as is seen in Fig. 4. Moderate cooperative coupling is observed in the (1, 4)-, (1, 5)-, and (1, 6)-torsion angle pair cross-correlation functions (see Figs. 5, 6, and 7, respectively). However, these couplings are much smaller than for the (1, 3)-torsion angle pairs.

Our calculations demonstrate the *ab initio* existence of selective cooperativity [47] among torsion angles as the simulation proceeds from β_0 to α_h. However, this particular cooperativity among the torsion angles is probably not general. We have also carried out MD simulations of the T_2G_2 to T_4 transition in syndiotactic polystyrene, sPS [48]. The structure of sPS is shown in Fig. 8. The stable T_4 and T_2G_2 conformations are shown in Figs. 9 and 10, respectively. Plots of torsion angle versus time for four adjacent backbone torsion angles, shown in Fig. 11, demonstrate the transition behavior of this chain molecule. The four torsion angles are initially in the $(\phi_5 \phi_6 \phi_7 \phi_8) =$ TGGT state, $(T = 180°, G = 300°)$. As the simulation proceeds, ϕ_6 and ϕ_7 convert over to the T state. However, these transitions have no influence on ϕ_5 and

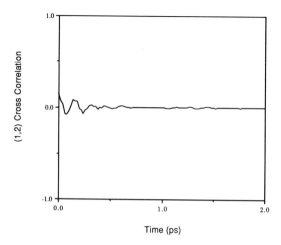

Figure 4   Cross-correlation function of all adjacent $(1, 2)$ torsion angle pairs.

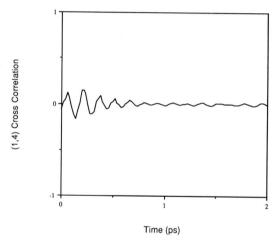

Figure 5   Cross-correlation function of all torsion angle pairs separated by two
$(1, 4)$ bonds.

ϕ_8, and, as it turns out, any other backbone torsion angles. Thus it appears that
transitions in ϕ_6 and ϕ_7 are uncoupled from the remainder of the molecule.

The question does arise: Are motions in ϕ_6 and ϕ_7 coupled? This can be
answered by inspecting Fig. 12, where the torsion angle trajectory is plotted in
$(\phi_6, \phi_7, \text{time})$-space. It can be seen that ϕ_6 first undergoes its torsion angle
transition before ϕ_7 begins its transition. Further, the ϕ_7 transition does not
influence the completed ϕ_6 transition (state T). In other words, the $G \rightarrow T$
transitions of ϕ_6 and ϕ_7 are decoupled from one another. More generally, at
this time it appears that conformational transition mechanisms of chain molecules can be quite different from one another.

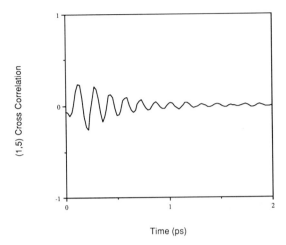

Figure 6 Cross-correlation function of all torsion angle pairs separated by three (1, 5) bonds.

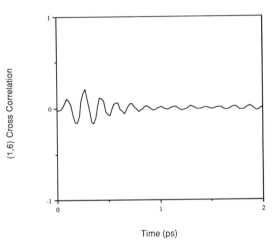

Figure 7 Cross-correlation function of all torsion angle pairs separated by four (1, 6) bonds.

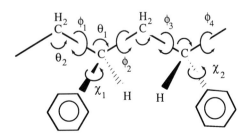

Figure 8 Structural repeat unit for sPS with backbone and torsion angles defined.

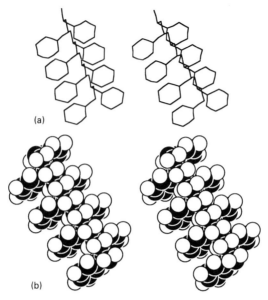

Figure 9   All-trans (T₄) conformation of sPS: (a) stick stereo, (b) space-filled stereo.

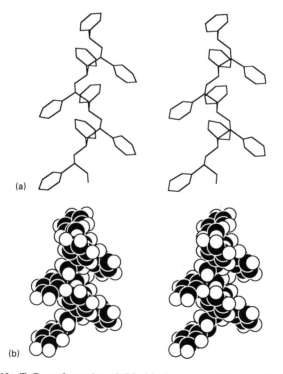

Figure 10   T₂G₂ conformation of sPS: (a) stick stereo, (b) space-filled stereo.

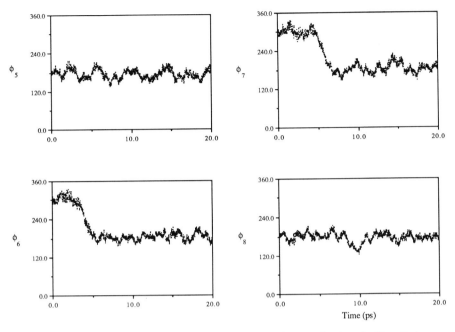

Figure 11 Torsion angle trajectory profiles for backbone angles ϕ_5 to ϕ_8. Note that ϕ_6 and ϕ_7 undergo a G^+-to-T transition.

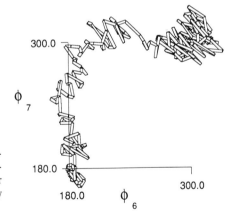

Figure 12 ϕ_7 versus ϕ_6 torsion angle trajectories during the GG (300°, 300°)-to-TT (180°, 180°) transition. Each cylinder represents 0.012 ps; the entire trajectory represents 3.0 ps.

We now return to the $\beta_0 \rightarrow \alpha_h$ C_{21}-alkane crystal transition. Figure 13 is a digitized plot of the matrix of the 18 torsion angles of one 21-carbon chain (vertical axis) versus time (horizontal axis). The darker blocks in angle-time space indicate values near 200°, above the trans state (180°); the lightest blocks indicate values near 160°, below the trans state. It can be seen that there exist diagonal lines of near-constant shading, that is, identical torsion angle deviations from trans, running from left to right. Two pronounced dark diagonals

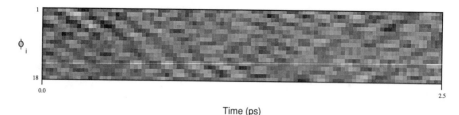

ϕ_i

1

18

0.0

2.5

Time (ps)

Figure 13 Torsion angle values of one chain (excluding end methyls) versus time for one chain during the MD trajectory. Darkest blocks represent torsion angle values near 200°; lightest blocks, near 160°.

and three pronounced light diagonals can be seen to start at the top of the plot near 0.6 to 0.8 ps and end at the bottom of the plot near 1.0 to 1.2 ps. Where such shaded bands exist, the plot suggests that a short time after one torsion angle reaches a maximum or minimum in its oscillation, the angle adjacent to it will do so also; and this behavior is repeated down the length of the chain. Thus Fig. 13 demonstrates that wavelike motions are propagated within the alkane chains, probably as a consequence of the $(1,3)$-torsion angle pair cooperativity. Similar cooperative behavior was noticed in many other chains also.

It is possible to estimate a wave velocity of about 4700 m/s from the slope of the diagonals in Fig. 13. We note that this is on the order of magnitude of the sound velocity through various crystalline media. Therefore, we believe that these motions may be related to the *damped torsional oscillator* (DTO) model, which has been used to explain acoustic relaxation data in alkane liquids [48].

Another type of cooperativity appears at the onset of the "rotator" (α_h) phase [6]. This phase is characterized by the chains rotating about their long axes into four equivalent positions ($\pm 45°, \pm 135°$). From the MD trajectory it is possible to calculate the setting angles made by each individual backbone bond vector projected into the *a-b* plane. Figure 14 represents the trajectories of the consecutive individual setting angles of one chain as it rotates from 135° to 45°. The darker blocks represent values near 135°; the lighter, near 45°. As can be seen from Fig. 14, cooperative motions contribute to an overall twisting of the chain as it rotates into its new position.

Overall, this work suggests many avenues of additional research in simulating phase transitions in chain molecules. One study we anticipate is simulating the TGTG' → T$_4$ transition in poly(vinylidene fluoride). It will be interesting to see how similar this transition's mechanism is to that of T$_2$G$_2$ → T$_4$ of sPS. The logical next study of C$_{21}$ alkane crystal behavior is to simulate the α_h → melt transition, using constant-pressure MD.

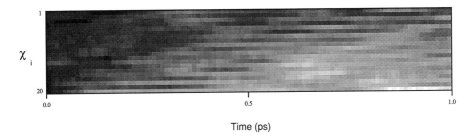

χ_i

Time (ps)

Figure 14 Chain setting angle values for individual bonds of one chain (as it underwent an overall rotation from 135° to 45°) versus time. Darkest blocks represent values near 135°; lightest blocks, near 45°.

ACKNOWLEDGMENTS

This work was supported through a Cooperative Research Grant from the Dow Chemical Company and by funds from the Laboratory of Computer-Aided Molecular Modeling and Design at UIC. We are also grateful to Cray Research, Inc., for a University Research and Development Grant for computer time at the National Center for Supercomputing Applications at UIUC. Conformational energy calculations were done by using the CHEMLAB-II molecular modeling package. The molecular dynamics calculations were done by using the MOLSIM system. Figures 9 and 10 were made by using the Micro-Chem molecular modeling software. Figures 13 and 14 were prepared using the ImageTool program from the National Center for Supercomputing Applications at UIUC. We believe that ImageTool has provided a very unique and useful way of representing this type of data. Figure 1 was made using the Molecular Viewer program developed by John Nicholas. We also thank John Nicholas, of our laboratory, for help in preparing the time correlation functions.

REFERENCES

1. J. A. McCammon and S. Harvey, *Dynamics of Proteins and Nucleic Acids,* Cambridge University Press, Cambridge, 1987.
2. A. J. Hopfinger, R. A. Pearlstein, P. L. Taylor, and F. P. Boyle, *J. Macromol. Sci. Phys.* **B26**(3), 359 (1987).
3. M. O. Steinitz, M. Kahrizi, J. Genossar, and R. G. Snyder, *J. Phys. Chem.* **91,** 3433 (1987), and references therein.
4. M. Maroncelli, S. P. Qi, H. L. Strauss, and R. G. Snyder, *J. Am. Chem. Soc.* **104,** 6237 (1982) and references therein.
5. R. G. Snyder, M. Maroncelli, S. P. Qi, and H. L. Strauss, *Science* **214,** 1586 (1981).

6. A. Müller, *Proc. R. Soc. London* **A138,** 514 (1932).

7. J. Doucet, I. Denicolo, A. Craievich, and A. Collet, *J. Chem. Phys.* **75,** 5125 (1981).

8. R. E. Dehl, *J. Chem. Phys.* **60,** 339 (1973).

9. H. G. Olf and A. Peterlin, *J. Polym. Sci.,* Part A-2, **8,** 791 (1970), and references therein.

10. D. H. Bonsor, J. F. Barry, M. W. Newberry, M. V. Smalley, E. Granzer, C. Koberger, P. H. Nedwed, and J. Scheidel, *Chem. Phys. Lett.* **62,** 576 (1979).

11. D. Bloor, D. H. Bonsor, D. N. Batchelder, and C. G. Windsor, *Mol. Phys.* **34,** 934 (1977).

12. J. D. Barnes, *J.Chem. Phys.* **58,** 5193 (1973).

13. J. D. Hoffman, *J. Chem. Phys.* **20,** 541 (1952).

14. T. Ishinabe, *J. Chem. Phys.* **72,** 353 (1980); M. Kobayashi, *J. Chem. Phys.* **68,** 145 (1977); D. W. McClure, *J.Chem. Phys.* **49,** 1830 (1968).

15. S. Blasenbrey and W. Pechhold, *Rheol. Acta* **6,** 174 (1967).

16. J. D. Barnes and B. M. Fanconi, *J. Chem. Phys.* **56,** 5190 (1972).

17. B. Ewen, G. R. Strobl, and D. Richter, *Faraday Discuss. Chem. Soc.* **69,** 19 (1980), and references therein.

18. T. Oyama, K. Takamizawa, and Y. Ogawa, *Kobunshi Ronbunshu* **37,** 711 (1980).

19. T. Oyama, K. Takamizawa, Y. Urabe, and Y. Ogawa, *Kyushu Daigaku Kogaku Shuho* **52,** 129 (1979).

20. G. Zerbi, R. Magni, M. Gussoni, K. H. Moritz, A. Bigotto, and S. Dirlikov, *J. Chem. Phys.* **75,** 3171 (1981).

21. R. G. Snyder, M. Maroncelli, H. L. Strauss, and V. M. Hallmark, *J. Phys. Chem.* **90,** 5623 (1986), and references therein.

22. T. F. Schatzki, *J. Polym. Sci.* **C14,** 139 (1966).

23. F. Boyer, *Rubber Chem. Technol.* **34,** 1303 (1963).

24. R. G. Snyder, N. E. Schlotter, R. Alamo, and L. Mandelkern, *Macromolecules* **19,** 621 (1986).

25. R. G. Snyder and S. L. Wunder, *Macromolecules* **19,** 496 (1986).

26. R. G. Snyder and H. L. Strauss, *J. Chem. Phys.* **87,** 3779 (1987).

27. A. J. Hopfinger, *Conformational Properties of Macromolecules,* Academic Press, New York, 1973.

28. M. Mezei, P. K. Mehotra, and D. L. Beveridge, *J. Am. Chem. Soc.* **107,** 2339 (1985).

29. For additional discussions, see (a) B. R. Brooks, R. E. Bruccoleri, B. D. Olafson, D. J. States, S. Swaminathan, and M. Karplus, *J. Comput. Chem.* **4,** 187 (1983); (b) T. A. Weber and E. Helfand, *J.Chem. Phys.* **71,** 4760 (1979).

30. See, for example, S. G. Whittington, J. E. G. Lipson, M. K. Wilkinson, and D. S. Gaunt, *Macromolecules* **19,** 1241 (1986).

31. M. Muthukumar, *Proceedings of the Workshop on Advanced Computation and Simulation of Complex Materials Phenomena,* San Diego Supercomputing Center, March 22–26, 1987.

32. H. Tadokoro, *Structure of Crystalline Polymers,* Wiley, New York, 1979.

33. B. B. Pant, J. Skolnick, and R. Yaris, *Macromolecules* **18,** 253 (1985).

34. D. N. Theodorou and U. W. Suter, *Macromolecules* **18,** 1467 (1985).

35. (a) E. Helfand, Z. R. Wasserman, and T. A. Weber, *J. Chem. Phys.* **70**(04), 2016

(1979); (b) E. Helfand, *Physica,* **A118,** 123 (1983); (c) E. Helfand, *Science* **226,** 647 (1984).

36. D. Rigby and R. J. Roe, *Abstracts of the American Physical Society Meeting,* New York, **32,** 423 (1987).

37. N. Karasawa and W. A. Goddard III, *J. Phys. Chem.* **92,** 5828 (1988).

38. J.-P. Ryckaert and A. Bellemans, *Faraday Discuss. Chem. Soc.* **66,** 95 (1978).

39. J.-P. Ryckaert and G. Ciccotti, *Mol. Phys.* **58,** 1125 (1986).

40. T. Yamamoto, *J. Chem. Phys.* **83,** 3790 (1985).

41. J.-P. Ryckaert, M. L. Klein, and I. R. McDonald, *Phys. Rev. Lett.* **58,** 698 (1987).

42. V. Burkert and N. L. Allinger, *Molecular Mechanics,* ACS Monograph 177, American Chemical Society, Washington, D.C., 1982.

43. See, for example, D. E. Williams, *J.Chem. Phys.* **47,** 4680 (1967).

44. D. Y. Yoon, P. R. Sundarajan, and P. J. Flory, *Macromolecules* **8,** 776, (1975).

45. (a) D. H. Reneker, *J. Polym. Sci.* **59,** 539 (1962); (b) W. Pechold, S. Blasenbrey, and S. Woerner, *Kolloid-Z.* **189,** 14 (1963).

46. E. Helfand, *Science* **226,** 647 (1984).

47. (a) E. Ising, *Z. Phys.* **31,** 253 (1925); (b) A. J. Hopfinger, *Conformational Properties of Macromolecules,* Academic Press, New York, 1973, Chap. 4.

48. (a) A. V. Tobolsky, and D. B. DuPre, *Adv. Polym. Sci.* **6,** 103 (1969); (b) M. A. Cochran, P. B. Jones, A. M. North, and R. A. Pethrick, *Trans. Faraday Soc.* **68,** 1719 (1972); (c) R. A. Pethrick, *J. Macromol. Sci., Rev. Macromol. Chem.* **C9**(1), 91 (1973).

49. H. J. C. Berendsen, J. P. M. Postma, W. F. van Gunsteren, A. Dinola, and J. R. Haak, *J. Chem. Phys.* **81,** 3684 (1984).

5

Use of an All-Atom Semiflexible Model in Molecular Dynamic Simulation of Long-Chain Paraffins

JEAN-PAUL RYCKAERT
Pool de Physique, C.P. 223
Universite Libre de Bruxelles,
1050 Bruxelles
Belgium

IAN R. McDONALD
Department of Physical Chemistry
University of Cambridge,
Cambridge CB2 1EP
United Kingdom

MICHAEL L. KLEIN
Department of Chemistry
University of Pennsylvania,
Philadelphia, Pennsylvania 19104

ABSTRACT

A method, based on the use of semiflexible chains, is described whereby all-atom molecular-dynamics calculations can be carried out for long-chain molecules. As an example, results are reported for two solid bilayer phases of the n-alkane tricosane ($C_{23}H_{48}$): the crystalline orthorhombic phase and the pseudohexagonal (R_I) rotator phase. In the disordered R_I phase, each chain has four well-defined orientations, the molecules exhibit enhanced longitudinal chain motion, and a significant number of conformational defects develop, predominantly at the chain ends. The results of the simulation are in excellent agreement with experiment.

INTRODUCTION

Rotator phases of paraffins have been thoroughly investigated in the last few years by a variety of experimental techniques [1–15]. It is now well established that the rotational disorder of the all-trans planar chains, around their long axis, is not the only form of disorder appearing at the rotator phase transition. Enhanced, though restricted longitudinal diffusion of the chains destroys somewhat the regular layer structure of the low-temperature crystal [1], and in addition, conformational defects (gauche bonds) are detected mainly at the chain ends [12]. Correlations between these various forms of disorder make it difficult to unravel, solely from experimental data, a coherent microscopic picture of these intriguing long-chain systems, which share many similarities with Langmuir–Blodgett films, micelles, and biological membranes. A portion of the phase diagram of long-chain paraffins is shown in Fig. 1. We report here the results of a realistic molecular-dynamics simulation [16–18] of the C_{23} n-alkane (tricosane) in its pseudohexagonal rotator phase R_I (see Fig. 1). The actual state point considered is indicated by an asterisk.

DETAILS OF THE CALCULATION

An all-atom, semiflexible chain model was employed. The intermolecular potential consisted of a sum of atom–atom pair interactions of the Buckingham (exponential-6) form parameterized to the low-temperature structures of alkanes [19]. This interaction is used not only to describe the intermolecular potential but also the intramolecular interactions between nonbonded atoms

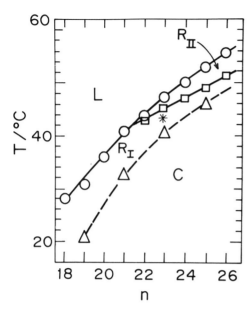

Figure 1 Phase diagram of paraffins $C_n H_{2n+2}$ in the $18 < n < 26$ region. The R_I phase investigated in the present work is pseudohexagonal.

belonging to methyl or methylene groups separated by at least three methylene units. The intramolecular potential further consists of a sum of CCC harmonic, bending potentials [18,20] and CCCC effective torsional potentials, of the butane type, for central CC bonds and of the propane type for end CC bonds [18,21]. All other vibrational modes are frozen by geometrical constraints [22–24]. The reason for this approach is that it indeed turns out to be easiest to treat the dynamics of such semiflexible chains in terms of atomic Cartesian coordinates [22,23]. The equations of motion must then contain constraint forces which can be made explicit once the constraint architecture is specified. First, all CC and CH chemical bonds are fixed at their desired length [19]. For methyl groups, the three H atoms are further constrained by (artificial) fixed bonds to the C atom in the β position along the chain; also, in principle, only two out of the three HH distances need to be explicitly fixed. In practice, however, it is more efficient if all three H atoms are constrained [25]. This only leaves a single degree of freedom describing the rotation of the methyl group around the last CC axis. The H atoms of the methylene groups are fully constrained to the carbon backbone by fixing M, the midpoint of the HH distance at $d_{CH} \cos(\phi/2)$ along the bisector of the adjacent C'CC' bending angle (see Fig. 2): $d_{CH} = 1.04$ Å is the constant distance between C and H atoms [19], while ϕ is the HCH angle (tetrahedral). Finally, to freeze the $>CH_2$ rocking mode, the HH axis is forced to remain perpendicular to the axis joining the C' atoms that are first neighbors with respect to the CH_2 group at which we are looking.

Such a model retains frequencies up to around 400 cm^{-1}, and hence a time step as large as $h = 3.3$ fs can be employed in the simulation, even at

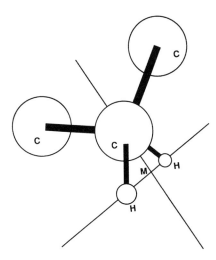

Figure 2 Schematic illustration of the methylene constraints.

$T = 300$ K. The calculation of the constraint forces does not affect, in a significant way, the efficiency of the MD program as, typically, it amounts to only 10% of the central processing unit (CPU) time per step.

We mention a few salient details on how a dynamical step is performed. A central difference algorithm is employed to obtain the positions at time $t + h$ from the positions at previous times t and $t - h$ and the forces at time t [26]. First, atomic positions at $t + h$, say $r(t + h)$, are obtained using the algorithm in which only the forces deriving from the potential energy are included (free propagation step). One then performs an additional move in which single constraints are treated individually in an iterative, sequential fashion, going many times (typically, 25 times) over the entire set of constraints until satisfactory convergence is achieved using a SHAKE routine [26]. It is worth pointing out that the order in which the constraints are implemented in the iterative sequence may affect the convergence characteristics [25]. A single constraint step consists of an updating of the appropriate atomic coordinates involved in either a rigid-bond constraint, one of the Cartesian projections of the vectorial constraint involving the location of the midpoint M, or the rocking constraint.

A bilayer of 30 tricosane chains per layer was set up initially with the experimental structure, for the low-temperature phase and for the rotator phase [3,7]. The all-trans chains were arranged in perfect layers with herringbone packing for the ordered phase and random distribution among four sites for the orientations of the zigzag planes in the rotator phase on the basis of previous predictions [7,16]. These systems were then brought close to the phase-transition temperature $T = 313$ K, using molecular dynamics combined with a thermalization procedure that acted on all degrees of freedom in an independent way [27,28].

RESULTS

The molecular dynamics simulations revealed quite drastic differences between the two systems. Typical snapshots taken from each experiment are shown in Fig. 3. The low-temperature phase appears to be stable and no disorder developed during an 8-ps run. On the contrary, conformational defects and longitudinal diffusion appear spontaneously in the rotator phase, which has been followed for a total of 130 ps.

A four-site angular distribution of the molecular planes appeared to be very stable during the entire run of the rotator phase (Fig. 4). The mean-residence time in a particular orientation is of the order of 8 ps. The stabilization of the distribution of local gauche defects requires about 20 ps. The result is a distribution that follows closely experimental data for the C_{21} *n*-alkane (Fig. 5). Kinks are observed in the chains, sometimes at their middles, but more frequently at the chain ends. Isolated gauche defects are detected not only around the first CCCC dihedral angle at each end of the chain, but also in the two adjacent positions. The mean residence time of the —CH_3 group is approximately 14 ps, whilst the mean time between gauche-trans or trans-gauche jumps increases (30 ps, 50 ps, 130 ps, . . .) for successive bonds along

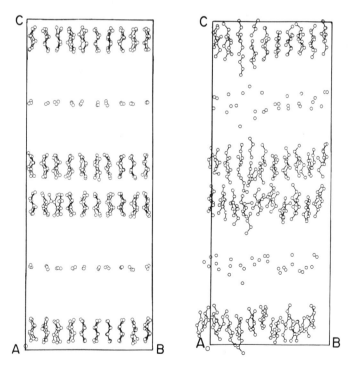

Figure 3 Snapshots taken from simulations of (a) the ordered and (b) the R_1 rotator phase as viewed down the *a*-axis of the orthorhombic cell.

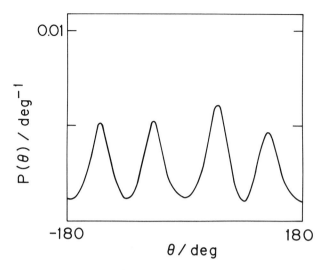

Figure 4   Stable four-site distribution of the zigzag planes in the R_I phase.

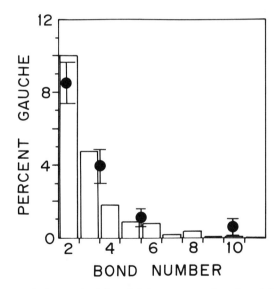

Figure 5   The circles are the inferred defect concentrations from infrared experiments on the C_{21} alkane [12], and the histogram, the simulation results for tricosane.

the chain [18]. Examples of the birth of gauche defects in randomly chosen chains are shown in Fig. 6. The figure illustrates that gauche-defect formation is a rather rapid process and that g^+ and g^- defects need not necessarily be born at the same time.

The longitudinal dispersion of the chain centers of mass first stabilizes close to its experimental value (about 1.5 Å) but later increases suddenly by

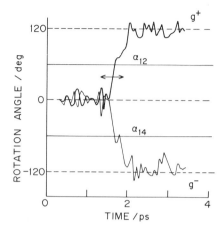

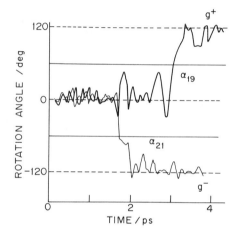

Figure 6 Gauche-defect formation in the R_I rotator phase of tricosane; α labels the torsion angle and the subscript its position. The upper curve is for defect formation near the chain end and the lower curve near the chain center.

about 20%, an observation that might be an artifact of the bilayer model [18]. Indeed, at the moment the longitudinal motion increases rapidly; one notes that the relative positions of the two layers in the simulation cell has changed in such a way that chains are collinear. Thus the observed enhanced diffusion should be viewed with caution.

Work is currently in progress on a six-layer system. We should then be able to analyze details such as (1) the formation of intramolecular defects and their possible migration along the chains, (2) the coupling between the overall rotation of the chains and their intramolecular dynamics, (3) the cooperativity of rotational jumps between neighboring chains, (4) domain structure of the orientational order, and (5) the effect of temperature and simulation of the $R_I \rightarrow R_{II}$ phase transition.

ACKNOWLEDGMENTS

The research outlined herein was supported by the Fonds National de la Recheche Scientifique (Belgique), the S.E.R.C. (U.K.), and the U.S. National Institutes of Health (GM-40712). The collaboration was encouraged by NATO Grant 0414/88.

REFERENCES

1. F. Guillaume, J. Doucet, C. Sourisseau, and A. J. Dianoux, *J. Chem. Phys.* **91**, 2555 (1989); A. Müller, *Proc. R. Soc. London* **A138**, 514 (1932).
2. E. Ewen, G. R. Stobl, and D. Richter, *Faraday Discuss. Chem. Soc.* **69**, 19 (1980).
3. G. Ungar, *J. Phys. Chem.* **87**, 689 (1983).
4. J. Doucet, I. Denicolo, A. F. Craievich, and A. Collet, *J. Chem. Phys.* **75**, 5125 (1981); J. Doucet, I. Denicolo, and A. F. Craievich, *J. Chem. Phys.* **75**, 1523 (1981); I. Denicolo, J. Doucet, and A. F. Craievich, *J. Chem. Phys.* **78**, 1465 (1983).
5. J. Doucet, I. Denicolo, A. F. Craievich, and C. Germain, *J. Chem. Phys.* **80**, 1647 (1984).
6. A. F. Craievich, I. Denicolo, and J. Doucet, *Phys. Rev.* **B30**, 4782 (1984).
7. G. Ungar, and N. Mašié, *J. Phys. Chem.* **89**, 1036 (1989).
8. G. Zerbi, R. Magni, M. Gussoni, K. H. Moritz, A. Bigotto, and S. Dirlikov, *J. Chem. Phys.* **75**, 3175 (1981).
9. H. L. Casal, H. H. Mantsch, D. G. Cameron, and R. G. Snyder, *J. Chem. Phys.* **77**, 2825 (1982).
10. M. Maroncelli, S. P. Qi, H. L. Strauss, and R. G. Snyder, *J. Am. Chem. Soc.* **104**, 6237 (1982).
11. H. L. Casal, D. G. Cameron, and H. H. Mantsch, *Can. J. Chem.* **61**, 1736 (1983).
12. M. Maroncelli, H. L. Strauss, and R. G. Snyder, *J. Chem. Phys.* **82**, 2811 (1985).
13. D. Bloor, D. H. Bonsor, D. H. Batchelder, and C. G. Windsor, *Mol. Phys.* **34**, 939 (1977).
14. J. Doucet, and A. J. Dianoux, *J. Chem. Phys.* **81**, 5043 (1984).
15. M. G. Taylor, E. C. Kelusky, I. C. P. Smith, H. L. Casal, and D. G. Cameron, *J. Chem. Phys.* **78**, 5108 (1983); E. Kelusky, I. C. P. Smith, C. Ellinger, and D. G. Cameron, *J. Am. Chem. Soc.* **106**, 2267 (1984).
16. J.-P. Ryckaert and M. L. Klein, *J. Chem. Phys.* **85**, 1613 (1986).
17. J.-P. Ryckaert, M. L. Klein, and I. R. McDonald, *Phys. Rev. Lett* **58**, 698 (1987).
18. J.-P. Ryckaert, I. R. McDonald, and M. L. Klein, *Mol. Phys.* **67**, 957 (1989).
19. D. E. Williams, *J. Chem. Phys.* **47**, 4680 (1967).
20. P. Van der Ploeg and H. J. C. Berendsen, *J. Chem. Phys.* **76**, 3271 (1978).
21. J.-P. Ryckaert and A. Bellemans, *J. Chem. Soc., Faraday Trans.* **66**, 95 (1978).
22. J.-P. Ryckaert, G. Ciccotti, and H. J. C. Berendsen, *J. Comput. Phys.* **23**, 327 (1977).
23. J.-P. Ryckaert, *Mol. Phys.* **55**, 549 (1985).
24. M. Ferrario and J.-P. Ryckaert, *Mol. Phys.*, **54**, 587 (1985).

25. K. D. Hammonds, I. R. McDonald, and J.-P. Ryckaert, 1989, *Comput. Phys. Commun.*

26. M. P. Allan and D. J. Tildesley, *Computer Simulation of Liquids,* Clarendon Press, Oxford, 1987.

27. H. C. Andersen, *J. Chem. Phys.* **72,** 2384 (1980).

28. J.-P. Ryckaert and G. Ciccotti, *Mol. Phys.* **58,** 1125 (1986).

6

Local Chain Motion Studied
by Molecular Dynamics Simulation
of Polymer Liquid and Glass

D. RIGBY and R. J. ROE
Department of Materials Science and Engineering
University of Cincinnati
Cincinnati, Ohio 45221

ABSTRACT

The dynamics of systems of chain molecules above and below the glass transition temperature have been studied by the molecular dynamics simulation technique, using a fully vibrational n-alkane model with included torsional and nonbonded intersegmental interactions. The two main quantities examined are the bond conformational barrier transition rates and the reorientation motion of individual backbone bonds. On the time scale of 10^2 to 10^3 ps of the MD simulations, conformational transitions cease around 130 K. The activation energy associated with the conformational transitions corresponds closely to a single trans-gauche barrier height and is nearly identical for butane and eicosane. Moreover, the effect of density, and hence intermolecular interactions, on the rates is found to be negligible. The bond reorientation distributions obtained are qualitatively similar to what is expected of a rigid rod undergoing rotational Brownian motion, but cannot be fitted by a theoretical reorientation distribution with a single rotational diffusion coefficient. As the temperature is lowered toward the glass transition temperature, the longer chains studied, decane and eicosane, show evidence of a discontinuous jump motion. This motion, which does not require conformational transitions, has an activation energy in the region of 6.5 kJ/mol, and persists even below the MD glass transition.

INTRODUCTION

The physical properties of glassy polymers reflect the structure frozen in at the T_g. A better understanding of the factors exerting influence on the physical behavior of both freshly prepared and aged glassy polymers can be obtained by elucidation of the static and dynamic aspects of chain molecule behavior in the glass transition region. Toward this aim, in recent work [1–3] we have applied the molecular dynamics simulation technique to study glass formation in rapidly quenched n-alkane-like systems. It has thus been observed that when forced to undergo stepwise cooling at constant pressure, a number of features characteristic of the laboratory glass transition occur within the model systems; for example, one finds discontinuities in the temperature coefficients of the density and internal energy at a temperature that increases with increasing chain length. At slightly higher temperatures, bond conformational transitions cease, leading to a freezing of the overall chain shapes, while at lower temperatures translational diffusion of polymer segments is arrested [1].

The local packing and development of orientation correlation within the systems in relation to the molecular dynamics glass transition has also been investigated via the radial distribution function and orientational correlation functions defined for subchains comprising a small number of backbone bonds [2]. Here it was found that the local packing shows a steady improvement as the temperature is lowered, as revealed by a gradual sharpening of the structure of the liquidlike radial distribution functions (RDFs). However, there was no sudden appearance of new features in the RDFs as the system passed through the glass transition region. Indeed, the splitting of the second peak in the RDF, observed around the T_g in MD simulations of simple liquids [4], was found to occur only at temperatures much lower than the T_g in these polymeric systems. The orientational correlation between subchains in neighboring molecules was also found to increase quite rapidly as the T_g was approached. At high cooling rates, this increase continued at a reduced rate below the T_g. However, when annealed just above the T_g, the systems were found to undergo a transition to an orientationally ordered nematic-type state.

In the present chapter we report a continuation of this program of analyzing the changes that occur as simulated alkane systems are cooled through the T_g, focusing on dynamic rather than static aspects of the behavior. Among the various dynamic behaviors of chain molecules that might be investigated, we concentrate on conformational and reorientational motions of individual bonds within the chain backbone rather than global chain motions (e.g., relaxation of the end-to-end vector). The relaxation times associated with the latter, even in the melt state, are longer than the time scale normally accessible to molecular dynamics simulations by at least two or three orders of magnitude.

COMPUTATIONAL METHOD

A detailed description of the chain model and simulation method has been given previously [1] and thus will be outlined only briefly here. Individual chains are modeled as sequences of particles each representing a CH_2 segment of mass equal to 14 amu. We use a fully vibrational treatment with quadratic potentials controlling bond lengths and valence angles and with a threefold torsional potential determining the conformational transition rates and preference for the trans state. A truncated and shifted Lennard-Jones potential is used for all nonbonded interactions between segments in different chains and for interactions between those separated by more than three bonds along the same chain. The forms of the bond stretching, bending, torsional, and nonbonded potentials may be summarized as follows:

$$E_b = \tfrac{1}{2} k_b (l - l_0)^2 \tag{1}$$

$$E_\theta = \tfrac{1}{2} k_\theta (\cos\theta - \cos\theta_0)^2 \tag{2}$$

$$E_\phi = \sum_{n=0}^{5} a_n \cos^n \phi \tag{3}$$

$$E_{nb} = \begin{cases} 4\epsilon^* \left[\left(\dfrac{r^*}{r} \right)^{12} - \left(\dfrac{r^*}{r} \right)^{6} \right] + \text{const.} & r \le 1.5r^* \\ 0 & r > 1.5r^* \end{cases} \tag{4}$$

Values used for the various constants in Eqs. (1) to (4) are those appropriate for an n-alkane system, and these are summarized in Table 1. Note, however, that the bond stretching force constant has been weakened by a factor of about 7 compared with a realistic value, in order to reduce the frequency of the stretching vibrations and permit use of a larger time step when integrating the equations of motion.

All calculations reported in the present work used a simulated system comprising a cubic box subject to the usual periodic boundary conditions and containing a total of 500 CH_2 groups. The majority of our calculations to date have been performed on systems containing 25 chains each of length 20 segments. To clarify some of the observed behavior, we have also performed a smaller number of simulations on systems containing chains of length, x, equal to 2, 4, and 10. In all cases, the Newtonian equations of motion for individual segments were integrated numerically using the Verlet algorithm [5] with a time step of 1.005×10^{-14} s, at constant energy and volume.

The stepwise cooling procedure is as used previously [2] and comprises simulations of duration 120 ps, followed by a stepwise reduction of temperature by 12° and adjustment of the density to yield the same pressure as the previous run. At the end of each 120-ps segment of the stepwise cooling runs, the system configuration was saved and used to start longer runs of duration between 500 and 2500 ps. The bond reorientation analysis was then performed using sets of coordinates output at intervals between 0.5 and 2.5 ps during these longer runs.

TABLE 1   List of Model Parameters

l_0	0.152 nm
$\cos \theta_0$	−0.3333
k_b	3.46×10^7 J/nm^2 per mole
k_θ	5×10^5 J/mol
k_ϕ	9000 J/mol
a_0	1.0
a_1	1.3100
a_2	−1.4140
a_3	−0.3297
a_4	2.8280
a_5	−3.3943
ΔE_{tg}	2842 J/mol
trans–gauche barrier	11,978 J/mol
cis barrier	43,444 J/mol

The reorientation distributions $W(\theta, t)$ give the probability that at some time t relative to an arbitrary time origin, a bond has rotated through angle θ from its orientation at time zero. In evaluating $W(\theta, t)$ from the present simulation data, the functions were averaged over all possible choices of time origin and over all bonds within the system. In selected cases, distributions of the terminal bonds were also determined separately.

The bond conformational transition rates have been evaluated by analysis of the total number of barrier crossings that occur during a given run, where the latter is estimated by monitoring the conformational states of all bonds at fixed intervals (set at 0.1 ps, except when stated otherwise). The resulting overall barrier crossing rate, R^{BC}, is comprised of contributions from the four processes $t \rightarrow g_+$, $t \rightarrow g_-$, $g_+ \rightarrow t$, and $g_- \rightarrow t$. The transition rate for the passage from trans to a gauche state may be obtained using the relation

$$R_{tg}^{BC} = \frac{R^{BC}}{4p_t} \qquad (5)$$

where p_t denotes the fractional population of trans conformers. This approach will overestimate the true transition rate due to the fact that a bond will generally cross and recross the transition state several times before settling into one of the adjoining energy wells. Typical values of this enhancement factor measured in rigorous analyses [6] of n-butane simulation data appear to be of the order of 3 to 4 although the effect of temperature remains to be investigated.

RESULTS AND DISCUSSION

Typical behavior of the density as a function of temperature observed during stepwise cooling at constant average pressure is shown in Fig. 1 for chains of 20 segments. By analogy with experimental practice, we define the molecular dynamics glass transition temperature as the temperature at which the discontinuity in slope occurs, in this case around 120 K. Also shown in Fig. 1 is the mean overall barrier crossing rate evaluated as described in the preceding section. Thus it is observed that on the time scale of a typical MD simulation run (10^{-10} to 10^{-9} s), barrier crossing would be extremely infrequent at temperatures less than, say, 130 K.

The associated trans-to-gauche barrier crossing rates, R_{tg}^{BC}, evaluated according to Eq. (5) for chains with 4 and 20 segments, are listed in Table 2 and are plotted in Arrhenius form in Fig. 2. The two straight lines are essentially parallel, with an associated activation energy of 11.6 kJ/mol. This result may be compared with two earlier sets of simulations performed using the same torsional potential. In the first, Brownian dynamics simulation was performed on a phantom ring molecule with 200 bonds [7]. The data were analyzed by the technique of hazard analysis [8] and were found to give an activation energy of 13.8 kJ/mol. In the second, MD simulations were

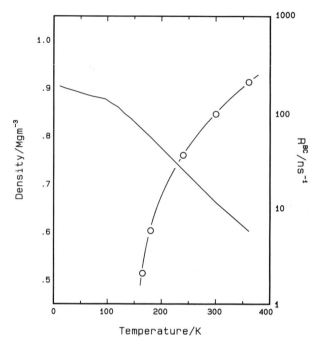

Figure 1   Behavior of density as a function of temperature observed during step-wise cooling simulations on a system containing twenty-five 20-segment chains. The circles show the overall barrier crossing rates measured during the stepwise cooling.

TABLE 2   Summary of Barrier Crossing Rates and Trans Conformer Populations for 4- and 20-Segment Chains Obtained During Simulations at Constant Pressure

Temperature (K)	Density (g/cm^3)	Transition Rate R_{tg}^{BC} (ns^{-1})	p_t
		$x = 4$	
361	0.3536	85.15	0.617
301	0.3956	46.12	0.634
240	0.4376	14.49	0.729
180	0.4952	2.166	0.766
		$x = 20$	
361	0.6046	72.53	0.678
301	0.6614	33.19	0.718
240	0.7276	11.62	0.762
180	0.7970	1.695	0.854
120	0.8230	0.5928	0.873

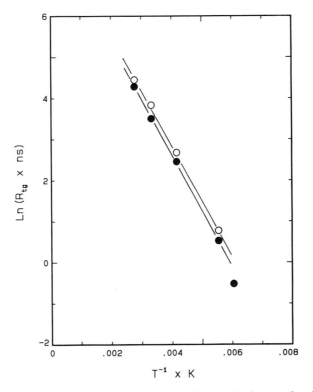

Figure 2 Arrhenius plots of the trans-gauche barrier crossing rate for chains of length $x = 4$ (o) and $x = 20(\cdot)$. The associated activation energy is 11.6 kJ/mol.

performed on butane with fixed bond lengths and angles [9]. When analyzed in terms of barrier crossing rates, the data give an activation energy of 10.0 kJ/mol.

It may also be noted that the absolute trans-to-gauche barrier crossing rates measured in the present work are about 25% higher than the rates found in the foregoing butane simulations and in subsequent simulations of a similar nature [6,10]. Since the apparent barrier crossing rate is sensitive to the sampling interval at which barrier crossings are checked, one might wonder whether this 25% difference is an artifact arising from small differences in the procedures used by the various authors. Figure 3 shows the effect of the sampling interval on the apparent barrier crossing rate. Symbols □ and ○ refer to those obtained at 240 K and 361 K at $x = 20$ in this work, while symbols ◇ are taken from the data of Ref. 6 for butane at 291.4 K. The sampling interval of 0.1 ps used in the present work is larger than the value of 0.02 ps used in Ref. 6. If a comparison were made on an identical basis, the difference in barrier crossing rates would actually be larger than 25%. Hence the difference seems to be significant and may possibly reflect differences in

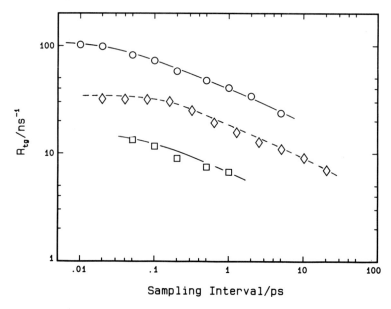

Figure 3　Effect of sampling interval on the apparent trans-gauche barrier crossing rate. O, 20-segment chains at 361 K; □, 20-segment chains at 240 K; ◇, 4-segment chains at 291 K (Ref. 6).

the dynamics of the constrained and unconstrained models. A previous Brownian dynamics simulation has shown that simultaneously increasing the strength of the bond stretching and bending force constants by a factor of $10^{0.5}$ leads to a decrease of about 35% in the trans-gauche conformational transition rate [11]. In a molecular dynamics simulation [12] of a protein, it was also found that the bond-length constraint introduces no significant modification of the dynamic properties of the molecule, while the imposition of both bond-length and bond-angle constraints markedly reduces conformational transition rates. This result thus suggests that the conformation transition rates evaluated in this study has in no way been affected adversely by the weakened bond stretching force constant used in our model.

　　The effect of densification on the barrier crossing rates, listed in Table 3

TABLE 3　Barrier Crossing Rates at Various Densities Measured During Simulations with 20-Segment Chains at 361 K

Density (g/cm^3)	Transition Rate R_{tg}^{BC} (ns^{-1})	p_t
0.6046	72.53	0.678
0.7276	70.47	0.684
0.7971	68.57	0.688
0.8603	71.00	0.685
0.9040	68.53	0.693

for the 20-segment chain at $T = 361$ K, is observed to be relatively minor when the density is varied over some considerable range from much below the actual density at 1 atm (about 0.743 g/cm^3; see Ref. 13) to a value achieved experimentally at pressures in excess of 10 kbar.

This seeming independence of bond conformational dynamics from density deserves further study, since it suggests the possibility that computationally less expensive studies of transitions in isolated molecules, by means of either molecular dynamics or the dynamic rotational isomeric state model [14], might be used to provide significant information on the transition behavior in the corresponding bulk systems.

Local chain motions have been investigated in this work by evaluating the bond reorientation distributions $W(\theta, t)$. Such distributions are much more informative than the autocorrelation functions which are more often presented for similar purposes. Recent advances using the two-dimensional NMR technique have also opened up the possibility [15] of experimental determination of such distributions against which the simulation results can be compared. Figure 4 illustrates typical behavior found at temperatures well above the T_g, in this case for $x = 20$ at 240 K and at times 1, 2, 5, 10, 20, and 50 ps. A gradual fanning out of the distributions is observed, as expected, as the elapsed time increases, eventually approaching $W(\theta, \infty) = \frac{1}{2} \sin \theta$, which one expects when the bond orientation correlation between the initial and final states is totally lost. The features exhibited in Fig. 4 are qualitatively similar to the behavior of a rigid rod undergoing rotational Brownian motion, or a rotational random

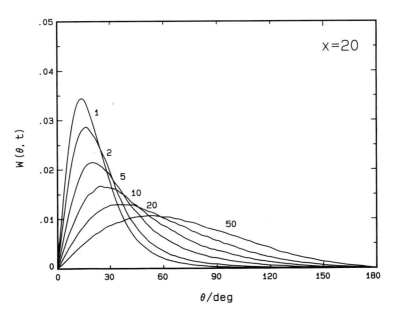

Figure 4   Reorientation distributions of chain backbone bonds within a 20-segment chain at 240 K at various times. Times given with curves are in picoseconds.

walk via a series of small angular steps. The distribution $W(\theta, t)$ for rotational diffusion is given [16–18] by

$$W(\theta, t) = \frac{1}{2} \sum_{n=0}^{\infty} (2n + 1) \exp[-n(n + 1)D_r t] P_n(\cos \theta) \sin \theta \qquad (6)$$

where D_r denotes the rotational diffusion coefficient and the P_n are the Legendre polynomials. Quantitatively, however, the observed behavior differs from rotational Brownian motion, which gives rise to much narrower reorientation distributions than those calculated from the MD simulations. This is illustrated in Fig. 5, which compares the "experimental" curves of Fig. 4 at 1 and 5 ps with distributions calculated according to Eq. (6), with D_r chosen to fit the experimental peak positions. A more detailed interpretation in terms of a distribution of rotational diffusion coefficients is in progress.

As the temperature is decreased to about $T_g + 40$ K and below, one observes a new feature in the reorientation distributions in the form of a second peak centered around an angle of about 70°. This is illustrated in Fig. 6, in which the distributions for the $x = 20$ chain after 50 ps are plotted at various temperatures. It is apparent that the bimodal feature, which first becomes visible around 165 K, persists at and below the T_g. (The feature is also observed at $T = 60$ K, but becomes noticeable only at times greater than about 500 ps.) The fact that the position of the second peak does not change with time suggests that it is associated with a jump of about 70° all at once, not

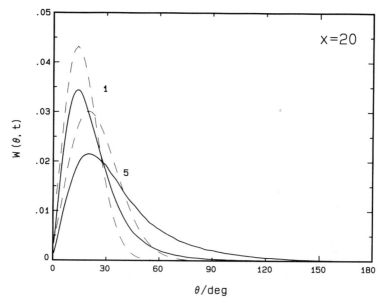

Figure 5 Comparison of reorientation distributions at 1 and 5 ps (from Fig. 4) with curves calculated according to Eq. (6). At $t = 1$ ps, the fitted curve was obtained using $D_r = 0.0291$ ps^{-1}. The corresponding D_r value at 5 ps is 0.0124 ps^{-1}.

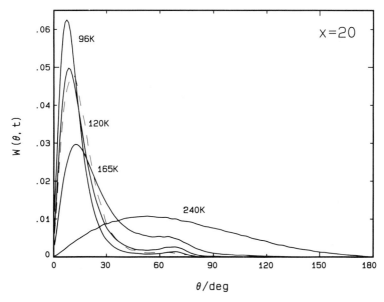

Figure 6   Effect of varying the temperature along the stepwise cooling curve of Fig. 1 on the reorientation distributions at $t = 50$ psi.

through a succession of small-angle reorientations. Its persistence at low temperatures indicates that it is not associated with conformational transitions, as further verified by noting the absence of any transitions during the simulations used to obtain the curves in Fig. 6. The feature also occurs in distributions evaluated for terminal bonds only, as shown by the dashed curve at $T = 165$ K.

A precise estimate of the activation energy for the process associated with the second peak would require deconvolution of the distributions arising from the diffusive-type motion and the discontinuous, 70°, motion. In the absence of a rigorous quantitative procedure for eliminating the diffusive component, we have obtained an approximate estimate of the activation energy by considering the reorientation distributions at $T = 165$, 120, and 96 K when the two components are relatively easy to distinguish (suitable curves are those at 10, 50, and 500 ps for $T = 165$, 120, and 96 K, respectively). The resulting Arrhenius plot indicates an activation energy in the range 6.5 ± 1 kJ/mol, or a little over one-half of the energy associated with conformational transitions.

Further information regarding the nature of the motion responsible for the bimodal reorientation feature has been obtained by performing simulations of chains of length 2, 4, and 10 segments. Reorientation distributions for these three systems just above their T_g values (30, 54, and 96 K) are shown in Figs. 7 to 9. It thus appears that the bimodal distributions are absent for $x = 2$ and $x = 4$, but that the feature becomes noticeable at $x = 10$, although less distinct

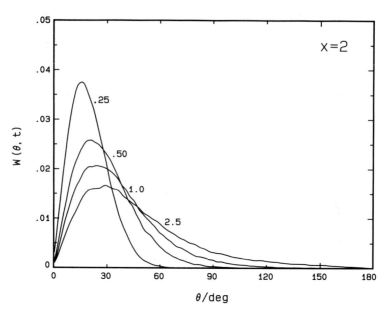

Figure 7   Reorientation distributions for dumbbell molecules ($x = 2$) just above the T_g (30 K). Times in picoseconds are given with curves.

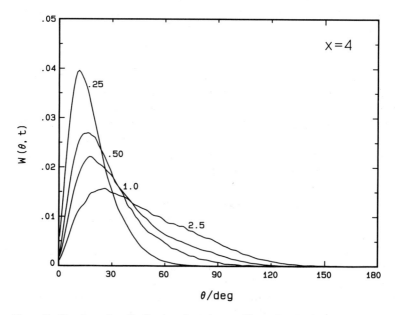

Figure 8   Reorientation distributions for n-butane-like molecules just above the T_g (54 K). Times in picoseconds are given with curves.

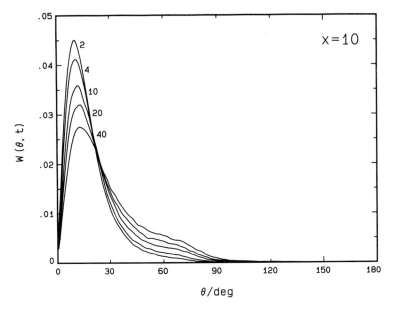

Figure 9 Reorientation distributions for *n*-decane-like molecules just above the T_g (96 K). Times in picoseconds are given with curves.

(perhaps masked more effectively by the diffusive component) than for the $x = 20$ calculations.

Some preliminary investigations of the physical nature of the discontinuous motions have been performed by identification of individual chains before and after the reorientation. Figure 10 illustrates the first two such events observed in the $x = 20$ simulations at 120 K. Both configurations shown contain extended trans-planar zigzag sequences, although neither are fully extended. In Fig. 10, translation of the chain along the axis of the planar

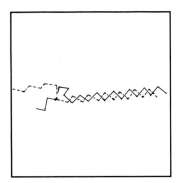
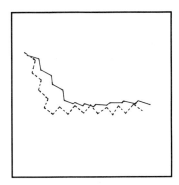

Figure 10 Sketch of two chains before and after undergoing a discontinuous jump motion. The internal configurations of each molecule are the same before and after the jump.

zigzag is apparently combined with rotation through 180°. Such a screw-type motion will result in fairly discrete reorientation of all the bonds in the planar zigzag portion of the chain by 71° ($=180° - 109°$). Possible movement of defects, called "dispiration loops," in crystalline polyethylene by a similar screw-like motion of chains has been modeled [19]. Such motions require relatively little cooperative displacements of surrounding chains, which are essentially immobilized by the crystalline lattice or the frozen glassy structure. The failure to detect such motions in systems of very short chains is presumably a consequence of the fact that they are associated with some lengths of planar zigzag conformations. It is not, at present, clear whether the existence of the discontinuous motion requires the presence of chain ends, and is therefore a characteristic feature only of molecules of intermediate length. Further studies of this motion, combined with simulations of longer chains, are thus desirable.

SUMMARY

In the present work we have performed a preliminary investigation of the nature of local chain motions as a supercooled system of chain molecules is cooled through the glass temperature using molecular dynamics simulation. Bond conformational transitions are observed to occur with an activation energy close to the trans-gauche barrier height as defined by the torsional potential for chains of length 4 and 20 segments. Moreover, for $x = 20$, the effect of densification on the rates is shown to be negligible. Hence, neither intermolecular segment–segment interactions, nor long-range intramolecular interactions, influence the activation energy for conformational transitions to any significant extent. The measured activation energies are quantitatively in close agreement with those deduced by Brownian dynamics simulation method for long chains, and by the constrained MD simulation method for butane. Comparison of absolute barrier crossing rates for butane simulated with and without bond length and angle constraints suggests that the fully vibrational model gives a rate at least 25% higher.

As the systems containing the longer molecules ($x = 10$ and $x = 20$) are cooled toward the T_g, a discontinuous 70° jump motion with an activation energy around one-half that associated with conformational transitions becomes apparent. The motion evidently is associated with nearly simultaneous reorientation of several bonds and takes place without the occurrence of conformational transitions.

ACKNOWLEDGMENT

This work was supported in part by NSF Grant DMR-8520921. The computation reported in this work was performed on a Cray Y-MP/832 at the Pittsburgh Supercomputing Center.

REFERENCES

1. D. Rigby and R. J. Roe, *J. Chem. Phys.* **87,** 7285 (1987).
2. D. Rigby and R. J. Roe, *J. Chem. Phys.* **89,** 5280 (1988).
3. D. Rigby and R. J. Roe, *Macromolecules* **22,** 2259 (1989).
4. J. H. R. Clarke, *J. Chem. Soc., Faraday Trans. 2* **75,** 1371 (1979).
5. L. Verlet, *Phys. Rev.* **159,** 98 (1967).
6. D. Brown and J. H. R. Clarke, preprint.
7. E. Helfand, Z. R. Wasserman, and T. A. Weber, *Macromolecules* **13,** 526 (1980).
8. K. V. Bury, *Statistical Models In Applied Science,* Wiley, New York, 1975.
9. R. Edberg, D. J. Evans, and G. P. Morriss, *J. Chem. Phys.* **84,** 6933 (1988).
10. R. Edberg, D. J. Evans, and G. P. Morriss, *J. Chem. Phys.* **87,** 5700 (1987).
11. E. Helfand, Z. R. Wasserman, T. A. Weber, J. Skolnick, and J. H. Runnels, *J. Chem. Phys.* **75,** 4441 (1981).
12. W. F. van Gunsteren and M. Karplus, *Macromolecules* **15,** 1528 (1982).
13. Texas A&M University System, *TRC Thermodynamic Tables: Hydrocarbons,* Texas A&M University, College Station, Tex. (1986).
14. I. Bahar and B. Erman, *Macromolecules* **20,** 1368, 2310 (1987).
15. S. Wefing, S. Kaufmann, and H. W. Spiess, *J. Chem. Phys.* **89,** 1234 (1988).
16. P. H. Roberts and H. D. Ursell, *Proc. Roy. Soc. London* **A252,** 317 (1960).
17. E. N. Ivanov, *Zh. Eksp. Teor. Fiz.* **45,** 1509 (1963) [*Sov. Phys. JETP* **18,** 1041 (1964)].
18. R. I. Cukier and K. Lakatos-Lindenberg, *J. Chem. Phys.* **57,** 3427 (1972).
19. D. H. Reneker and J. Mazur, *Polymer* **29,** 3 (1988).

7

Simulation of Polymer Liquids and Glasses

RICHARD H. BOYD and KRISHNA PANT
Department of Materials Science and Engineering
and Department of Chemical Engineering
University of Utah
Salt Lake City, Utah 84112

ABSTRACT

The use of Monte Carlo statistical mechanical methods in realistically simulating the structure of simple polymeric liquids is discussed. Off-lattice constant-pressure simulations where the system volume and thermal expansion are predicted rather than imposed can be carried out. In addition to PVT predictions, in conjunction with vibrational analysis, the constant-pressure heat capacity can be calculated. An equilibrated liquid prepared by simulation forms an excellent starting point for simulating polymeric glasses. The structures of glasses prepared by energy minimization starting from the liquid are described. The distribution of free volume as holes defined as tetrahedral interstitial sites is discussed.

INTRODUCTION

The realistic simulation of the structure and properties of polymeric liquids and glasses has proven to be an extraordinarily difficult task. The molecules are intimately intertwined, yet individually pursue courses that are conformationally largely oblivious of other chains. In using equilibrium statistical mechanical Monte Carlo methods, this makes it difficult to devise moves that are efficient in producing the intramolecular torsional sampling in the presence of the intermolecular constraints. Simply put, random variation of torsional angles produces molecular excursions that are almost certain to produce serious intermolecular interferences. If the torsional variations are small, an unacceptably large number of moves are required to sample the torsional space. The same considerations render the physical time scales of molecular motion long enough that direct integration of the equations of motion [molecular dynamics (MD)] is limited to trajectories (100 to 1000 ps) that are relatively short compared to the time scale of structure formation for many processes of interest, such as vitrification.

There are promising developments, however. In the Monte Carlo case, lattice simulations have found that moves by reptation are effective in sampling configuration space even when the lattice site occupancy is high (but not, of course, fully occupied) [1–4]. Most important, in the context of simulations incorporating realism, the reptational move can be effected off-lattice in a continuous space [5,6]. Molecular dynamics is proving valuable as well since considerable useful information can be extracted from the trajectories available [7]. In the present work, recent use of the reptational Monte Carlo method in simulating simple polymeric liquids, with sufficient realism that comparisons with experiment may be undertaken, is described.

A major element of realism that can be injected into the simulation concerns the condition of constant pressure as opposed to constant volume. Most simulations are carried out with molecules confined to a periodic box of fixed dimensions. However, this invokes the supposition that the density of the system being studied is known. In fact, the response of polymer volume to temperature and pressure constitutes one of the more important properties to be predicted. It has been found that it is possible to carry out an off-lattice constant-pressure simulation, one where the volume adjusts to an equilibrium value at a given temperature and pressure along with other properties [8].

A realistic simulation of an equilibrated liquid, especially one carried out at constant pressure, forms an ideal starting point for the simulation of polymeric glasses. Thermal expansion and the change of free volume with temperature and pressure form the basis of many models for the formation of glasses. Thus a simulation that incorporates volume change with temperature and thus thermal expansion is particularly appropriate [9]. The structure of glasses constructed from such liquids is also described here.

MONTE CARLO SIMULATION OF POLYMERIC MELTS

Reptational Monte Carlo Moves

As used here, *reptation* means the moving of a chain atom from one end of the chain to the other, cutting a bond at one end (end 1) and making a new one at the other (end 2) in the process (Fig. 1). As a result, the chain has moved by one chain atom along its contour. In placing the atom at its new position, the torsional angle ($\phi_{end\,2}$) is selected at random. If another atom from end 1 later moves to end 2, the torsional angle already there, $\phi_{end\,2}$, moves one position to the interior, and so on. After a large number of success-ful moves where the ends are selected at random, a given torsional angle placed at an end will move diffusively through the chain. In any given snapshot it can be in any position in the chain. In essence, the reptational move is useful because by creating the torsional angle change at the chain end, the problem of large chain stem excursions caused by a large torsional angle change at an interior bond is eliminated. In practice, the reptation moves are supplemented by moves where an attempt is made to reset the torsional angle at an end without moving the atom to the other end (i.e., a purely "bond rotational" move). This speeds up the torsional angle equilibration.

To be useful, there has to be sufficient room around the chain ends that an acceptable number of placement attempts are successful. It appears to be the case that in real homologous series of liquids such as the n-alkanes, there is in fact a large density or packing defect associated with the chain ends. For

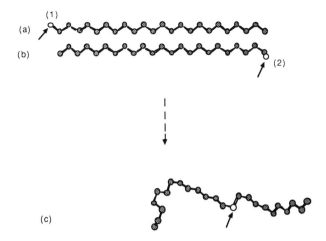

Figure 1 Reptational moves. (a) An end at (1) is removed and (b) placed at the other end (2) with a random torsional angle. (c) After many such successful attempts at either end, the original bead moves diffusively through the chain and the torsional angles throughout the chain become disordered (and the chain itself moves along its contour).

example, the group molar volume at 20°C of the —CH_3 group in the liquid n-alkanes is 33 cm³/mol but that for the —CH_2— group is 16 cm³/mol. In the simulations this translates into an acceptable success rate. In any event, the method appears to be ergodic [6]. Any chain configuration can be reached and in a reasonable number of steps.

There are disadvantages. The migration of the torsional angles, originally set at the chain ends, through the chain is diffusive; that is, of the order of N^2 moves are required to effect movement of an end-set torsional angle through N chain bonds. In practice, this places a restriction to relatively short chains in off-lattice simulations. Another obvious restriction is that in order to have sufficient room near the chain end for a successful move, the group being moved must be small. This probably means that single atoms or single atoms plus substituent hydrogen atoms should be moved. However a number of important polymers can be reptated by moves decomposed into events of this type. Polypropylene, for example, can be reptated by moving the —CH_3, >CH—, and —CH_2— groups as separate events.

The use of the reptation move in Monte Carlo simulation, as discussed here, is a means of achieving equilibrated torsional angles (i.e., sampling configuration space) and is not meant or intended to be a physically realistic move. It has no bearing one way or another on the question of whether polymer molecules actually move physically along their chain contours. There is a view which states that when incorporated with other moves, the reptational Monte Carlo method does approximate a master equation for the chain motion and can form the basis for a *dynamic* Monte Carlo simulation of the time evolution of chain motion and thus explore the question of reptation as a mode of physical chain motion [2,3]. However, in the use described here, this question is not meant to be addressed.

Constant-Pressure Simulation

The methodology for carrying out a constant-pressure simulation is well known and is based on the distribution function, $f(\mathbf{x}, V)$, for the classical isobaric, isothermal ensemble [10]:

$$f(\mathbf{x}, V) = \frac{e^{-\beta(U(\mathbf{x}, V) + pV)}}{Z}$$

$$Z = \int_0^\infty dV \int d\mathbf{x}\, e^{-\beta(U(\mathbf{x}, V) + pV)} \tag{1}$$

where $\beta = (k_B T)^{-1}$, the potential energy; $U(\mathbf{x}, V)$ is a function of the atomic coordinates $\mathbf{x}$ for a given volume V; and the pV term in the exponential is the pressure p times the system volume. In other words, the particle coordinates are to be integrated over (sampled) for a given volume and the volume is to be integrated over (sampled). A simulation proceeds (1) by carrying out a number of reptation–rotation moves for a fixed volume, and (2) occasionally, a volume fluctuation is attempted under a pressure p introduced by the pV term. In part (1) the usual ΔU evaluation is made and considered via the Metropolis rules. In part (2) a volume fluctuation is attempted and the Metropolis consideration is based on $\Delta U + p\,\Delta V = U''(\mathbf{x}'', V'') - U(\mathbf{x}', V') - p(V'' - V')$.

Consideration must be given to how an unacceptable rejection rate for the volume fluctuation can be avoided. That is, the trial atom configuration after the volume change, $\mathbf{x}''$, must be chosen as well as the trial new volume, V''. Because the atoms are partitioned into valence-bond-connected sets (i.e., a number of molecules), it is easy to imagine that a practical procedure for choosing the new atom configuration so that some volume changes are accepted would be difficult to devise. Certainly, the coordinates are not to be individually scaled by the volume. That would result in always unacceptable changes in bond lengths and angles. However, a very simple procedure that gives an adequate acceptance rate is the following. One end of each chain is picked arbitrarily and moved affinely with the proposed volume change (Fig. 2). The rest of the molecule is moved rigidly by this translation. The periodic boundary conditions are, of course, reapplied after the translations. Although most of the volume changes do result in chain movements that give rise to rejection because of the generation of serious intermolecular steric interferences, an acceptable number are successful.

A complete recipe for such a Monte Carlo simulation could consist of the following [8]:†

---

† This is the procedure used in Ref. 8 except for deletion of a feature of making several attempts at reptation at the same end. This change has had no observable effect on results but was made to remove any possible uncertainty concerning the symmetry of transition probabilities.

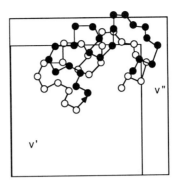

Figure 2 Volume fluctuation. The atom coordinates of a chain are placed in the new box (filled circles) of volume V'' by moving the starting atom of the chain in the original box (open circles) of volume V' to a new position determined affinely by $V' \rightarrow V''$. The rest of the chain is moved rigidly by this translation.

1. Select an end at random.
2. Part of the time, move the bead to the other end with a randomly selected torsional angle. Part of the time, keep the bead at the selected end but change the torsional angle randomly.
3. Evaluate the energy change in step 2; if ≤ 0, accept; if > 0, accept with probability $\exp[-\beta \Delta U(\mathbf{x}, V)]$ (i.e., if a random number, $0 \rightarrow 1$, is $\leq \exp[-\beta \Delta U(\mathbf{x}, V)]$ accept).
4. If the move suggested in step 2 is not accepted, start over at step 1.
5. After $N_{\text{rep/rot}}$ cycles starting at step 1, attempt a volume variation by changing the periodic box size. Evaluate the energy change, but base acceptance or rejection on $\Delta U(\mathbf{x}, V) + p \, \Delta V$.
6. Begin over at step 1 for $N_{\text{rep/rot}}$ more attempts.

The selection in step 2 between reptation and rotation without reptation has been made randomly with probability $= 0.5$. The number of reptation or rotation attempts, $N_{\text{rep/rot}}$, before a volume fluctuation is attempted has been taken as 1000.

Results on Liquids

Figure 3 shows the results obtained previously for the PVT behavior of polymethylene [8]. In that simulation 32 chains of 24 methylene units each were used. To save computer time, the hydrogens were not explicitly included but a "united atom" Lennard-Jones 6-12 potential appropriate for interactions between CH_2 units was employed. Bond lengths and valence angles were held fixed. Agreement between the simulation and experimental [11] values of the specific volume for tetracosane ($C_{24}H_{50}$) was good. The compressibility calculated compared well with experimental values for polyethylene (Fig. 4).

The heat capacity is a property that would be very useful to calculate. It cannot be obtained directly because the heat capacity is dominated by the internal molecular vibrations and the bond lengths and valence angles are fixed. Furthermore, the simulation is a classical one and the vibrations are

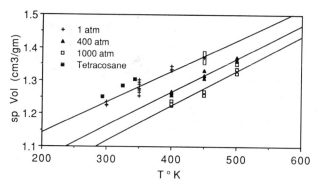

Figure 3 Specific volume of liquid C_{24} polymethylene versus temperature along 3 isobars (from Ref. 8). Points are the results of individual simulation runs except at 450 K and 1 atm, where the vertical dimension of the box indicates the standard deviation of 11 runs. The lines are least-square fits. The points shown for tetracosane, $C_{24}H_{50}$, are experimental (1 atm).

highly quantized. However, the internal vibrations are relatively insensitive to environment and can be calculated on typical individual molecules isolated from the melt simulation. The interesting contribution is that from the change in intermolecular packing energy with temperature (and the attendant volume change). There is also a contribution from the change in the gauche/trans occupational ratio with temperature. Figure 5 shows the calculated heat capacity resolved into its components and compared with experiment [12].

Other quantities that can be calculated include the mean-square end-to-end distance and the effect of pressure on the gauche/trans population ratio. The former is found [6,8] to be in good agreement with predictions of rotational isomer state calculations and thus confirms the idea that the melt is a

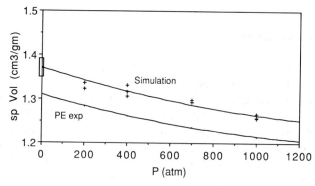

Figure 4 Specific volume of C_{24} polymethylene versus pressure at 450 K (from Ref. 8). The points are individual simulation runs except at 1 atm, where the vertical dimension of the box indicates the standard deviation of 11 runs. The curve without points is the experimental behavior of PE (from Ref. 11).

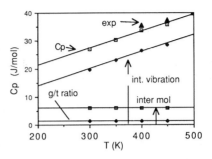

Figure 5 Heat capacity of the liquid C_{24} polymethylene, labeled C_p, from simulation. Experimental values (from Ref. 12) are also shown. The contributions to C_p from the internal vibrations, the temperature dependence of the gauche/trans ratio, and the intermolecular packing energy are also shown.

theta solvent for its own chains. No change in gauche/trans ratio with pressure was found.

POLYMERIC GLASSES

As described, for the liquid simulation above, thermal expansion at a given pressure follows from the modeling rather than being imposed. This feature is especially important in assessing the atomistic packing changes that take place in the liquid with lowering temperature and precede vitrification. However, being a Monte Carlo method, the simulation can only represent the system at equilibrium and the study of glass formation per se cannot be addressed. An equilibrated liquid does constitute an attractive starting point for preparing a glass via other simulation methods. There are only two, but closely related, choices for doing this at present. One is molecular dynamics. Since the MD time trajectories in very long runs correspond to the order of a nanosecond, forming a glass from the liquid corresponds to an extremely fast quench to a finite temperature. The assessment of volume changes at constant pressure is indirect. A somewhat simpler but related alternative is energy minimization. If energy minimization is thought of as a gradient, steepest descent method, it corresponds to MD carried out at 0 K (zero kinetic energy and vanished acceleration term). The incorporation of variable volume in energy minimization is relatively straightforward. If liquids prepared by Monte Carlo simulation are energy minimized, physically, the process simulated can be thought of as an extremely fast quench to 0 K from an equilibrated liquid. It differs from previous atomistic simulations by energy minimization [13] in that by starting from the liquid the role of energy minimization is computationally relatively minor and in that a physical process for the glass formation is fairly well defined.

Calculations

The conformational model used for glass preparation was the same as for liquid simulation [8] except that bond stretching and bending terms using previously adopted force constants [14] were included. The energy minimizations were carried out using a conjugate gradient quasi-Newton technique. The system volume was included as a variable along with the bead coordinates in the minimization. This involved including the box size as a variable in the formulation of the periodic boundary conditions.

Considerable attention attaches to the concept of free volume and its distribution in the glass. No one definition is accepted for this quantity. In order to assess the packing in terms of a concept that resembles the idea of "holes," the following scheme was implemented. The concept of interstitial sites is much used in the study of crystals. A tetrahedral interstitial site is described by inscribing the largest sphere at a site (Fig. 6); that is, its center is defined by its being equidistant from four points and its radius by subtracting the radii of the objects at the four points. The center is found by the mutual intersection of three planes that perpendicularly bisect lines joining the four points.

In defining what the objects are that constitute the filled space, a simple definition is to place a sphere at each methylene unit. The radius of the filling objects can be defined several ways, but one reasonable definition is to take it as half the Lennard-Jones "repulsion distance," the distance where the potential crosses zero. Then it is a relatively easy task to find an "open" or unfilled place. Points can be selected at random and the atom coordinates scanned to generate a near-neighbor list. If the point lies outside the radii of all the neighboring points, the space at the randomly selected point is open. Drawing the largest sphere nearby is not straightforward, however, because there can be a relatively large number of neighboring objects. If four are selected to place an interstitial point equidistant, the inscribed sphere may intercept nearby objects not of the four used to create the interstitial sphere. Complete tesselation of the space to find all the interstitial spaces accurately is very time consuming. An approximate procedure used by us is to select the four *nearest* objects to a random point in unfilled space and to find the interstitial sphere (i.e., the point equidistant from the four objects). Then the interstitial sphere is checked to see if it is valid in not intercepting other objects. If not, it is accepted. About two-thirds of the attempts to locate a tetrahedral interstitial site in this manner are successful.

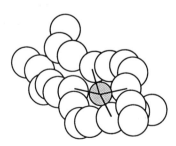

Figure 6 Depiction of interstitial sites (after Ref. 9).

Results on Glasses

A question of interest is the degree to which the glass formed in energy minimization depends on the state of the liquid from which it was prepared. Minimizations have been carried out on liquids generated at 300, 350, and 400 K, all at 1 atm. At each of the temperatures, three snapshots were selected for minimization to assess the variability of minimized structure with the starting coordinates. It may be seen in Fig. 7 that there is essentially no difference in the specific volumes of the energy-minimized 300- and 350-K liquids. There may be some tendency for the 400-K liquid to exhibit a slightly larger minimized 0-K volume.

Figure 8 shows the distribution of tetrahedral interstitial sites for the liquid and compares it with that for the glass. The distribution is more sharply peaked and has a somewhat smaller peak-value radius in the glass.

Another point of interest concerns the relative role of intermolecular packing energy versus intramolecular energy in going from the liquid to the glass. Table 1 shows the change in packing energy (nonbonded energy) along with the intramolecular bond stretching, bending, and twisting energy on going from the *energy-minimized system constrained at the liquid volume* to the system minimized with respect to volume (0-K glass). It is seen that the

Figure 7 Specific volume versus temperature from Monte Carlo liquid simulation (from Ref. 8) and of three 0-K glasses made by energy minimization of 300, 350, 400 K liquids (from Ref. 9). Three snapshots at each starting temperature were minimized. The arrows connect the starting temperatures with the energy-minimized points. The latter are offset from 0 K by different amounts for clarity.

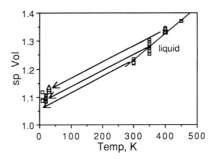

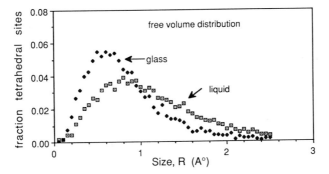

Figure 8 Free-volume distribution expressed as tetrahedrally coordinated vacant sites in the liquid at 300 K compared to energy minimized 0 K glass (from Ref. 9).

TABLE 1 Partitioning of Energy Change on Going from Energy-Minimized System Constrained at Liquid Volume to Energy-Minimized 0-K Glass[a]

Packing Energy	Intramolecular Energy				
	Bond Stretch	Bond Bend	Torsion	Nonbond	Total
−26.4	+1.5	+5.8	+7.4	+5.0	+19.7

[a] Units are kJ/system box.

better packing experienced in the volume-minimized system is achieved at the expense of intramolecular valence distortions, and the energy minimum with volume participating is a balance between inter- and intramolecular contributions. In fact, the net lowering in energy on volume contraction is considerably smaller than the energy increase in valence distortions considered separately.

ACKNOWLEDGMENTS

The authors are indebted to the Polymers Program, Division of Materials Research of the National Science Foundation for financial support, and to the San Diego Super Computer Center for a grant of computer time.

REFERENCES

1. A. Baumgaertner, *J. Chem. Phys.* **84,** 1905 (1986); a number of references to previous work are contained here also.
2. A. Kolinski and J. Skolnick, and R. Yaris, *Macromolecules* **19,** 2550, 2560 (1986).
3. A. Kolinski, J. Skolnick, and R. Yaris, *J. Chem. Phys.* **86,** 1567, 7164 (1987).
4. K. Kremer, *Macromolecules* **16,** 1632 (1983).
5. M. Bishop, D. Ceperly, H. L. Frisch, and M. H. Kalos, *J. Chem. Phys.* **72,** 3228 (1980).
6. M. Vacatello, G. Avitabile, P. Corradini, and A. J. Tuzi, *J. Chem. Phys.* **73,** 548 (1980).
7. D. Rigby and R. J. Roe, *J. Chem. Phys.* **87,** 7285 (1987); *J. Chem. Phys.* **89,** 5280 (1988); *Macromolecules* **22,** 2259 (1989).
8. R. H. Boyd, *Macromolecules* **22,** 2477 (1989).
9. R. H. Boyd and K. Pant, *Polym. Prep., Am. Chem. Soc., Div, Polym. Chem.* **30**(2), 30 (1989); full manuscript in preparation.
10. T. Hill, *Statistical Mechanics,* McGraw-Hill, New York, 1956.
11. D. P. Maloney and J. M. Prausnitz, *J. Polym. Sci., Polym. Phys. Ed.* **18,** 2703 (1974).
12. U. Gaur and B. Wunderlich, *J. Phys. Chem. Ref. Data* **10,** 119 (1981).
13. D. N. Theodoru and U. Suter, *Macromolecules* **18,** 1467 (1985).
14. R. H. Boyd and S. M. Breitling, *Macromolecules* **7,** 855 (1974).

Investigation by Atomistic Simulation of Structural and Dynamic Differences in the Glassy and Liquid States of Atactic Poly(propylene)

M. F. SYLVESTER
Program in Polymer Science and Technology
and Department of Mechanical Engineering
Massachusetts Institute of Technology
Cambridge, Massachusetts 02139

S. YIP
Department of Nuclear Engineering
Massachusetts Institute of Technology
Cambridge, Massachusetts 02139

A. S. ARGON
Department of Mechanical Engineering
Massachusetts Institute of Technology
Cambridge, Massachusetts 02139

ABSTRACT

The molecular dynamics simulation technique (canonical, isobaric ensemble) has been used to investigate the structural and dynamic characteristics of a model for atactic poly(propylene). Relaxed structures at temperatures above and below the experimental T_g were generated by means of a sequence of heating and stress relaxation steps starting from configurations of static energy minimized RIS polymer chains. Calculated properties of the simulated polymer were found to be in reasonable agreement with experiment. The detailed microstructure of the simulated polymer was studied by use of Voronoi tessellation of the simulation volume. Results indicated there are substantial structural differences between the glass and liquid state. In particular, evidence was found for the development of spatial heterogeneity in the distribution of volume available to chain segments below T_g. Preliminary data also revealed significant dissimilarity between the dynamics of the simulated polymer above and below the glass transition.

INTRODUCTION

Atomistic simulation techniques, in conjunction with experimental and theoretical approaches, have proven themselves to be useful tools for the study of materials. Recent years have seen the first application of a variety of simulation methods to investigating the phenomena occurring in bulk polymers [1–5]. Among these, possibly the most important and challenging is the glass transition. Despite its significance, the glass transition is still poorly characterized except in a phenomenological manner. For example, comprehension of how the structure of a polymer glass differs from that of the liquid is limited to the somewhat vaguely defined concept of "free volume" [6,7]. It is not known how the volume associated with atoms might be distributed within the polymer or how changes in the distribution are related to the property changes seen at the glass transition. In addition, what other structural changes occur on passing between the liquid and glass states are not known. Similarly, the change in chain dynamics that occurs on passage through the glass transition is not understood beyond the idea that a shift from long-range chain motion to local, correlated motion must take place. The nature of these motions or the definition of the terms "long range" and "local" remains to be established.

These questions involve the local atomic-level structure and motion of individual or small groups of atoms. Available experimental and theoretical methods are not well suited to investigate these details. Therefore, simulation is the only way to obtain physical insight into these problems at the present time. Reported here are initial results obtained from a series of such computer simulations carried out on a simple vinyl polymer, atactic poly(propylene).

SIMULATION METHODOLOGY

Among simulation techniques, molecular dynamics (MD) is most appropriate for directly probing a time-dependent phenomenon such as the glass transition. In MD, the system of atoms or molecules is represented by point masses obeying the laws of classical mechanics. The "atoms" interact through semi-empirical interatomic potentials that mimic the true quantum mechanical nature of the system. The time evolution of this system is followed and properties of interest are calculated using the precepts of classical statistical mechanics.

The equations of motion governing the phase-space trajectory of the simulated atoms are found by the solution of Lagrange's equation, once a system Lagrangian, $\mathcal{L}$, is defined. The choice of $\mathcal{L}$ establishes the character of the statistical mechanical ensemble being simulated. For comparison with experimental results, it is most desirable to simulate the isothermal–isobaric ensemble. This can be accomplished by using the extended system methods of Nosé and Hoover [8,9] for the isothermal ensemble and Andersen [10], as modified by Nosé and Klein [11], for the isobaric ensemble. The combined ensemble Lagrangian is

$$\mathcal{L} = \frac{1}{2} \sum_{i=1}^{N} m_i \xi^2 L^2 \dot{\mathbf{s}}_i^T \mathbf{g}^T \mathbf{g} \dot{\mathbf{s}}_i - \sum_{i=1}^{N} \sum_{j>i} \phi_{ij}(\mathbf{r}_i, \mathbf{r}_j) - \sum_{i=1}^{N} \sum_{j \neq i} \sum_{k \neq j \neq i} \phi_{ijk}(\mathbf{r}_i, \mathbf{r}_j, \mathbf{r}_k)$$

$$- \sum_{i=1}^{N} \sum_{j \neq i} \sum_{k \neq j \neq i} \sum_{l \neq k \neq j \neq i} \phi_{ijkl}(\mathbf{r}_i, \mathbf{r}_j, \mathbf{r}_k, \mathbf{r}_l) + \frac{1}{2} W \dot{L}^2 - p L^3 |\mathbf{g}|$$

$$+ \frac{1}{2} Q \dot{\xi}^2 - NkT \ln \xi$$

where $\mathbf{s}_i$ is the position of the ith particle within the simulation cell, m_i its mass, $\mathbf{g}$ a 3×3 constant matrix describing the shape of the simulation cell (Fig. 1a), L the added degree of freedom (DOF) which controls the dilation/compression of the simulation cell, ϕ's the particle potential energies which are a function of the particle positions ($\mathbf{r}_i = L \mathbf{g} \mathbf{s}_i$), W the "mass" of the cell DOF, p the externally applied hydrostatic pressure, ξ the thermostating DOF, Q the "mass" of the thermostat DOF, T the desired system temperature, k the Boltzmann constant, and N the number of particles in the simulation.

The first term in the Lagrangian represents the kinetic energy of the particles ("atoms") in the simulation cell, and the next three terms, the contribution to the particle potential energy arising from two-, three-, and four-body interactions, respectively. The fifth term is the kinetic energy of the simulation

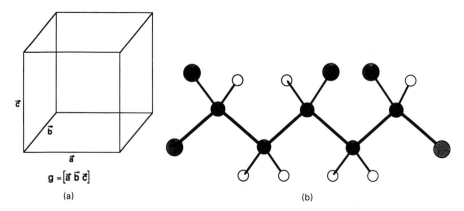

Figure 1   (a) The initial shape and size of the simulation cell is defined by the three vectors **a, b,** and **c.** The shape matrix **g** is formed using **a, b,** and **c** as its columns. The volume dilation variable L scales the components of **g** evenly. The real space positions of the atoms in the system are therefore defined by their position in a unit cube, s_i, times **g** times L. (b) The model polymer chain consists of carbon atoms (black), hydrogen atoms (white), and methyl groups (gray). Shown is a chain with degree of polymerization 3 and a terminal methyl group.

cell degree of freedom, and the sixth is the potential energy due to an alteration of the cell volume. The last two terms are the kinetic and potential energy contributions from the thermostat degree of freedom.

The identity of the material being simulated is set by the choice of the particle masses and their interaction potential energy functions, ϕ. For the case of atactic poly(propylene), three particles are used to build the simulated polymer, following the work of Theodorou and Suter [1]. Explicit atoms are used for the backbone carbon (12 amu) and pendant hydrogens (1.007 amu). However, the three hydrogens and carbon of the pendant methyl groups are represented by a single particle of mass 15.021 amu. Figure 1b shows a schematic picture of a short segment of such an atactic poly(propylene) chain. This "unified atom" approximation should not have a significant effect on the behavior of the model, as the internal motions of the methyl group are thought not to effect the properties of the polymer except at very low temperatures. Computationally, it reduces the number of degrees of freedom in the simulated system considerably.

To simulate the polymer bulk while keeping the computational problem to a manageable size, the technique first utilized by Theodorou and Suter has been adopted. The bulk is imitated by a single polymer chain packed in the simulation cell through the use of periodic boundary conditions and image chains. Figure 2 shows one such chain in the simulation cell and unpacked. As can be seen, whenever the chain exits through one wall of the cell, an image chain enters through the opposite wall. The use of this approximation certainly affects phenomena that occur over long times and distances (e.g., chain

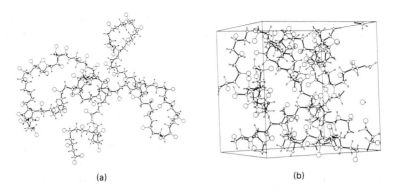

(a)   (b)

Figure 2 The system is constructed by packing a chain such as that shown in (a) into the simulation cell through the use of periodic boundary conditions. Twenty-six image chains are generated by translation of the chain along the **a, b,** and **c** axes of the cell by permutations of $[(-1, 0, 1), (-1, 0, 1), (-1, 0, 1)]$. This results in a cell, shown in (b), which contains portions of both the original chain and the image chains. Whenever a chain exits through one side of the cell, another chain enters from the opposite side.

diffusion). It should not affect the more local behavior we are interested in, as long as the persistence length of the chain is small compared to the contour length of the chain and smaller than the simulation cell.

Potential-energy functions are used to describe the interactions of the particles in the simulation. For a relatively simple nonpolar polymer such as poly(propylene), potentials are required only to describe covalent bonds, intrinsic torsional barriers, and dispersion forces. While complex potentials have been developed for these [12], simpler forms are preferable for the polymer model to minimize the computational burden. The well-known Lennard-Jones 6-12 function is used for the dispersion forces,

$$\phi_{ij}(\mathbf{r}_i, \mathbf{r}_j) = 4\epsilon_{ij}\left[\left(\frac{\sigma_{ij}}{|\mathbf{r}_{ij}|}\right)^{12} - \left(\frac{\sigma_{ij}}{|\mathbf{r}_{ij}|}\right)^{6}\right]$$

where $|\mathbf{r}_{ij}|$ is the distance between particles i and j and ϵ_{ij} and σ_{ij} are constants for each pair of particle types. Following the usual procedure, the potential function is truncated for all $|\mathbf{r}_{ij}| > 2.33\sigma_{ij}$. In addition, a quintic spline in the range $1.45\sigma_{ij} > |\mathbf{r}_{ij}| > 2.33\sigma_{ij}$ was used to force the potential and its derivatives to go smoothly to zero at the cutoff.

The covalent structure of the model polymer is formed through the use of two simple harmonic potentials, one in the covalent bond length and the other in the valence angle,

$$\phi_{ij}(\mathbf{r}_i, \mathbf{r}_j) = C_{ij}^{bl}(|\mathbf{r}_{ij}| - r_{ij}^0)^2$$
$$\phi_{ijk}(\mathbf{r}_i, \mathbf{r}_j, \mathbf{r}_k) = C_{ijk}^{\theta}(\theta_{ijk} - \theta_{ijk}^0)^2$$

where C_{ij}^{bl} and r_{ij}^0 are constants for each type of covalent bond, C_{ijk}^{θ} and θ_{ijk}^0 are constants for each type of valence angle, $|\mathbf{r}_{ij}|$ is again the distance between particles i and j, and θ_{ijk} the angle formed by particles i, j, and k. Finally, the intrinsic barrier to rotation about backbone carbon–carbon bonds is incorporated using a threefold cosine function,

$$\phi_{ijkl}(\mathbf{r}_i, \mathbf{r}_j, \mathbf{r}_k, \mathbf{r}_l) = \frac{1}{2}C^{\omega}_{ijkl}[\cos(3\omega_{ijkl}) + 1]$$

where C^{ω}_{ijkl} is a constant and ω_{ijkl} is the torsional angle defined by the four backbone particles i, j, k, and l.

The values used for the parameters in the potential functions are given in Table 1. These values were obtained from several sources. Those for the Lennard-Jones potential and the intrinsic torsional potential were taken from Theodorou and Suter [1]. The values for C^{θ}_{ijk} were taken from the work of Suter and Flory [13], while the θ^0_{ijk}'s were found by trial and error such that the average equilibrium valence angles agreed with the fixed values used by Theodorou and Suter. Similarly, the parameters for the bond stretch potential were repeatedly adjusted until the equilibrium bond lengths agreed with the fixed values of Theodorou and Suter and frequencies of the bond stretch vibrations roughly agreed with the known experimental values.

Two final parameters used in the model are W and Q, the "masses" of the degrees of freedom that control the temperature and pressure. Their value controls the inertia of these DOFs and hence the response time of the system

TABLE 1   Simulation Parameters

	HH	CH	RH	CC	CR	RR
ϵ_{ij} (kJ/mol)	0.31935	0.30176	0.37086	0.35197	0.44719	0.58112
σ_{ij} (Å)	2.3163	2.7617	2.9400	3.2072	3.3854	3.5636

	CH	CC	CR
C^{bl}_{ij} (kJ/Å^2)	3349.44	1674.72	1674.72
r^0_{ij} (Å)	1.0990	1.4712	1.5065

	HCH	a-HCC[a]	c-HCC[a]	HCR	a-CCC[a]	c-CCC[a]	RCC	RCR
C^{θ}_{ijk} (kJ/rad^2)	164.96	199.29	199.29	199.29	302.29	302.29	302.29	302.29
θ^0_{ijk} (rad)	1.9216	1.9190	1.8863	1.9246	1.8778	1.9380	1.9465	2.0559

C^{ω}_{ijkl} (kJ/mol)	11.723
$W\left(ps^2 - \dfrac{kJ}{mol}\right)$	1081.635
$Q\left(ps^2 - \dfrac{kJ}{mol}\right)$	0.216327

[a] a-HCC designates H-C-C valence angles where the central carbon is achiral; c-HCC designates H-C-C valence angles where the central carbon is chiral; a-CCC designates C-C-C valence angles where the central carbon is achiral; c-CCC designates C-C-C valence angles where the central carbon is chiral.

to changes in T or p. On the other hand, average, or equilibrium, properties do not depend on what W and Q are. Values for W and Q (Table 1) were chosen that were large enough to provide the model with numerical stability and small enough to allow the response of the model to be controlled by the response of the simulated polymer chain rather than the additional DOFs.

Preparation of the polymer structures used as input to the MD simulation was carried out using the technique developed by Theodorou and Suter [1]. This process consists of generation of an initial guess for the chain configuration using a rotational isomeric states/Monte Carlo scheme followed by an energy minimization to relax the structure to mechanical equilibrium. These initial structures have fixed covalent bond lengths and valence angles. Furthermore, they are completely static, equivalent to a structure at 0 K. Temperature enters the procedure only in setting the system density and statistical weights used in the RIS generation. To use these structures in the molecular dynamics simulation, it is necessary to equilibrate them, introducing thermal motion and allowing the bond lengths and valence angles to take on a distribution of values about the desired ones.

To do this, a series of startup simulations were used. Based on trials with the model, it was found that an effective schedule is (1) to give the particles in the system a random Maxwellian velocity distribution appropriate for the desired temperature; (2) to run the model in a microcanonical ensemble (constant total energy, constant volume) for 5 ps of simulated time with occasional rescaling of the particle velocities to add kinetic energy to the system, which allows the distributions on bond lengths and angles to form and some partitioning of the kinetic energy; and (3) to run the simulation for 40 ps in the canonical ensemble (constant temperature, constant volume). This permits continued partitioning of the kinetic energy among the modes available to the system. During this time partial relaxation of the system toward a new mechanical equilibrium consistent with the additional degrees of freedom added to the system by both the particle momenta and the release of the fixed bond lengths and angles also occurs. Finally, run the simulation for a further 60 ps in the isobaric–isothermal ensemble to allow the volume to relax to an equilibrium value for the system. At the end of this process the system is in full mechanical equilibrium, as indicated by an essentially zero internal pressure ($< \sim \pm 5$ atm) and the kinetic energy appears to be partitioned among the modes available to the system.

To carry out the study of how the structure and dynamics of the polymer glass and liquid differ, simulations were carried out at six temperatures: 393, 343, 293, 233, 213, and 173 K. Three are above the experimental T_g (253 K) and three below. At each temperature, three initial structures of degree of polymerization 76 (455 total atoms) were created at the experimental density for each temperature. These initial structures were equilibrated as described above. The phase-space trajectories of each structure were recorded over the last 20 ps of the equilibration process and used for analysis. The use of the three initial structures for each temperature was necessitated by the sluggishness of the simulation. While ergodic, the time required to allow a single structure to probe a wide volume of phase space is extremely long. Therefore, it was far more efficient to use the multiple initial structures to start in widely separated regions of phase space. Similarly, it was more efficient to prepare statically equilibrated structures at each temperature than, for instance, to prepare the low-temperature simulations by quenching from higher temperatures. Using present-day computers, molecular dynamics is capable of studying only phenomena that occur on a time scale of under 1 ns. The time required for a polymer system to fully respond to a large temperature change is many orders of magnitude greater than this.

To investigate the structure of the simulated polymer, use was made of the technique of Voronoi tessellation of space [14,15]. Voronoi tessellation is a method of subdividing a volume with space-filling convex polyhedra around a given set of reference points within the volume. In this instance the reference points are the locations of the atoms in the simulation. Especially useful for the purposes of this work is the property of Voronoi polyhedra that every point within a polyhedron is closer to the enclosed reference point than to any other reference point. Therefore, Voronoi tessellation allows the unambiguous apportionment of the volume of the system among each of the atoms in the simulation. The recorded positions of the atoms were used to tessellate the simulation volume at regular time intervals over the final 20 ps of each simulation. Since it would be expected that the atoms making up each achiral CH_2 or chiral CHR group would make most significant motions as a unit, it was more appropriate to look at the volume associated with each group, or "chain segment," rather than each atom separately. Consequently, the polyhedra for the three atoms in each chain segment were joined together into a "superpolyhedron." For each simulation, the volume distribution of the superpolyhedra was calculated, as were the distributions of volumes for neighbors of particular polyhedra.

RESULTS AND DISCUSSION

It is desirable to compare the output from a computer simulation to experimental results whenever possible. Experimental data for atactic poly(propylene) is skimpy, but Figs. 3 through 5 present three examples of such comparisons. Figure 3 shows the calculated and experimental [16] x-ray structure factor for atactic poly(propylene) in the low-Q region. The agreement in peak positions and relative heights is excellent and indicates the model polymer reproduces the intermolecular structure of the actual material. In the high-Q region (not shown) agreement was less good, particularly in peak heights. The intensity variation in the high-Q area mostly reflects intramolecular spatial correlations between atoms. Hence the lack of agreement in this region is thought to be

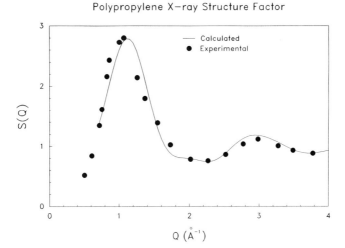

Figure 3 Low-Q x-ray static structure factor obtained experimentally (●) and calculated from the radial distribution functions determined from simulation (———). The experimental results are taken from the work of Maeda [16] and represent the structure factor of atactic poly(propylene) at an indeterminate temperature near or above the glass transition. The calculated results were obtained from simulations at 293 K using the method described by Theodorou and Suter [1].

due primarily to the loss of the contributions from the hydrogens and carbons in the pendant methyl groups, which are ignored in the model polymer.

The calculated generalized vibrational frequency spectrum at 233 K and 393 K is presented in Fig. 4. The tickmarks along the abscissa indicate the experimental frequencies of the major optical absorption peaks reported for the polymer [17]. Agreement between the location of these peaks and the major peaks in the calculated vibrational frequency spectrum is acceptable given the approximations associated with the simple model potentials. It may be noted that fine features such as the splitting of the methylene C—H stretching (2800 to 3100 cm^{-1}) into symmetric and antisymmetric modes are apparently reproduced by the model. Extra peaks may be seen in the calculated frequency spectrum, particularly those in the region 1200 to 1500 cm^{-1}. The generalized frequency spectrum is a distribution of vibrational modes weighted by the square of the vibrational amplitude. It does not take into account optical selection rules and indicates what vibrational modes are present that could absorb optical energy rather than which ones do. Hence some discrepancy is expected. In addition, the use of a single particle to represent the methyl groups will result in new vibrational modes not found in the actual polymer and a loss of the stretching, bending, and torsional modes internal to the methyl group.

Vibrational Density of States

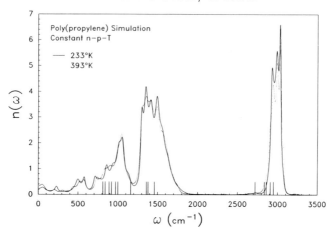

Figure 4 The generalized frequency spectrum for atactic poly(propylene) at 233 K (———) and 393 K ($\cdots\cdots$) was calculated as the Fourier transform of the velocity autocorrelation function, $c(t) = \langle \dot{\mathbf{r}}(t) \cdot \dot{\mathbf{r}}(0)\rangle/\langle \dot{\mathbf{r}}(0) \cdot \dot{\mathbf{r}}(0)\rangle$, where the average is taken over all atoms and $0 \leq t \leq 2.5$ ps. The tickmarks along the abscissa indicate the locations of infrared optical absorptions in an experimental spectrum from Ref. 17.

Figure 5 shows the specific volume of the simulated polymer versus temperature. Also shown are the experimental values [18] used as the starting points of the simulations at the same temperatures. It can be seen that the simulation duplicates the experimental values to within 5%. Furthermore, the simulated volume–temperature curve displays the change in slope conventionally taken to signify the glass transition. The transition temperature for the simulation is 277 K, in contrast to the experimental value of 253 K. The calculated volumetric expansion coefficients also agree quite well with the measured values. Although the agreement of the experimental and calculated volume expansion curve is quite good, it must be emphasized that the calculated curve was not produced by quenching the simulated polymer system from the liquid state, in contrast to the experimental procedure. As described earlier, the simulations at each temperature were started at the experimentally determined specific volume for that temperature and allowed to evolve to mechanically stable states. The residual time-average pressure at these states was quite low, under $\sim\pm5$ atm for all simulations. From these results it can be concluded that the molecular dynamics model is a physically reasonable description of the actual polymer. Further analysis of the structure and dynamics of the simulated polymer is therefore warranted.

Figures 6 and 7 depict the distribution of Voronoi volume associated with all chain segments, achiral (CH_2) chain segments, and chiral (CHR) chain segments at 233 K and 393 K. These distributions are representative of the results for the other temperatures above and below the glass transition. Several

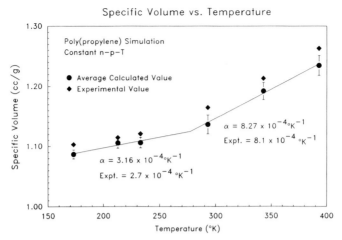

Figure 5 The calculated (——●——) specific volume versus temperature curve was obtained by allowing static polymer structures prepared by the method of Theodorou [1] at each temperature and the experimental specific volume (◆) to relax during molecular dynamics simulations to new mechanical equilibria following the removal of fixed bond lengths and angles and introduction of thermal motion. Each calculated point represents the average of three simulations, while the error bars are the averaged standard deviations over 20 ps. Regression lines through the three points above and below the experimental glass transition (253 K) intersect at 277 K, while the slopes provide constant-pressure thermal expansion coefficients that are close to the experimental values [18].

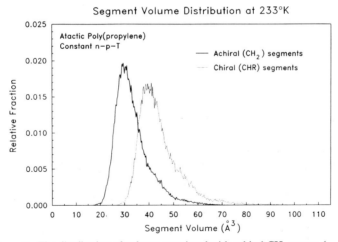

Figure 6 The distribution of volume associated with achiral CH_2 groups (——————) and chiral CHR groups (······) at 233 K was calculated by carrying out a Voronoi tessellation of space based on the recorded positions of the atoms in the system over the last 20 ps of each simulation. The Voronoi polyhedra for the three atoms in each group were joined together and the volume of the group computed. Each curve represents the average from tessellation of 250 "snapshots" for each of three simulations.

Segment Volume Distribution at 393°K

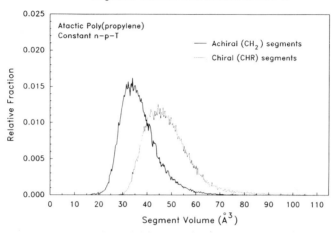

Figure 7 The distribution of volume associated with achiral CH_2 groups (———) and chiral CHR groups ($\cdots\cdots$) at 393 K was calculated as described in Fig. 6. The distributions at the higher temperature have a higher average value and are significantly broader than those at 233 K (Fig. 6).

things may be noted. First, the chiral groups occupy more volume than the achiral groups. This might be expected since the chiral segments contain a bulky methyl group. Second, the distribution of segment volumes is found to be quite wide with a pronounced large volume tail, particularly for the chiral groups. A small fraction of segments have far more volume associated with them than their minimum van der Waals volume. Presumably, segments with more than minimum volume represent excess or free volume. The breadth of the distributions implies that the excess volume in the system is not shared evenly among the chain segments, but rather, tends to be localized. Third, some differences can be noted between the distributions above and below the glass transition. There is an overall upward shift in volume at the higher temperature, as would be expected from Fig. 5. In addition, the distributions at 393 K seem broader and flatter than those at 233 K.

A more quantitative characterization of these perceptions is presented in Figs. 8 and 9, which display plots of the average volume and the standard deviation of the volume distributions against temperature. The average volume of the achiral and chiral segments increases uniformly with temperature and a slight difference in the slope of the two curves above and below the glass transition is discernible (Fig. 8). As the two curves are essentially parallel, it can be deduced that both chiral and achiral groups contribute equally to the overall volume expansion. From Fig. 9 it appears that there is a slight increase in the breadth of the distribution above the glass transition. The average standard deviation of the achiral segments in the structures above T_g was 23% higher than that of the segments below T_g. For the chiral segments the increase

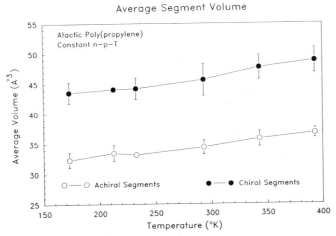

Figure 8   The average volume for the achiral (—○—) and chiral (—●—) groups versus temperature shows the same behavior as is seen for the overall volume expansion (Fig. 5). No evidence that one type of group contributes more to the expansion than the other can be seen.

was found to be only 15%. Examination of the distributions revealed that most of the increase was due to an increase in the number of segments with significantly larger than average volume (i.e., an enhancement of the large volume tails of the distributions). There were, however, still a sizable fraction of chain segments with a surprisingly small volume, even at temperatures well into the liquid region.

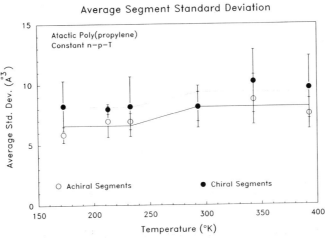

Figure 9   The average standard deviation for the achiral (—○—) and chiral (—●—) group volume distributions versus temperature shows an increase of 15 to 25% in the region of the apparent glass transition temperature for the polymer model.

Taken together, these facts indicate that the overall volume expansion with increasing temperature does not occur solely by a uniform dilation of the volume associated with all segments. Instead, it takes place, in part, by a process in which the population of chain segments with considerable excess volume is increased. This change appears to occur most noticeably around the glass transition temperature. Since Figs. 6 to 9 indicate that the segment volume is not distributed homogeneously, an important question is what the nature of the spatial distribution of segment volume might be. Figures 10 and 11 provide information that begins to answer this. Both show, first, the distribution of volume associated with all polymer chain segments, and second, the distribution of volume associated with all segments that are nearest neighbors to an achiral or chiral segment with a volume in the top 5%.

At 393 K (Fig. 10), it can be seen that the distribution of volume for neighbors of segments with large excess volume is very close to the distribution of volume for chain segments as a whole. A slight shift to higher volume is seen in the distribution for neighbors, but this is attributable to the process of Voronoi tessellation used. The shape of the distribution is essentially unchanged. At 233 K (Fig. 11), the difference in the distribution for neighbors of segments with large excess volume and the general segment volume distribution is more pronounced. The neighbor distribution is not only shifted upward but is seen to have a different shape. A sizable enlargement in the number of segments with larger than average volume over the general segment population is found. This suggests that the neighbors of chain segments with signifi-

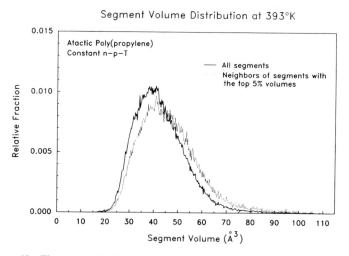

Figure 10 The volume distribution of all segments at 393 K (————) is compared to the distribution for segments which are nearest neighbors to any segment that is in the top 5% of the achiral or chiral segment distributions (······) (Fig. 7). The two curves have similar shapes, although the distributions for the neighbors of large segments is shifted upward slightly.

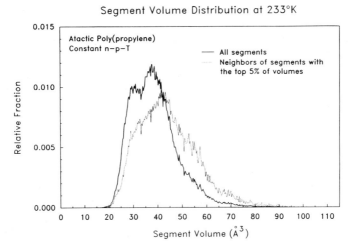

Figure 11   The distribution of all segments at 233 K (———) is compared to that for those segments that are neighbors of the largest 5% of chiral or achiral segments (······) (Fig. 6). The shape of the distribution for the neighbors of large segments is distorted considerably toward larger volumes, in addition to an upward shift.

cantly larger than average volume are themselves significantly larger than the norm. This implies that in the liquid state the segments with large excess volume are distributed more or less uniformly through the simulation volume. Below the glass transition, the tendency is more for the largest segments to be clustered with other segments of large volume. Thus the length scale of the inhomogeneity in the liquid is on the order of the size of the segments— individual segments have different volumes associated with them, but larger regions look more alike. In the glass, the inhomogeneity extends beyond the segments and the picture of a glass becomes one of pockets of relatively low density (and presumably higher mobility) material embedded in a matrix of higher density (lower mobility) polymer. The development of such inhomo- geneity has been reported during simulations of the "amorphization" of crystals by irradiation [19].

One explanation of the glass transition that has been proposed on the basis of theory and simulations of simple atomic liquids and glasses is that the transition from glass to liquid occurs when local regions of material with ex- cess volume ("liquid-like") increase in number and/or size until a percolation threshold is reached [7, 20]. A continuous network of "liquidlike" material then exists and the macroscopic properties take on the characteristics of a liquid. A more detailed analysis than that described here is necessary to de- termine if this intuitively appealing concept can be applied to the polymer glass transition; it will be reported on in a later publication.

The study of the dynamics of the simulated polymer is not yet complete. However, the data of Fig. 4 do indicate that some differences in the dynamics of the glass and liquid do exist. The generalized frequency spectrum in the region below 500 cm^{-1} is due mostly to backbone torsional modes. As can be seen, there are differences in this region between the spectra calculated for 233 K and 393 K. The lowest-frequency peak, at 50 to 60 cm^{-1}, is enhanced in the high-temperature spectra. Presumably, this is due to an increase in the amount of torsional motion taking place in this low-frequency range. The low-frequency regime is amenable to probing through the use of inelastic neutron scattering. Studies of atactic poly(propylene) utilizing this technique are currently in progress. In a future report we will discuss the results of this and associated simulation work to identify the motions taking place and how they affect the macroscopic properties of the polymer.

In summary, the development of a realistic molecular dynamics model for a simple vinyl polymer is reported. The model replicates the behavior of the actual polymer to an acceptable degree. Using the model, a study of how the structure and dynamics of the polymer differ in the liquid and glass states has been undertaken. Initial results obtained by examining the volume associated with the chain segments indicate that there is a significant broadening of the volume distribution above T_g. In addition, there is evidence that in the glassy state, chain segments with much larger than average volume tend to segregate together, while in the liquid they are more uniformly dispersed. Very preliminary data on the dynamics of the system indicate that the model does show differences between simulations conducted at different temperatures. Work in progress with the molecular dynamic model will shed light on the finer details of the structural and dynamic differences between the liquid and glassy states. Additional work being carried out with a Monte Carlo model of atactic poly(propylene) will provide information on the kinematic pathways by which both one state can be reached from the other and the process of physical aging takes place.

ACKNOWLEDGMENTS

This work was supported by the DARPA University Research Initiative through the Office of Naval Research under Contract N00014-86-K-0768. The San Diego Supercomputer Center is acknowledged for a generous grant of time on a Cray X-MP/48 on which some simulations were performed.

REFERENCES

1. D. N. Theodorou and U. W. Suter, *Macromolecules* **18,** 1467 (1985).
2. K. Kremer, G. S. Grest, and I. Carmesin, *Phys. Rev. Lett.* **61,** 566 (1988).
3. J. H. R. Clarke and D. Brown, *Mol. Simul.* **3,** 27 (1989).
4. R. H. Boyd and K. Pant, *Polym. Prepr. Am. Chem. Soc., Div. Polym. Chem.* **30**(2), 30 (1989).

5. D. Berman and J. H. Weiner, *J. Chem. Phys.* **83,** 1311 (1985).
6. D. Turnbull and M. H. Cohen, *J. Chem. Phys.* **34,** 120 (1961).
7. M. H. Cohen and G. S. Grest, *Phys. Rev.* **B20,** 1077 (1979).
8. S. Nosé, *J. Chem. Phys.* **81,** 511 (1984).
9. W. G. Hoover, *Phys. Rev.* **A31,** 1695 (1985).
10. H. C. Andersen, *J. Chem. Phys.* **74,** 2384 (1980).
11. S. Nosé and M. L. Klein, *Mol. Phys.* **50,** 1055 (1983).
12. S. Lifson and A. Warshel, *J. Chem. Phys.* **49,** 5116 (1968).
13. U. W. Suter and P. J. Flory, *Macromolecules* **8,** 765 (1975).
14. G. F. Voronoi, *Z. Reine Angew. Math.* **134,** 198 (1908).
15. M. Tanemura, T. Ogawa, and N. Ogita, *J. Comput. Phys.* **51,** 191 (1983).
16. T. Maeda, "Structural Analysis of Atactic Polypropylene: A Study of the Glass Transition by X-ray Diffraction," Master's thesis, Department of Materials Science and Engineering, University of Pennsylvania, Philadelphia, August 1987.
17. D. O. Hummel, *Atlas of Polymer and Plastic Analysis,* Vol. 1, 2nd ed., Hanser Verlag, Munich, 1979.
18. H. S. Kaufmann and J. J. Falcetta, *Introduction to Polymer Science and Technology,* Wiley, New York, 1977.
19. H. Hsieh and S. Yip, *Phys. Rev.* **B39,** 7476 (1989).
20. D. Deng, A. S. Argon, and S. Yip, *Philos. Trans. R. Soc. London* **A329,** 549 (1989).

Molecular Mechanics Simulation of Glassy Polymers at Interfaces

K. F. MANSFIELD and D. N. THEODOROU
Department of Chemical Engineering
University of California, Berkeley
and Center for Advanced Materials
Lawrence Berkeley Laboratory
Berkeley, CA 94720

ABSTRACT

A general technique for the detailed atomistic computer simulation of interfacial systems containing *glassy* amorphous polymers is described. The technique is implemented for two distinct types of systems: (1) a thin film of glassy atactic polypropylene exposed to vacuum on both sides, and (2) a film of glassy atactic polypropylene sandwiched between two semi-infinite phases of graphite. Each of these two systems is modeled as a collection of microstates in detailed mechanical equilibrium. The microstates, characterized by periodic boundary conditions in two dimensions, are created by a combination of Monte Carlo and energy minimization techniques. Polypropylene is represented in terms of individual carbon, hydrogen, and methyl centers, interacting via Lennard-Jones potentials. In case 2, polymer–graphite interatomic interactions are included explicitly, using a Fourier series summation of the Lennard-Jones potential field. The simulations are used to extract interfacial thermodynamic properties and a wealth of structural information about the glassy polymer surface and the glassy polymer–solid interface. A formulation for the prediction of the internal energy contribution to surface tension, adhesion tension, and work of adhesion is developed and shown to give very satisfactory results. The interfacial thickness, defined on the basis of the density distribution of segments, is predicted to be on the order of 10 Å. Density at the free surface falls along a sigmoidal profile; near the solid phase, local density is enhanced relative to the bulk. Polymer backbone bonds exhibit a tendency to orient parallel to the phase boundary in the interfacial region. The overall shape, position, and orientational distribution of chains is characterized by computing chain centers of mass, spans, and radius of gyration tensors.

INTRODUCTION

The surface and interfacial properties of amorphous polymers in the condensed, solvent-free state, are relevant in many applications. Polymers are used to control the wettability and biocompatibility of surfaces. Polymer–solid interfaces are encountered in the technologies of adhesives and coatings. Interfaces between *glassy* polymers and solids govern the mechanical behavior of composite materials. To select judiciously and efficiently design materials in these and similar areas, we need to understand polymer interfacial behavior *at the molecular level*. Interfaces are intrinsically inhomogeneous and anisotropic. There is an asymmetry between cohesive interactions within the polymer and adhesive interactions between the polymer and the adjacent phase. In

123

addition, chain conformations experience strong entropic constraints at a phase boundary, against which they cannot propagate. These factors cause the organization and shape of macromolecules at the interface to be quite different from those in the bulk. Structural characteristics, in turn, govern all macroscopic properties that we wish to control and manipulate.

The importance of polymer melts at interfaces has recently stimulated considerable simulation work in this area using lattice Monte Carlo [1–3], off-lattice Monte Carlo [4–6], and even molecular dynamics techniques [7]. The objective of this work is to apply computer simulation to well-defined glassy polymer–vacuum and glassy polymer–solid interfaces, in order to understand and predict (1) microscopic structural features, such as density distribution, bond orientation, chain shape, and conformation in the interfacial region, and more important, (2) macroscopic thermodynamic properties such as surface tension, adhesion tension, and work of adhesion. Our approach is distinguished from other work in this area in that we use a detailed atomistic representation of the polymer and solid phases that corresponds as closely as possible to the real systems under examination. In addition, our work focuses on polymers in the glassy state, as opposed to polymer melts. We lay emphasis on the prediction of actual values for surface thermodynamic properties, which can be tested quantitatively against experiment. Finally, we strive to design our approach as efficiently as possible so as to keep computations tractable.

The simulation method we describe in the following is perfectly general and applicable, in principle, to any glassy polymer and solid phase, provided that their molecular structures are known and appropriate interatomic interaction potentials are available. The underlying picture is that of a polymer glass created by slow vitrification of the melt while in contact with the non-polymeric phases. We have chosen glassy atactic polypropylene at −40°C and basal planes of graphite as suitable materials to demonstrate our technique. Atactic polypropylene has many advantages as a model polymer; it is non-crystallizable, it possesses a simple alkane chemical constitution that can be described well in terms of simple interatomic potential functions, and its conformational statistics in the unperturbed state are well understood. In addition, extensive simulations of this polymer in the bulk are available [8] against which we can compare our interfacial predictions. Graphite was chosen as the solid substrate because of its relevance in composite applications. Its surface properties have been heavily studied experimentally; also, its interactions with polypropylene are largely of a London dispersion type and can be described satisfactorily in terms of existing potential expressions [9].

THE MODEL

Our molecular model rests on the following assumptions. The polymeric glass is viewed as a collection of mutually inaccessible microstates, each characterized by "frozen in" liquidlike disorder. We will assume that the polymer is glassy throughout the entire extent of our interfacial systems. Each microstate is in detailed mechanical equilibrium, constituting a *local* minimum of the total potential energy in configuration space. Microstates are static; temperature enters only indirectly through specification of the bulk density. We obtain estimates of macroscopic properties of the glass by *arithmetically* averaging the properties of individual microstates. We use a realistic representation of molecular geometry and energetics. We treat bond lengths and bond angles as classical springs of infinite stiffness, according to the "classical flexible" model of Gō and Scheraga [10]. Molecular rearrangement occurs through overall chain translation and rotation, and through torsion around skeletal bonds. Skeletal carbons, pendant hydrogens, and methyl groups are treated explicitly. They interact with one another through Lennard-Jones potentials, whose parameters are taken from the recommendations of Flory and Bondi [8,11]. In addition, there is an intrinsic torsional potential, associated with skeletal bonds.

Through a statistical mechanical analysis of a microstate "locked" in the vicinity of a local minimum of potential energy, one can arrive at a useful result: The derivative of the internal energy U with respect to any space-related thermodynamic quantity ξ, such as volume, strain, or surface area, can be well estimated by the corresponding derivative of potential energy $\mathcal{V}^0$ at the local minimum:

$$\left.\frac{\partial^m U}{\partial \xi^m}\right|_{T,\xi'} = \left.\frac{\partial^m \mathcal{V}^0}{\partial \xi^m}\right|_{\xi'} \tag{1}$$

Equation (1) allows one to deduce the internal energy contribution to thermodynamic quantities, such as surface tension and work of adhesion, from our static computer simulations.

Our simulation method, as mentioned above, is applicable to both glassy polymer–vacuum and glassy polymer–solid interfaces. We will refer to these two geometries as "free surface" and "composite" systems, respectively. The free-surface simulation, which is computationally simpler, is discussed first, to point out the essential features of the method. Special strategies adopted for the composite simulation are discussed subsequently.

FREE SURFACE SIMULATION

Our technique [12–14] requires the creation of an ensemble of model microstates, such as the one shown in Fig. 1. A microstate is an orthorhombic cell, containing segments from several "parent" chains and characterized by periodic boundary conditions in the x and y directions. It can be viewed as part of a thin film, extending to infinity in these directions. The x and y dimensions must be chosen larger than twice the range of polymer–polymer interatomic interactions, and ideally should be considerably larger than the radius of

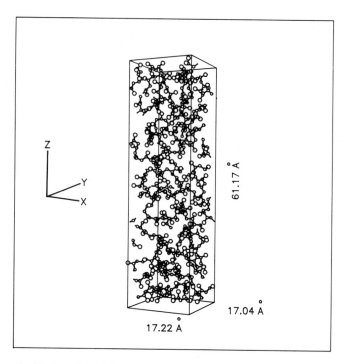

Figure 1   "Relaxed" model structure of a thin film of a glassy polymer, composed of several H—[CH$_2$(CH$_3$)$_{76}$$^{-CH_3}$ chains, exposed to vacuum at its upper and lower surfaces. Periodic boundary conditions are incorporated in the x and y directions. The smaller and larger spheres correspond to carbons and methyls, respectively.

gyration $\langle s^2 \rangle^{1/2}$ of constituent chains. The top and bottom faces of the cell are exposed to vacuum. They must be separated by a distance large enough to make the middle region behave as bulk polymer, yet small enough to keep computations tractable. A good rule to follow is to use a separation distance (film thickness) larger than $4\langle s^2 \rangle^{1/2}$ in the z dimension. We have chosen the x, y, and z dimensions of our model cell as 17.22, 17.04, and 61.17 Å, respectively. In addition, one must specify the number and molecular weight M of the chains. This is done so that the average density in the middle region of the cell corresponds to the bulk density of the experimental system of interest. In our base-case simulations, we use three "parent" chains per cell, each of degree of polymerization $x = 76$ ($M = 3214$). This corresponds to a density of 0.892 g/cm^3, the experimental value for atactic polypropylene at $-40°$C, roughly 20°C below its glass transition temperature. The unperturbed radius of gyration of these chains in the bulk [8] is 17.1 ± 2.1 Å.

Packing chains at such a glassy density, in order to create a microstructure, is a challenging problem. We perform this in two steps, as in our previous bulk work [8]. First, an initial guess microstructure is generated. Subsequently, the total potential energy of the microstructure is minimized with respect to all microscopic degrees of freedom.

The initial guess generation step aims at the formation of a "liquidlike" microstructure of realistic density and sufficiently low energy. This means that excluded volume effects must be accounted for, both inter- and intramolecularly. Briefly, the Monte Carlo generation of an initial guess proceeds as follows [8, 13]. First we decide on the tacticity of the chains. Next, we randomly place the terminal methyl group of each chain into the cell, and also the first two skeletal bonds, in random orientation. Subsequently, we grow all our chains in parallel, bond by bond, through a procedure that resembles polymerization, until we reach their last segments. During this procedure it is important that one prevents chains from exiting through the top and bottom faces of the cell into the surrounding vacuum. To accomplish this, we place a fictitious, steeply repulsive "soft-wall" potential at each polymer–vacuum boundary. In the course of the conformation generation, we use a discrete representation of chain conformations, based on Suter and Flory's five-state rotational isomeric state model (RIS) for polypropylene [15]. The conditional probabilities of our Monte Carlo scheme blend the short-range conformational energetics of the RIS model with long-range interactions so as not to allow chains to cross themselves, each other, or the walls. In the presence of the walls, we have found it necessary to introduce a new two-bond scheme for the conformation generation. The rotational state of bond i is decided upon after considering the energetic consequences of placing the bonds i and $i + 1$ in all pairs of states accessible to them. The new scheme allows the chains more freedom to redirect themselves when approaching a steep obstacle.

The initial guess generation is followed by potential-energy minimization with respect to all microscopic degrees of freedom. These are: the start coordinates of all parent chains, referred to those of the first chain; three Eulerian angles per chain, specifying overall orientation with respect to the cell frame of reference; and the sequence of torsion angles for all rotatable bonds of each chain, which define conformation. Overall, a microstructure has 465 degrees of freedom. As in our bulk work [8], we use a three-stage optimization strategy. We start with a purely repulsive, "soft sphere" potential and reduced atomic radii (stage I), proceed using the soft sphere potential and radii of actual size (stage II), and finally incorporate the attractive part of the potential, arriving at a fully relaxed microstructure (stage III). The constraining walls are removed during the last stage of the minimization, so that the polymer film relaxes solely under its own cohesive interactions. Minimization in the presence of the soft walls (i.e., during stages I and II) requires an additional degree of freedom, namely the z coordinate of the first group of the first parent chain.

The minimization is performed with the powerful quasi-Newton BFGS algorithm [16] using analytical derivatives. Our method lends itself to extensive vectorization, making it possible to arrive at a fully relaxed microstructure of atactic polypropylene within 2 CPU (central processing unit) hours on a Cray X-MP/48. The number of microstates one needs to generate depends on the "computer experimental" error one is willing to tolerate in the results. The convergence of structural features at the level of individual segments (e.g., density profile) is faster than that of thermodynamic estimates. We have found 70 microstructures to give adequate predictions of surface thermodynamic properties, which we are most interested in.

The internal energy contribution is predicted from out static microstructures based on a thermodynamic analysis of the glassy polymer film. The fundamental equation for the film, whose middle (bulk) region is isotropic, is

$$dU = T\,dS - P\,dV + \gamma\,da + \mu\,dn \tag{2}$$

The surface tension γ is the derivative of Helmholtz energy A with respect to surface area a at constant temperature T, volume V, and number of moles n. It can be separated into an internal energy contribution γ^U and entropic contribution γ^S as follows:

$$\gamma = \left.\frac{\partial A}{\partial a}\right|_{T,\,V,\,n} = \left.\frac{\partial U}{\partial a}\right|_{T,\,V,\,n} - T\left.\frac{\partial S}{\partial a}\right|_{T,\,V,\,n} = \gamma^U + \gamma^S \tag{3}$$

Through a Maxwell relation, γ^S can be expressed simply in terms of the product of temperature and the temperature gradient of surface tension; γ^U then becomes

$$\gamma^U = \left.\frac{\partial U}{\partial a}\right|_{T,\,V,\,n} = \gamma - T\left.\frac{\partial \gamma}{\partial T}\right|_{V,\,n} \tag{4}$$

Thus γ^U is an experimentally measurable quantity. On the other hand, γ^U can be obtained directly from our simulations. One way of estimating γ^U is to subject each microstructure to a small deformation that dilates the surface area a, relax anew, and monitor the change in potential energy $\mathcal{V}^0$ at the minimum. By Eqs. (1) and (4), the finite-difference ratio $\Delta\mathcal{V}^0/\Delta a$ equals γ^U. In practice, we have found that our estimate remains practically unchanged if we instead subject the film to an *infinitesimal affine* deformation that dilates the surface area while keeping the total volume constant. The associated change in potential energy is calculated as the work done on all atoms of the film during the affine deformation. An advantage of the affine deformation method is that γ^U is obtained directly through an analytical expression evaluated at the undeformed minimum energy microstructure. This expression, Eq. (5), is cast in terms of interatomic forces, including bonded forces, and interatomic distances; it is essentially a static analog of the virial theorem for surface tension [17].

$$\gamma^U = -\frac{1}{2a}\sum_i\sum_{j\neq i}\left[\frac{1}{2}F_{ij,x}(x_i - x_j) + \frac{1}{2}F_{ij,y}(y_i - y_j) - F_{ij,z}(z_i - z_j)\right] \tag{5}$$

It is very important to include the potential "tail" contribution, due to attractive interactions at long range, to any thermodynamic property predicted by computer simulation. We have included this contribution by direct integration, after "smearing" each atom in each microstructure in the x and y directions [13]. Our prediction for γ^U of atactic polypropylene, which is based on averaging over 70 minimum energy microstructures, is 43 ± 33 mN/m (errors reported here constitute 95% confidence limits from averaging over all microstructures). Compared to the experimental [18] value of $\gamma^U = 46 \pm 0.2$ mN/m, this indicates agreement within 7%. Nevertheless, we consider our predicted value quite satisfactory given the fact that we do not adjust any parameters in our simulation.

In addition to thermodynamic properties, structural information is also of interest. Structural features were analyzed by superimposing all our model microstructures, so that their centers of mass coincide. The information resulting from this superposition was divided into bins, by drawing planes parallel to the film surface; average structural features were computed within each bin. To improve statistics, we took advantage of the fact that the film is macroscopically symmetric with respect to its midplane. Structural information can be analyzed on the length scale of individual segments, as well as on the length scale of entire chains. In the figures discussed below, position $z = 0$ corresponds to the film midplane, while vacuum lies at large z values. Error bars indicate 95% confidence limits.

To examine the chain-length dependence of structural features, a series of simulations was undertaken using nine parent chains of degree of polymerization $x = 25$ (molecular weight 1068) in a cell of the same dimensions. The radius of gyration in the bulk region of this shorter chain system is 9.03 ± 0.25 Å. Results from this system are presented together with those from the $x = 76$ system below.

Figure 2 displays the mass density distribution across the film. Over most of the film, the density is constant and practically indistinguishable from its bulk value. At the surface, density falls along a sigmoidal profile, which spans a distance smaller than 10 Å. In this respect the polymer surface resembles the free surfaces of simple liquids [19]. The individual components of density due to carbon, methyl, and hydrogen atoms are also shown. The surface is enriched in methyls at the extremities of the polymer. Further, skeletal carbons and pendant hydrogens exhibit shallow peaks. Density profiles from the short chain ($x = 25$) system are quantitatively indistinguishable from those obtained for $x = 76$ and are therefore not displayed here.

Local orientational tendencies of bonds are examined in Fig. 3. We define a bond order parameter $S_B = 0.5[3\langle\cos^2\theta\rangle - 1]$ in terms of the angle θ formed between a bond and the direction normal to the surface. S_B would assume a value of -0.5, 0.0, or 1.0, respectively, for bonds characterized by perfectly parallel, random, and perpendicular orientation with respect to the

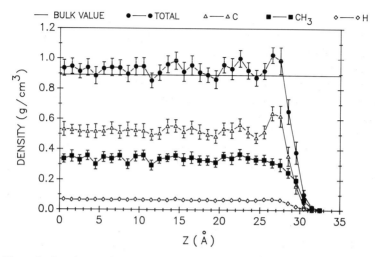

Figure 2 Local mass density distribution at a glassy polymer–vacuum interface. Vacuum is located at the extreme right, and the region near $z = 0$ is representative of unperturbed bulk polymer. The solid line represents the experimentally observed macroscopic density at the simulation temperature. Individual contributions to the mass density profile from skeletal carbon, hydrogen, and methyl groups are also shown.

surface. We find a weak tendency for the skeletal bonds to lie parallel to the surface in the interfacial region. On the contrary, pendant bonds exhibit a weak perpendicular orientation there. Bond orientational tendencies are again insensitive to chain length.

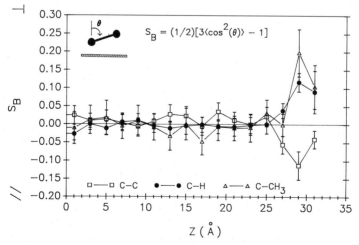

Figure 3 Local order parameter for C—C, C—H, and C—CH₃ bonds as a function of distance from the midplane of the model structures. A positive (negative) order parameter indicates an orientational tendency that is perpendicular (parallel) to the surface planes.

The spatial distribution of chain centers of mass, shown in Fig. 4, characterizes structure at the level of entire chains. There is a depletion of chain centers of mass at the surface: Since chains are finite objects, their centers of mass cannot come arbitrarily close to the surface. Associated with this exclusion effect, there is a peak in the distribution, observed at a distance slightly less than one radius of gyration from the edge of the film in the $x = 76$ system (Fig. 4a). Clearly, this peak appears closer to the surface in the shorter chain system (Fig. 4b), its position scaling roughly in proportion to the radius of gyration. The distribution in the $x = 76$ system shows more structure than that of the $x = 25$ system. This is largely a result of the small size of our model cell.

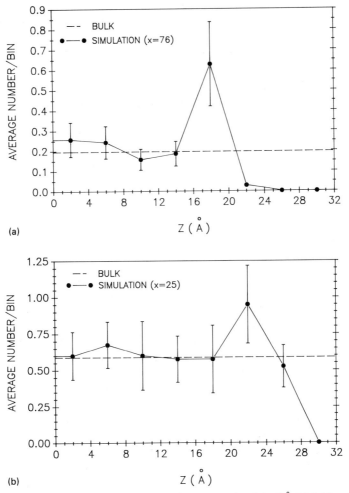

(a)

(b)

Figure 4 Chain center-of-mass distribution, accumulated in 4-Å-thick bins. (a) Model system with degree of polymerization $x = 76$ (152 skeletal bonds per chain). (b) Degree of polymerization $x = 25$ (50 skeletal bonds per chain).

The x and y dimensions of the cell are roughly equal to the radius of gyration in the larger-chain system; this induces some artificial layering of chains, which is absent from the shorter-chain system.

We explored the overall shape of chains by determining their spans and eigenvalues of the radius of gyration tensors. Spans are defined following Rubin and Mazur [20] as the dimensions of the smallest orthorhombic box that can completely enclose the chain segment cloud. Our analysis led to the conclusion that overall chain shape, when looked at in the reference frame of the principal axes, remains virtually unaffected by the presence of a phase boundary. How chain segment clouds tend to *orient* in the interfacial region is a different question. We explored this question by computing order parameters for chain spans. The definition of the order parameter for spans is analogous to that used for the bonds. Plotted in Fig. 5a are the order parameters for the shortest and longest spans in the $x = 76$ system as functions of chain center-of-mass position. In the surface region we see a strong tendency for the longest span to orient parallel to the surface and for the shortest span to orient perpendicular to the surface. This tendency is observed in both the long- and short-chain systems. Moreover, it remains unchanged if, instead of spans, one uses the principal axes of inertia (i.e., the eigenvectors of the radius of gyration tensor) as measures of shape (Fig. 5b). In the long-chain system, one detects, in addition, some tendency for perpendicular orientation of the largest spans at roughly $1.5\langle s^2 \rangle^{1/2}$ from the edge of the film (Fig. 5a). This is a special organization effect, induced by the small x and y dimensions of our model box. It is absent from the short-chain system, as seen in Fig. 5b.

COMPOSITE SIMULATION

The methodology described in the preceding section is straightforwardly extended to polymer–solid interfaces. In our "composite" geometry, a glassy atactic polypropylene film adheres to basal planes of graphite on both sides (Fig. 6). The dimensions of the model box in the x and y directions, equal to those used in the free surface case, are commensurate with the graphite lattice; two-dimensional periodic boundary conditions are thus observed. The top layers of carbons of the two basal planes are 66.27 Å apart. The graphite phases extend infinitely on either side of the polymer. They are modeled as sets of Lennard-Jones carbon centers, at their crystallographically correct positions. The solid structure is assumed to remain unperturbed by the adsorption of polymer. For each atom in the polymer, an interaction potential due to all atoms in the solid phase is included in the total potential-energy function. To calculate this potential, we implement the efficient and accurate Fourier summation method designed by Steele [9] which takes advantage of the symmetry in the crystalline substrate.

Again, a model microstructure is obtained through the Monte Carlo

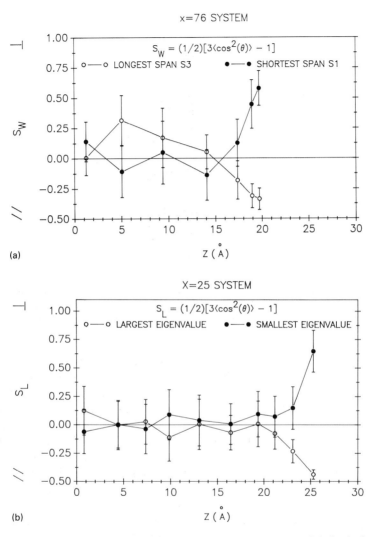

Figure 5 (a) Order parameters for the longest and shortest spans of chains in the $x = 76$ system as functions of center-of-mass position. (b) Order parameters for the longest and shortest principal axes of chains in the $x = 25$ system as functions of center-of-mass position.

generation of an initial guess, followed by total potential-energy minimization in three stages. In the initial guess generation and the first two (soft sphere) stages of the minimization, soft walls are used to constrain the polymer. These stages are carried out exactly as in the free surface case, with the soft walls placed at exactly the same positions. It does not pay to introduce the actual texture of the substrate surfaces until the third stage of the minimization. Here the soft walls are turned off and replaced by explicit graphite phases. The

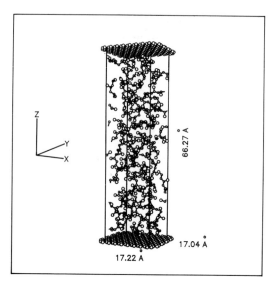

Figure 6 "Relaxed" model structure of a thin film of glassy atactic polypropylene sandwiched between two basal planes of graphite. Periodic boundary conditions are incorporated in the x and y directions. This system is referred to in the text as a "composite" system.

carbon centers in the first layer of each graphite phase are placed 2.55 Å out from the original position of the soft walls, so as to keep the density in the middle of the film equal to its bulk value. Since interactions of the polymer with the entire semiinfinite graphite phases are included, the "tail contributions" from long-range cohesive interactions within the polymer must also be incorporated in the minimization. This is done by modeling the polymer at large distances from a given atom as a background of uniform density in all three atomic species present. The number of degrees of freedom during stage 3 of the composite system minimization is 468; now, the start coordinates of the first chain enter as degrees of freedom.

At the present time, 35 composite microstates have been fully minimized. It is noted that structural features converge more quickly in the composite system than they do in the free-film system. To analyze structure, all microstates are superposed so that the graphite phases coincide.

The methodology for predicting interfacial thermodynamic properties in the composite system are similar to that used in the free surface case. In place of the surface tension we now have the difference $\gamma_{SP} - \gamma_S$ between the solid–polymer interfacial tension γ_{SP} and the surface tension of the pure solid γ_S. The negative of this quantity is sometimes referred to as the "adhesion tension." The internal energy part of the adhesion tension is related to the derivative of the internal energy of the film with respect to surface area, at constant temperature, volume, and number of chains:

$$(\gamma_{SP} - \gamma_S)^U = (\gamma_{SP} - \gamma_S) - T\frac{\partial(\gamma_{SP} - \gamma_S)}{\partial T}\bigg|_{V,na} = \frac{\partial U}{\partial a}\bigg|_{TV,n} \approx \frac{\partial \mathcal{V}^0}{\partial a}\bigg|_{V,n} \tag{6}$$

$\partial\mathcal{V}^0/\partial a|_{V,n}$ is estimated from our simulations by subjecting each composite microstructure to an *infinitesimal affine* deformation, which displaces the graphite phases inward while at the same time flattening the film. This deformation leaves the graphite phases undistorted. It increases the area of interfacial contact but preserves the volume of the film. The associated change in potential energy is calculated from the work done on all polymer atoms. The final expression,

$$(\gamma_{SP} - \gamma_S)^U = -\frac{1}{2a}\Bigg\{ \sum_{\substack{i \\ \text{polymer}}} \sum_{\substack{j \neq i \\ \text{polymer}}} \left[\frac{1}{2}F_{ij,x}(x_i - x_j) + \frac{1}{2}F_{ij,y}(y_i - y_j) - F_{ij,z}(z_i - z_j)\right]$$

$$+ \sum_{\substack{i \\ \text{polymer}}} \sum_{\substack{k \\ \text{upper} \\ \text{solid}}} \left[F_{ik,x}x_i + F_{ik,y}y_i - 2F_{ik,z}\left(z_i - \frac{h}{2}\right)\right] \tag{7}$$

$$+ \sum_{\substack{i \\ \text{polymer}}} \sum_{\substack{l \\ \text{lower} \\ \text{solid}}} \left[F_{il,x}x_i + F_{il,y}y_i - 2F_{il,z}\left(z_i + \frac{h}{2}\right)\right]\Bigg\}$$

is cast in terms of the interatomic forces and distances within the polymer, as well as all polymer–solid forces. It involves the coordinates of all film atoms and the film thickness h. One can readily prove that this expression is invariant to the coordinate system used for a system in mechanical equilibrium.

Experimental data on interfacial thermodynamic properties are not readily available. Fowkes [21] recommends a Girifalco and Good-type geometric mean expression for estimating the work of adhesion from the dispersion force contributions to the surface tensions of the two phases involved. Furthermore, he suggests that the work of adhesion is not very sensitive to temperature. Experimental values of the dispersion force contribution to the surface tensions of solid polypropylene and graphite, from equilibrium spreading pressure measurements, are given by Fowkes as $\gamma^d = 28.5$ mN/m and $\gamma_S^d = 123$ mN/m, respectively. Based on these, the best experimental estimate of the work of adhesion is

$$W_A = \gamma + \gamma_S - \gamma_{SP} \approx 2(\gamma^d \gamma_S^d)^{1/2} = 118 \text{ mN/m} \tag{8}$$

By subtracting the surface tension of polypropylene, one obtains an experimental estimate of 90 mN/m for the adhesion tension. No evidence is available on the temperature dependence of this quantity. Our model predictions for the internal energy contribution to the work of adhesion and to the adhesion tension at $-40°C$ are 145 ± 52 mN/m (23% error) and $102\pm$ mN/m (44%

error), respectively. We should emphasize, however, that the validity of this comparison is limited by the unavailability of experimental evidence on the temperature dependence of these quantities.

Structural features at the level of segments are distinctly different from what is observed at the free surface. The mass density profile is shown in Fig. 7 together with its individual components. In the middle region of the film, the total density assumes its bulk value. Density displays a strong maximum next to the highly attractive graphite surfaces. Chains adsorb on the surface through their pendant groups. There is a sharp peak in the methyl density profile at a distance from the basal plane roughly equal to the sum of the van der Waals radii of methyls and graphite carbons. An analogous shoulder peak can be discerned in the hydrogen density profile. Skeletal carbons peak at a somewhat larger distance from the graphite surface, because of the constraints imposed on them by pendant bonds. There are no traces of oscillatory behavior in the density profile. Oscillatory features have been observed in simulations of simple, freely jointed bead-and-spring chains in the liquid state [4–7]. Apparently, the presence of substituents of different sizes in polypropylene has a structure-breaking effect on the density profile, which causes any oscillations to disappear.

Bond orientational characteristics in the composite system are displayed in Fig. 8. Again there is a tendency for skeletal carbon–carbon bonds to orient parallel to the solid substrate at the interface. Perhaps the most interesting feature in these results is the strongly perpendicular orientation of pendant

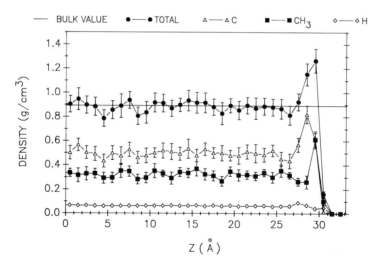

Figure 7 Local mass density distribution at a glassy atactic polypropylene–graphite interface. The first layer of carbon atoms of the solid is located at $z = 33.1$ Å. Individual contributions to the mass density profile from carbon, hydrogen, and methyl groups are also shown.

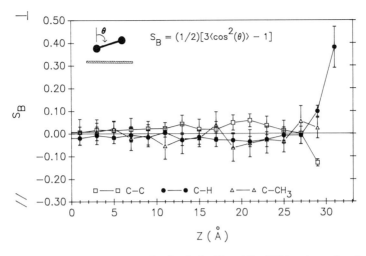

Figure 8 Local order parameter for C—C, C—H, and C—CH₃ bonds as a function of distance from the midplane of the model composite structures.

carbon–hydrogen bonds next to the surface. This suggests some registry between adsorbed hydrogens and the surface.

To explore this idea we accumulated two-dimensional surface density maps for adsorbed hydrogens. We find, as shown in Fig. 9, that the hydrogens show a strong tendency to locate themselves above the centers of the graphite hexagons when adsorbed on the surface. Such a tendency is observed to a lesser extent for adsorbed methyls.

At the level of entire chains, one again sees a strong tendency for the segment cloud to orient itself with its maximal dimension parallel to the phase boundary (Fig. 10). The overall shape of the segment cloud in its principal axis frame of reference is not strongly perturbed. The organization of chains at the interface relaxes down to its bulk characteristics over a distance commensurate with the radius of gyration.

CONCLUSIONS

We have used a detailed, atomistic molecular mechanics approach to simulate the free surface of a glassy polymer, as well as a glassy polymer–solid interface. Although our implementation of this approach focused on the system atactic polypropylene–graphite, the method is quite general and can be applied to more complex systems. Our predictions for surface thermodynamic properties—the internal energy contribution to surface tension, adhesion tension, and the work of adhesion—are within 7 to 23% of available experimental values. The density profile at the free surface is sigmoidal. On the contrary, density is strongly enhanced near the graphite surface in the composite sys-

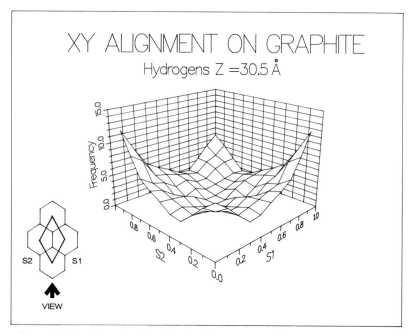

Figure 9 Two-dimensional surface density map for pendant hydrogens adsorbed on the graphite surface. Only hydrogens within a 1-Å-thick bin region, centered 2.5 Å from each graphite basal plane, have been considered in constructing this map. The $(s1, s2)$ domain of the graph constitutes a side view onto the surface unit cell of the graphite basal planes, as indicated in the inset. The plotted function is proportional to the probability that an adsorbed hydrogen sits above a certain point of the surface unit cell. Clearly, hydrogens exhibit a siting preference for the centers of hexagons in the graphite surface honeycomb.

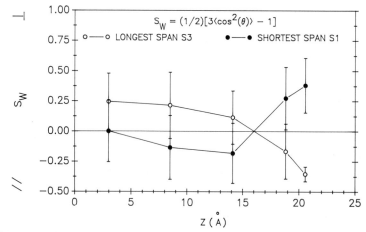

Figure 10 Order parameters for the longest and shortest spans of chains as functions of center-of-mass position for the composite system. Each point represents an average over 15 model chains.

tem. Perturbations in local segment density and bond orientation from isotropy are confined to a narrow region at the interface, approximately 10 Å thick. Structure at the level of entire chains displays some organization over a region whose thickness scales with the radius of gyration. The chain center of mass distribution exhibits a well-defined peak. At both free surfaces and polymer–solid interfaces, chains tend to lie flat, with their longest dimension parallel to the surface.

ACKNOWLEDGMENTS

This work was supported by the Director, Office of Energy Research, Office of Basic Energy Sciences, Materials Science Division of the U.S. Department of Energy under Contract DE-AC03-76SF00098. We also wish to thank BP America and the Union Carbide Corporation for generous gifts. We are grateful to Cray Research, Inc. for making our computations possible through a University Research and Development Grant during 1987 and 1988. We thank the San Diego Supercomputer Center, where all computations were performed, for excellent technical support. D.N.T. expresses gratitude to the National Science Foundation for a 1988 P.Y.I. Award, No. DMR-8857659.

REFERENCES

1. W. G. Madden, *J. Chem. Phys.* **87**, 1405 (1987).
2. G. ten Brinke, D. Ausserré, and G. Hadziioannou, *J. Chem. Phys.* **89**, 4374 (1988).
3. K. F. Mansfield and D. N. Theodorou, *Macromolecules* **22**, 3143 (1989).
4. S. K. Kumar, M. Vacatello, and D. Y. Yoon, *J. Chem. Phys.* **89**, 5206 (1988).
5. C. Lastoskie and W. G. Madden, *Polym. Prepr., Am. Chem. Soc., Div. Polym. Chem.* **30**(2), 39 (1989).
6. A. Yethiraj and C. K. Hall, *Macromolecules* **23**, 1865 (1990).
7. I. Bitsanis and G. Hadziioannou, *Polym. Prepr., Am. Chem. Soc., Div. Polym. Chem.* **30**(2), 78 (1989).
8. D. N. Theodorou and U. W. Suter, *Macromolecules* **18**, 1467 (1985); **19**, 139 (1986); **19**, 379 (1986).
9. W. A. Steele, *Surf. Sci.* **36**, 317 (1973).
10. N. Gō and H. Scheraga, *Macromolecules* **9**, 535 (1976).
11. A. Bondi, *Physical Properties of Molecular Crystals, Liquids and Glasses,* Wiley, New York, 1968.
12. K. F. Mansfield and D. N. Theodorou, *Polym. Prepr., Am. Chem. Soc., Div. Polym. Chem.* **29**(2), 402 (1988).
13. K. F. Mansfield and D. N. Theodorou, *Macromolecules* (in press).
14. K. F. Mansfield and D. N. Theodorou, *Polym. Prepr., Am. Chem. Soc., Div. Polym. Chem.* **30**(2), 76 (1989).
15. U. W. Suter and P. J. Flory, *Macromolecules* **8**, 765 (1975).

16. K. Hillstrom, *Nonlinear Optimization Routines in AMDLIB,* Technical Memorandum No. 297, Argonne National Laboratory, Applied Mathematics Division, 1976: Subroutine GQBFGS in AMDLIB, Argonne, Ill., 1976.
17. R. J. Swenson, *Am. J. Phys.* **51,** 940 (1983).
18. S. Wu, *Polymer Interface and Adhesion,* Dekker, New York, 1982.
19. F. F. Abraham, D. E. Schreiber, and J. A. Baker, *J. Chem. Phys.* **62,** 1958 (1975); G. A. Chapela, G. Saville, S. M. Thompson, and J. S. Rowlinson, *J. Chem. Soc., Faraday Trans. 2* **73,** 1133 (1977).
20. J. R. Rubin and J. Mazur, *Macromolecules* **10,** 139 (1977).
21. F. M. Fowkes, *Ind. Eng. Chem.* **56**(12), 40 (1964).

Interaction Between Grafted Polymeric Brushes

MICHAEL MURAT and GARY S. GREST
Corporate Research Science Laboratory
Exxon Research and Engineering Company
Annandale, New Jersey 08801

ABSTRACT

Equilibrium properties of polymeric brushes consisting of linear polymers grafted at one end onto a repulsive surface as well as the interaction between two brushes are studied using molecular dynamics simulations. Polymers consisting of 10 to 150 monomers are grafted randomly at an average surface coverage of ρ_a onto one of the two parallel plates immersed in a good solvent. ρ_a is taken to be large enough to induce overlap and consequent stretching of the chains. For infinite separation of the plates and intermediate values of ρ_a, we find a monomer density profile in reasonable agreement with the parabolic form predicted by the self-consistent field theory of Milner et al. We also find a nonzero density of the free ends everywhere within the brush, in agreement with that theory. For separations smaller than the contact separation, we monitor these properties as well as the interpenetration of the brushes and the force between the plates. We find that the amount of interpenetration can be described by a simple scaling function for different values of N and ρ_a. The force versus separation curves agree well with the results of the experiments performed on such systems.

INTRODUCTION

The configuration of end-grafted polymers, the so-called polymeric brushes, and the interactions between such brushes have been the subject of recent interest. Experimental studies of such systems have dealt with the direct measurement of the force between surfaces onto which polymers have been terminally attached [1–4]. Theoretical treatments have utilized scaling arguments [5–7] and self-consistent field (SCF) theory [8,9]. Numerical SCF [10,11], Monte Carlo [10,12], and molecular dynamics [13,14] studies of the system have also been performed.

The system considered in this chapter is one consisting of linear polymer chains with N monomers (statistical units) grafted randomly onto an otherwise repulsive surface through an end group located at only one of the ends. The number density of the chains on the surface (surface coverage) is donated by ρ_a. When the average distance between the anchoring points of the chains $d = \rho_a^{-1/2}$ is smaller than their Flory radius, $R_F \sim N^{3/5}$, the chains will tend to stretch in the direction normal to the grafting surface. Assuming the density of the monomers to be constant from the surface up to a maximum height, h^*, Alexander [5] used scaling arguments to show that $h^* \sim N\rho_a^{1/3}$. Milner et al.

[8,9] solved the SCF equations analytically in the limit of very large N and found a parabolic monomer density which vanishes at a distance h_{par}, which has the same scaling form as h^*. Furthermore, they found that the free ends of the chains have a nonzero probability of being at any distance from the surface up to h_{par} and gave an analytical expression for that probability.

Both approaches have been extended to study the interactions resulting from the compression of a polymeric brush to another. Both in the scaling [7] and SCF [15] analyses one assumes that there is no interpenetration of the chains from the two brushes so that the calculation of the interaction energy is reduced to calculating the change in the free energy of a single brush restricted to have a height $h < h_{eq}$, where h_{eq} is either h^* or h_{par}. Both calculations give a repulsive interaction energy at all separations, with different functional forms.

Model experiments of such systems have been performed using end-functionalised polystyrene chains of different molecular weights grafted on mica surfaces [1–3]. In these experiments, the surface was covered up to the saturation density corresponding to each molecular weight, so that N and ρ_a were not varied independently. Force versus separation profiles were measured and found to be well fitted qualitatively by both the scaling and SCF predictions. It should be emphasized, however, that the fit to the SCF theory requires no adjustable parameters, while for the fit to the scaling predictions one extracts the brush height and the absolute magnitude of the forces from the experiments themselves. Furthermore, by identifying the separation at which the force first becomes detectable as twice the brush height, the experiments have verified its scaling as predicted by both theories provided that the interdependency between N and ρ_a is taken into account.

In this study we present the results of a molecular dynamics simulation of these systems at several values of N and ρ_a. This method allows us to simulate systems with a high local density of monomers, unlike the lattice Monte Carlo algorithms, which are very inefficient at such densities. We find that for intermediate values of ρ_a, the monomer density profile is in reasonable agreement with the parabolic form suggested by the SCF theory. Different measures for the brush height all converge to the scaling form predicted by both scaling and SCF theories as the scaling variable $N\rho_a^{1/3}$ increases. The density profile of the free ends is bound to be nonzero anywhere within the brush and its functional form agrees with the SCF predictions, not with the simple constant density profile used by Alexander [5] in his original scaling analysis. We find that when two brushes are brought to contact with each other, the chains from the brushes interpenetrate. The amount of interpenetration is found to decrease with increasing N and ρ_a. We propose a simple scaling form for the amount of interpenetration as a function of the separation between the surfaces. We also calculate the force as a function of the separation and compare our results with experiments and with the theoretical predictions. These results have partially been reported in two recent publications [13,14].

METHOD

The simulations are performed using a molecular dynamics method in which each monomer is coupled to a heat bath [16]. This method has already been applied successfully to several problems related to static and dynamic properties of polymers [16–18]. The monomers are treated as beads of mass m which interact through a shifted short-range Lennard-Jones potential given by

$$U^0(r) = \begin{cases} 4\epsilon\left[\left(\dfrac{\sigma}{r}\right)^{12} - \left(\dfrac{\sigma}{r}\right)^6 + \dfrac{1}{4}\right] & \text{if } r \leq r_c \\ 0 & r > r_c \end{cases} \tag{1}$$

with $r_c = 2^{1/6}\sigma$. As this potential is purely repulsive, our simulations are in the good solvent regime. In addition, there is an attractive interaction between neighboring monomers along the chains and a strongly repulsive one of range $\sigma/2$ between the wall and the monomers. Parameters of these potentials are the same as in Ref. 13. Denoting the total potential of monomer i by U_i, the equation of motion for monomer i is given by

$$m\frac{d^2\mathbf{r}_i}{dt^2} = -\nabla\cdot U_i - m\Gamma\frac{d\mathbf{r}_i}{dt} + \mathbf{W}_i(t) \tag{2}$$

Here Γ is the bead friction which acts to couple the monomers to the heat bath. $\mathbf{W}_i(t)$ describes the random force acting on each bead. It can be written as a Gaussian white noise with

$$\langle \mathbf{W}_i(t)\cdot\mathbf{W}_j(t')\rangle = 6k_B Tm\Gamma\delta_{ij}\delta(t - t') \tag{3}$$

where T is the temperature and k_B is the Boltzmann constant. We have used $\Gamma = 0.5\tau^{-1}$ and $k_B T = 1.2\epsilon$. With this choice of parameters, the average bond length between neighboring beads along the chains is found to be 0.97σ. The equations of motion are then solved using a predictor–corrector algorithm with a time step $\Delta t = 0.006\tau$. Here $\tau = \sigma(m/\epsilon)^{1/2}$. Further details of the method can be found elsewhere [13,16].

    Using this method, we simulated systems of flat surfaces onto which M polymers of $N + 1$ monomers are anchored at one end. Each plate has a surface area S in units of σ^2, giving a grafting density of $\rho_a = M/S$. Periodic boundary conditions are used in the two directions parallel to the plane of the plates. To study the interaction between two brushes, we simulated two such systems facing each other at a distance D. We varied N in the range 20 to 150 and ρ_a in the range 0.01 to 0.2. In the study of the individual brushes two different sets of initial conditions were used to ensure that our final results were equilibrated. In the first, one end of each polymer was grafted at a random point on the surface, after which N monomers at a distance σ apart were added along a straight line perpendicular to the plane, in a fully stretched configuration. In the other one, the N monomers performed a self-avoiding walk in the half-space $z > 0$, where z is the distance along the axis perpendicular to the grafting surface. In the latter case, two monomers from different chains often overlapped, and it was necessary to run the simulations for a few hundred steps with a softer potential until they did not overlap any more before switching on the repulsive Lennard-Jones potential. After equilibration we found that both starting states gave the same results. Simulations involving the interactions of two brushes were carried out in the following manner: We start with the plates at a separation D slightly above the contact distance $D_c = 2h_{ext}$, where h_{ext} is the maximum extent of a single brush [13] and equilibrate the system. After equilibration, we continue the simulation while reducing D by a very small amount at each time step. The configurations at several selected values of D are saved. These are then used as starting states for subsequent longer runs at those values of D, from which the quantities of interest are calculated. We run the simulations for times much longer than the configurational relaxation times of the chains to assure that proper averaging is achieved.

RESULTS AND DISCUSSION

Figure 1 shows the monomer density profiles for chains of length $N = 50$ for two values of ρ_a. For both cases one can see a layering near the wall, a feature that is not exhibited by the scaling or the SCF approaches. For $\rho_a = 0.2$ the profile is very close to the step function profile assumed by Alexander [5]. However, the situation is quite different for $\rho_a = 0.03$. The profile clearly has a parabolic form, with a maximum slightly away from the wall. This last feature has also been observed in the numerical SCF calculations of Cosgrove et al. [10]. Simulations at other values of the surface coverage in the range 0.01 to

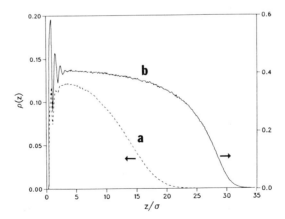

Figure 1   Monomer density profile for polymers of length $N = 50$ at (a) $\rho_a = 0.03$ and $\rho_a = 0.2$. Note the different scales of the vertical axes for the two cases.

0.1 also show similar behavior. We also find that different measures of the height of brush, such as the first moment of the density profile or the average z-component of the free ends converge to the scaling form predicted by both theories as the scaling variable $N\rho_a^{1/3}$ increases. To make the comparison with the parabolic profile more quantitative, we plot in Fig. 2 the density profile as a function of $(z/N)^2$ at $\rho_a = 0.03$ for $N = 20, 50, 100$, and 150. The parabolic fit clearly improves with increasing N. We have also numerically fitted the density profiles to a parabolic form and compared [13] the resulting parameters to the predictions of the SCF theory. The agreement is reasonable for the intermediate values of ρ_a.

Other properties of the brushes, such as the distribution of the free ends and the average orientation of the nth bond of each chain agree with the predictions of the SCF theory for intermediate values of ρ_a. For higher values of ρ_a ($\rho_a = 0.1$ and 0.2), we find that our results cannot be reconciled with that theory. This, however, is not surprising since the theory is applicable to the case of weakly interacting chains. This cannot be achieved with the purely repulsive monomer–monomer interaction that we have used in our simulations.

We now consider the results for the case of two such brushes held at a distance D between the grafting surfaces, with $D < 2h_{\text{ext}}$. Figure 3 shows the results for the monomer density profile for $N = 50$, $\rho_a = 0.03$ at several separations. $h_{\text{ext}} = 21$ for this system. As the separation is decreased, the density

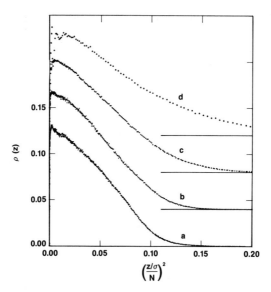

Figure 2 Monomer density as a function of the square of the distance from the grafting surface for chains of length (a) $N = 150$, (b) $N = 100$, (c) $N = 50$, (d) $N = 20$ for a surface coverage of $\rho_a = 0.03$. Each curve is displaced vertically for clarity. Note that z is scaled by N.

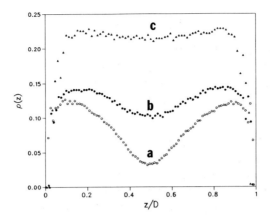

Figure 3   Monomer density profile $\rho(z)$ for polymers of length $N = 50$ at $\rho_a = 0.03$.
(○) $D = 35$; (●) $D = 25$; (Δ) $D = 15$.

increases and becomes almost uniform. This result resembles the one obtained by Muthukumar and Ho [11] using a numerical SCF calculation. When we separate the contribution of each brush to the overall monomer density (Fig. 4), however, we see that the parabolic profile of each brush is conserved. One can also see that as D is decreased, the number of monomers from each brush found at $z > D/2$ increases, indicating larger interpenetration. To quantify the amount of interpenetration at separation D, we determine $I(D)$, given by

$$I(D) = \frac{\displaystyle\int_{D/2}^{D} \rho_1(z)\, dz}{\displaystyle\int_{0}^{D} \rho_1(z)\, dz} \tag{4}$$

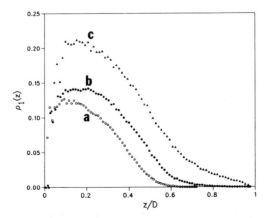

Figure 4   Contribution to $\rho(z)$ from one of the brushes for the case of $N = 50$, $\rho_a = 0.03$. The symbols have the same meaning as in Fig. 3.

Here $\rho_1(z)$ is the contribution of each brush to the overall density. We can use the results of the SCF theory of Milner et al. [8] to develop a scaling form for $I(D)$ with N and ρ_a. For a brush compressed to a height $h = D/2 < h_{par}$, the monomer density profile is still a parabolic one up to $z = h$ and vanishes abruptly for larger values of z. The density at h is then given by

$$\rho(h) = \frac{\rho_a N}{h}\left[1 - \left(\frac{h}{h_{par}}\right)^3\right] \tag{5}$$

Assuming that the bulk of the monomers is at $z < h$ with an exponentially decaying tail of length δ beyond h and noting that the integration over the whole density profile of a single brush yields $\rho_a N$, we find that the amount of interpenetration can be approximated as $I(D) \simeq 2\delta/D$. Recently, Witten et al. [19] suggested an estimate for δ in the context of lamellar copolymer phases. Their estimate is applicable to our problem since in both cases the polymers are in identical configurations. Equating the gain in translational entropy as a result of interpenetration with the accompanying penalty of interaction energy, they obtained $\delta \sim (h/N^2)^{-1/3}$. Using this relation in the estimate for $I(D)$, we obtain

$$I(D)N^{-2/3} h_{par}^{4/3} \sim x^{-4/3}(1 - x^3) \tag{6}$$

where $x = D/2h_{par}$. Figure 5 shows the left-hand side of this equation as a function of x for all the cases studied. The data from the different systems collapse reasonably well for small compressions. We also observe that the points for $\rho_a = 0.2$ start to deviate from this curve first (at the lowest values of

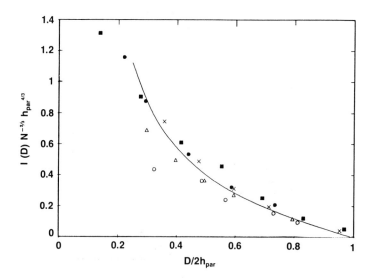

Figure 5 Interpenetration $I(D)$ scaled as explained in the text for the cases studied. ($\bullet$) $N = 100$, $\rho_a = 0.03$; ($\blacksquare$) $N = 50$, $\rho_a = 0.03$ ($\times$) $N = 50$, $\rho_a = 0.05$; ($\triangle$) $N = 50$, $\rho_a = 0.1$; ($\circ$) $N = 50$, $\rho_a = 0.2$.

compression), followed by those for $\rho_a = 0.1$. This can be qualitatively under-stood by noting that when the overall monomer density approaches unity, the density profile is almost uniform, and further decrease of D causes a uniform compression rather than an increase in the interpenetration. The scaling re-sults for $I(D)$ deviate sooner the higher the surface coverage since a smaller compression is needed to reach such high densities. As such high densities are achieved at smaller compressions the higher the surface coverage, $I(D)$ curves for those cases saturate first.

Figures 6 and 7 shows the probability of finding the free end of each chain at a distance z from the surface to which the chain is attached for $N = 100$, $\rho_a = 0.03$ and $N = 50$, $\rho_a = 0.2$ respectively. The respective h_{ext} for these sys-tems are 42 and 33. When the separation between the plates is slightly above D_c, the end density profile is very similar to the case of the single brush [13], with a very small probability of penetration into the region $z > D/2$. As D decreases, the density profile broadens and the probability of finding the end in the region $z > D/2$ significantly increases. For the higher compressions one finds a relative increase in this density near the walls where the overall mono-mer density is low.

At each separation we also calculated the force between the plates. This is done by calculating the pressure using the virial [20]

$$PV = NMk_B T - \frac{1}{3} \sum_{i=1}^{NM} \mathbf{r}_i \cdot \nabla U_i \qquad (7)$$

where U_i is the total potential energy of monomer i due to other monomers and V is the volume of the system. The force per unit area, $f(D)$, between the two plates is then found by subtracting from this pressure the pressure at $D = D_c$. We find that this force relaxes very rapidly; runs of lengths of only a few τ are sufficient to obtain $f(D)$ quite accurately. To compare our result with

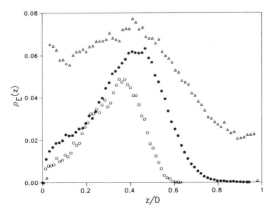

Figure 6 Probability of finding the free end of each polymer at a distance z from the grafting surface for $N = 100$ at $\rho_a = 0.03$. (○) $D = 70$; (●) $D = 40$; (△) $D = 20$.

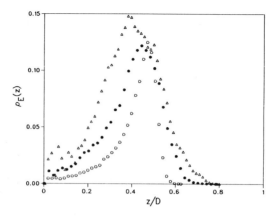

Figure 7 Probability of finding the free end of each polymer at a distance z from the grafting surface for $N = 50$ at $\rho_a = 0.2$. (○) $D = 50$; (●) $D = 30$; (Δ) $D = 20$.

recent experiments, which measure the interaction energy per unit area, we calculated

$$E(D) = \int_{D_c}^{D} f(D')dD' \tag{8}$$

Both scaling arguments [7] and SCF theory [9] predict that this energy should scale as $E(D) \sim h\rho_a^x \bar{E}(D/2h)$, with $x = \frac{3}{2}$ and $\frac{4}{3}$, respectively. Here h is the height of the brush as predicted by either theory. The difference originates from the dependence of the osmotic pressure on the monomer density in the two theories. The two theories also predict different scaling functions $\bar{E}(x)$. In Figure 8 we show $E(D)$ scaled as suggested by the scaling theory. We find that the scaling works much better when we substitute h_{ext} rather than h_{par} for h. The energies for $N = 50$, $\rho_a = 0.01$, and the two different values of N at $\rho_a = 0.03$ collapse to a single curve. The other cases converge to this curve only for $D/2h_{ext} \to 1$, although all the curves have the same basic form. The insert in Fig. 8 shows the experimental results of Taunton et al. [3] from experiments with terminally attached polystyrene chains of various molecular weight. The solid line is a smooth curve passing through *our* data points for $\rho_a = 0.03$. Since our energy scale is arbitrary, this line is vertically shifted to lie on the experimental points for comparison. The overall agreement between the simulation and the experimental results is excellent. We emphasize that the fit is achieved using no adjustable parameters, except the vertical shift due to the arbitrary energy scale of our simulations. The length scale used in the scaling of D is taken from an independent simulation.

As is the case with the experiments [3], our data for $\rho_a = 0.03$ and 0.01 comply equally well with the scaling functions predicted by both theories. We expect the agreement to break down at high compressions when Φ, the average number density of the monomers between the surfaces, approaches unity. At this limit, the osmotic pressure, which is the leading contribution to the

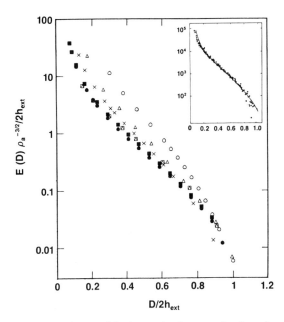

Figure 8 Energy per unit area of the interaction between the plates for all the cases studied. (●) $N = 100$, $\rho_a = 0.03$; (△) $N = 50$, $\rho_a = 0.01$; (■) $N = 50$, $\rho_a = 0.03$; (×) $N = 50$, $\rho_a = 0.05$; (△) $N = 50$, $\rho_a = 0.1$; (○) $N = 50$, $\rho_a = 0.2$. The insert shows a comparison between the experimental results of Taunton et al. [3] (data points) and our results (full line obtained by joining our data points for $N = 100$, $\rho_a = 0.03$. It is vertically shifted by an arbitrary amount). The experimental points include results from measurements with polymers of various molecular weights at corresponding grafting densities.

force, diverges logarithmically with $1 - \Phi$ rather than increasing as a power of Φ, as predicted by both theories. In Fig. 8 we indeed see this deviation for $\rho_a = 0.1$ and 0.2 at $D/2h_{\text{ext}} \approx 0.6$ and 0.8, respectively, as this dense regime is already reached at these compressions. Such a trend, however, is not seen in the experiments. Systems with polymers of various lengths (and consequently, at different surface coverages, as these two are not varied independently in the experiments [3]) yield profiles that collapse on the same scaling curve. This difference between the simulations and the experiments can be explained by observing that in the experiments the average monomer number density in the uncompressed system for all the cases (except possibly those with the lowest N) is well within the semidilute regime. Densities of order unity are achieved only at the lowest separation values. Any possible deviation from the theoretical behavior would be hidden in the experimental noise. The predicted deviation from the scaling (or SCF) behavior could be seen experimentally if one used an anchoring unit which attaches to the surface with a much higher sticking energy, which would increase the average density. It can easily be shown that if the sticking energy is $\alpha k_B T$, the average density scales with α at constant N as $\alpha^{4/5}$. Thus if α is increased by a factor of 3 to 5, one should observe these systematic deviations at intermediate compressions.

CONCLUSIONS

In conclusion, we have performed a molecular dynamics study of the system of end-grafted polymers using a simple bead-spring model for the polymers. For a single brush, we confirm the predictions of the SCF of Milner et al. [8] for intermediate values of the surface coverage. When two such brushes are brought to contact, we find that the interaction is purely repulsive and that it sets in as soon as the brushes touch each other before any interpenetration occurs. The interaction versus separation profiles agree extremely well with the experimental curves. We also find that these profiles for intermediate values of surface coverage are satisfactorily described by both scaling and SCF theories if one scales the separation by the maximum extent of the brushes. We attribute the deviations from the theoretical behavior at higher surface coverages to the fact that the resulting monomer density is much higher than the regime of validity of both theories. We also monitor the amount of interpenetration in the compressed brushes and show that it can be described by a simple scaling form.

ACKNOWLEDGMENTS

We thank J. Klein, T. A. Witten, and S. T. Milner for helpful discussions. Further thanks are due to J. Klein for permitting the use of the experimental data shown in the insert to Fig. 8 prior to publication. MM is a recipient of the Chaim Weizmann postdoctoral fellowship.

REFERENCES

1. H. J. Taunton, C. Toprakcioglu, and J. Klein, *Macromolecules* **21**, 3336 (1988).
2. H. J. Taunton, C. Toprakcioglu, L. J. Fetters, and J. Klein, *Nature* **332**, 712 (1988).
3. H. J. Taunton, C. Toprakcioglu, L. J. Fetters, and J. Klein, *Macromolecules* **23**, 571 (1990).
4. G. Hadziioannou, S. Patel, S. Granick, and M. Tirrell, *J. Am. Chem. Soc.* **108**, 2869 (1986).
5. S. Alexander, *J. Phys. (Paris)* **38**, 983 (1977).
6. P.-G. de Gennes, *Macromolecules* **13**, 1069 (1980).
7. P.-G. de Gennes, *C. R. Acad. Sci. Paris* **300**, 839 (1985).
8. S. T. Milner, T. A. Witten, and M. E. Cates, *Macromolecules* **21**, 2610 (1988).
9. S. T. Milner, T. A. Witten, and M. E. Cates, *Europhys. Lett.* **5**, 413 (1988).

10. T. Cosgrove, T. Heath, B. van Lent, F. Leermakers, and J. Scheutjens, *Macromolecules* **20,** 1692 (1987).

11. M. Muthukumar and J.-S. Ho, *Macromolecules* **22,** 965 (1989).

12. A. Chakrabarti and R. Toral, preprint (1989).

13. M. Murat and G. S. Grest, *Macromolecules* **22,** 4054 (1989).

14. M. Murat and G. S.Grest, *Phys. Rev. Lett.* **63,** 1074 (1989).

15. S. T. Milner, *Europhys. Lett.* **7,** 695 (1988).

16. G. S. Grest and K. Kremer, *Phys. Rev.* **A33,** 3628 (1986).

17. K. Kremer, G. S. Grest, and I. Carmesin, *Phys. Rev. Lett.* **61,** 566 (1988).

18. K. Kremer and G. S. Grest, Chapter 12, this volume.

19. T. A. Witten, L. Leibler, and P. A. Pincus, *Macromolecules* **23,** 824 (1990).

20. M. P. Allen and D. J. Tildesley, *Computer Simulation of Liquids,* Clarendon Press, Oxford, 1987, p. 47.

Brownian Dynamics Simulations of Local Polymer Motions in Polyisoprene: Comparison to NMR Experiments

DAVID B. ADOLF and M. D. EDIGER
Department of Chemistry
University of Wisconsin
Madison, Wisconsin 53706

ABSTRACT

Local segmental motions of polyisoprene have been analyzed via Brownian dynamics computer simulations. The simulations were performed in the high-friction limit using independent torsional potentials and some structural approximations. The simulations indicate that the vector connecting the methine carbon to its bonded proton has a correlation time 40% longer than that of a CH vector on the neighboring methylene group. Comparison of experimental correlation times obtained from nuclear magnetic resonance (NMR) measurements to the analogous correlation times produced by the simulations reveals semiquantitative agreement.

INTRODUCTION

In recent years a wide variety of experimental techniques have been used to investigate the local segmental dynamics of polymer chains. Our fundamental understanding of how these dynamics emerge from structure and potentials has not advanced at the same rate as the experimental measurements. Several analytical treatments of this problem have pointed out essential features of local dynamics in polymer chains [1]. Yet most of these models have not incorporated sufficiently realistic features of the polymer structure and potentials to explain the different local dynamics observed for different polymers. Computer simulations afford a method of incorporating more realistic structures and potentials and offer the possibility of understanding the nonuniversal nature of local polymer motions.

The simulation of local polymer dynamics in dilute solution presents some unique challenges. The simulated chain must be long enough that end effects and end-over-end motions do not dominate the observed dynamics. Since the isolated chain is in a random coil configuration, the volume of the simulated system must be large. Most of the interactions in this large volume will be solvent–solvent interactions which are only weakly coupled to the dynamics of the polymer chain. Thus the application of molecular dynamics simulations to this problem would require large amounts of computer time, with most of the time spent on the least interesting aspects of the system.

An alternative approach to molecular dynamics is a Brownian dynamics simulation in which the solvent molecules are not treated explicitly. In a Brownian dynamics simulation, solvent molecules are replaced by a viscous continuum. Stochastic forces act on each chain atom, mimicking the effects of solvent–polymer collisions and allowing energy exchange between the chain and the bath. Each chain atom has a friction coefficient associated with it which accounts for the viscosity of the solvent continuum. The solvent continuum approximation is supported by experimental evidence which shows that specific polymer–solvent interactions do not play an important role in solutions of polyisoprene [2,3].

Stochastic techniques such as Brownian dynamics have a long history. A review of the fundamental principles of these methods is found in Ref. 4. The work of van Gunsteren and co-workers [5], Karplus and coworkers [6], and Weiner and Pear [7] illustrates the application of Brownian dynamics to chain-like molecules. Helfand and coworkers [8,9] have simulated the conformational dynamics of polyethylene using this approach. The simulation described in this chapter extends the approach used for polyethylene to polyisoprene. Polyisoprene is an interesting polymer to simulate because of the wide range of experimental techniques that have been used to study its local segmental dynamics. In this chapter, we restrict our attention to comparisons with ^{13}C NMR experiments. We demonstrate that Brownian dynamics simulations can provide a semiquantitative description of the experimental results and thus show considerable promise for a fundamental understanding of local segmental dynamics in this system. In a future publication we will discuss the simulation results in terms of specific conformational transition processes [10].

DESCRIPTION OF SIMULATIONS

The starting point for Brownian dynamics simulations is the Langevin equation. In the high-friction limit, inertial terms can be ignored and the equation of motion written as

$$\frac{dx_i}{dt} = \frac{1}{\zeta_i} \nabla_{\mathbf{x}_i} V + \mathbf{B}_i(t) \tag{1}$$

The subscript i runs over all the particles that comprise the polymer chain used in this project. ζ_i is the friction coefficient for particle i and the potential function is denoted by V. $\mathbf{B}_i(t)$ is the Brownian noise term with mean zero and variance

$$\langle \mathbf{B}_i(t)\mathbf{B}_j(t') \rangle = \frac{2k_B T}{\zeta_i} \delta_{ij} \delta(t - t') \tag{2}$$

In Eq. (2), k_B is Boltzmann's constant and T is the bath temperature.

The upper part of Fig. 1 shows the structure of *cis*-polyisoprene, and the lower part shows the structure used in the simulations. As is customary in Brownian dynamics simulations, united atoms were used to represent methylene groups. In the same spirit, the entire $CHCCH_3$ group has also been collapsed into a single particle, as indicated. Relative friction coefficients for

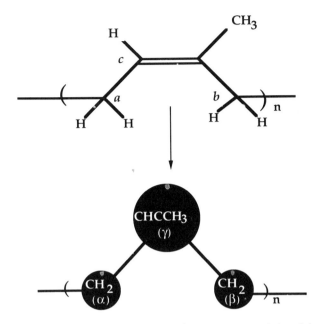

Figure 1   The upper part of the figure shows the true structure of *cis*-polyisoprene and labels the locations of the three carbons referred to in the text. The lower part of the figure shows the structural approximation used in the simulations.

the two types of particles were fixed by applying Stokes' law to the volumes of the groups they represent. In the discussion below we use English characters to refer to real atoms and Greek characters to indicate the united atoms. The simulated chain had 67 monomer units, each comprised of three particles.

The potential V in Eq. (1) is a sum of three terms. The motion of each particle is affected by two bond stretching forces (V_b), three valence angle bending forces (V_θ), and four dihedral angle torsion forces (V_ϕ). These terms take the form

$$V_{b,i} = 0.5\gamma_b(b_i - b_0)^2 \tag{3}$$

$$V_{\theta,i} = 0.5\gamma_\theta(\cos\theta_i - \cos\theta_0)^2 \tag{4}$$

$$V_{\phi,i} = 0.5\gamma_\Phi \sum_{n=0}^{5} a_n \cos^n\phi_i \tag{5}$$

The stretching and bending potentials were harmonically constrained to reasonable b_0 and θ_0 values. Following Helfand, we chose the force constants for bond stretching and bending to be about three times looser than a realistic force constant in order to allow larger integration time steps [8]. The torsional potentials [Eq. (5)] were obtained by using Allinger's MM2 program [11] on small molecules resembling pieces of a polyisoprene polymer. Three distinct torsional potentials were utilized, two of which are illustrated in Fig. 2. The

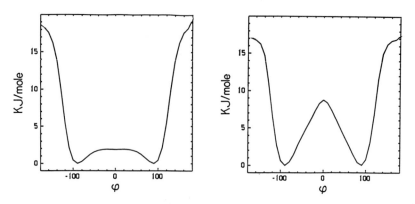

Figure 2   Torsional potential-energy curves used for dihedral angles in the simulation. The left curve refers to rotation about the $\alpha\gamma$ bond and the right curve refers to the $\gamma\beta$ bond. The conformation shown in Fig. 1 corresponds to $\phi = 0$ for both dihedral angles.

left part of the figure shows the potential for revolution about the $\alpha\gamma$ bond. It was obtained by running MM2 on 2-methyl-2-pentene. The central torsional angle was fixed at various angles, while all other degrees of freedom were energy minimized. The right part of the figure shows the torsional potential for rotation about the $\gamma\beta$ bond obtained with the same procedure using 3-methyl-2-pentene as the model compound. The results from the MM2 calculations were fit to a six-term cosine series. The torsional potential for rotation about the $\beta\alpha$ bond is very similar to that for butane [8] and is not shown. The barrier height for this potential is about 14 kJ/mol. No torsional coupling was utilized in these simulations. Table 1 contains structural and potential parameters used in these simulations. In addition, $\gamma_b = 5.81 \times 10^{-17}$ kg/ns^2, $\gamma_\theta = 3.32 \times 10^{-21}$ J, and $\gamma_\Phi = 1.54 \times 10^{-20}$ J.

The second-order integration scheme used for these simulations was developed by Helfand and Greenside [12]. We verified that a step size of 5 fs was sufficiently small, and this value was used throughout. We initially performed short runs on a VAX 8650. A 1-ns trajectory required 10 hours of central processing unit (CPU) time. All major runs were performed on an IBM 3090 with a vector processor. On this machine, 1 ns of trajectory required about 1 hour of CPU time. The runs discussed in this chapter are shown in Table 2.

Periodic boundary conditions were used to avoid end effects. The end-to-end vector was constrained to be 4.3 nm, which is close to the average value for the simulated chain. Random starting configurations were generated using a Boltzmann weighting of conformations. A harmonic potential was then applied to obtain the desired end-to-end distance. The end-to-end distance was fixed and a 5-ps trajectory was run to ensure that the chain was equilibrated with the bath. All 201 coordinates were stored every 250 fs after this initialization procedure.

TABLE 1 Structural and Potential Parameters

Particle Properties			
	α	β	γ	Units
ζ/m	1.0×10^5	48,265	1.0×10^5	ns^{-1}
m	0.014	0.040	0.014	kg/mol

Bond Lengths			
	$\gamma\alpha$	$\alpha\beta$	$\gamma\beta$	Units
b_0	0.153	0.287	0.287	nm

Bond Angles			
	$<\beta\alpha\gamma$	$<\alpha\gamma\beta$	$<\gamma\beta\alpha$	Units
θ_0	70.5	120.0	70.5	degrees

Torsional Potentials		
	$\alpha\gamma$	$\gamma\beta$	$\beta\alpha$
a_0	10.137	12.496	18.567
a_1	−0.102	−0.350	−21.028
a_2	17.215	22.253	−6.354
a_3	−19.625	−19.233	27.280
a_4	−6.534	−9.343	7.451
a_5	11.381	15.562	3.531

As a check on our simulation code and procedures, we ran a 10-ns trajectory of polyethylene at 425 K using the parameters given by Weber and Helfand [9]. We calculated bond, chord, bisector, and out-of-plane correlation functions and compared them with the fits reported in their paper. The agreement was quantitative.

CORRELATION FUNCTIONS

The major focus of this chapter is the comparison of the simulated dynamics of polyisoprene with the actual solution dynamics as determined by ^{13}C NMR T_1 measurements. These experiments are sensitive to P_2 orientation autocorrela-

TABLE 2 Simulation Runs

Run	T (K)	Trajectory Length (ns)
I	425	7.5
II	425	10.0
III	273	20.0

tion functions associated with the motion of various CH vectors in the chain backbone. A publication by Glowinkowski, Gisser, and Ediger reported NMR measurements on dilute solutions of polyisoprene in 10 solvents [3]. Correlation times were reported for the motion of CH vectors on the a, b, and c carbons (see Fig. 1). For the methylene carbons, the average correlation time for the two CH bonds is observed. The second-order orientation autocorrelation function for a particular CH bond vector is given by

$$CF(t) = \langle P_2(\mathbf{x}(0)\cdot\mathbf{x}(t))\rangle = 0.5\langle 3(\mathbf{x}(0)\cdot\mathbf{x}(t))^2 - 1\rangle \qquad (6)$$

where P_2 is the second Legendre polynomial and $\mathbf{x}(t)$ is a unit vector in the direction of the CH bond vector at time t. In the extreme narrowing region, the NMR experiments allow the correlation time τ_c associated with a particular correlation function to be determined. The correlation time is the integral of the correlation function:

$$\tau_c = \int_0^\infty CF(t)\, dt \qquad (7)$$

To compare the simulation results with the NMR experiments, we used the following procedure. First, vectors with the appropriate orientation were constructed with respect to the simulated chain structure. Second, correlation functions associated with each of these vectors were calculated, averaging over all monomer units in the chain. Finally, the correlation functions were integrated in order to obtain the correlation times. This integration was performed both by direct numerical integration and by analytically integrating the fit to a sum of exponentials (obtained by using Provencher's SPLMOD exponential fitting routine) [13]. Good agreement was obtained between the two methods and the differences between the two methods are included in the uncertainties given with the results presented below.

The orientation of vector $\mathbf{C}$ (connecting methine carbon c to its bonded proton) on a particular unit was calculated according to the following procedure:

$$\mathbf{D} = \boldsymbol{\gamma\alpha} \times (\boldsymbol{\beta\gamma} \times \boldsymbol{\gamma\alpha}) \qquad (8)$$

$$\mathbf{C} = 0.866\frac{\mathbf{D}}{|\mathbf{D}|} - 0.5\frac{\boldsymbol{\gamma\alpha}}{|\boldsymbol{\gamma\alpha}|} \qquad (9)$$

In these equations, $\boldsymbol{\beta\gamma}$, for example, represents a vector connecting beads β and γ. The orientation of vector $\mathbf{B}$ (connecting methylene carbon b to one of its bonded protons) is given by

$$\mathbf{E} = \boldsymbol{\alpha\beta} \times \boldsymbol{\beta\gamma} \qquad (10)$$

$$\mathbf{F} = \boldsymbol{\alpha\beta} - \boldsymbol{\beta\gamma} \qquad (11)$$

$$\mathbf{B} = 0.816\frac{\mathbf{E}}{|\mathbf{E}|} + 0.577\frac{\mathbf{F}}{|\mathbf{F}|} \qquad (12)$$

The expression for the orientation of vector $\mathbf{A}$ is analogous.

The correlation functions calculated from the simulations showed relaxation times spanning more than two decades in time. To characterize the correlation function on all relevant time scales we calculated the correlation functions from a number of positions within the trajectory using three different time increments (0.25 ps, 2.5 ps, and 25 ps). In this way we did not require the large amount of CPU time needed to calculate the correlation function over the entire trajectory with the finest time increment. In Fig. 3 we show a typical correlation function for vector **C** in polyisoprene at 425 K. The calculated values for the finest time increment are not shown.

NMR results on polyisoprene in solution indicate that some very low amplitude (<0.5%), long time components (>10 ns) are present in the experimental correlation functions in addition to the components that decay on the time scale of 100 ps or less [3]. These long time components presumably are associated with large-scale chain motions which must occur in order to relax completely the orientation of a given CH vector. Contributions from these components are not included in the NMR correlation times to which we compare the simulations in the next section. In this sense the NMR experiments measure an effective correlation time. In practice, these long components are also excluded from the correlation times that we calculate from the simulated correlation functions. This is because our trajectories are not long enough to

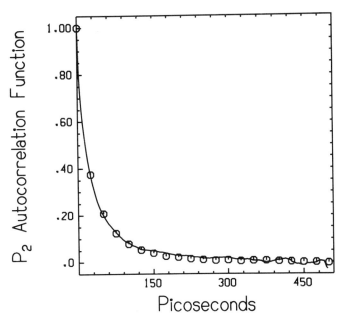

Figure 3 Correlation function for the vector connecting the methine carbon to its bonded proton (425 K). The line represents results for a 500-ps correlation with a time increment of 2.5 ps. The open circles indicate results for a 5000-ps correlation with a time increment of 25 ps.

sample large-scale motions accurately and because the calculated correlation functions have decayed to zero (within statistical noise) in a few hundred picoseconds, as shown in Fig. 3.

RESULTS AND DISCUSSION

Breakdown of the nT_1 Rule

One of the most striking features of NMR results on polyisoprene is that different carbon–hydrogen vectors in the chain backbone have significantly different correlation times (thus the "nT_1 rule" does not hold) [3,14–16]. The vector connecting carbon c to its bonded proton has the longest correlation time, while the vector connecting carbon a to its protons has the shortest correlation times. Figure 4 shows the simulated correlation functions for 425 K, corresponding to the three different vectors sensed in the NMR experiments. The correlation function decays are ordered as expected from the NMR experiments. Table 3 shows a quantitative comparison of the ratios of the various correlation times. The NMR results represent average values for

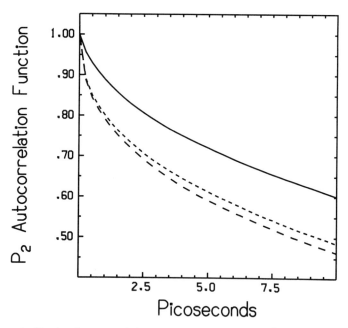

Figure 4   Simulated autocorrelation functions associated with ^{13}C NMR T_1 relaxation measurements. The curves show correlation functions for CH vectors on carbon c, ———; carbon b, – – –; and carbon a, — — —. The simulation temperature was 425 K.

TABLE 3 Ratio of Correlation Times for Methine and Methylene Carbons[a]

	τ_b/τ_a	τ_c/τ_a
Simulation		
Run I	1.13	1.37
Run II	1.21	1.50
Run III	1.16	1.43
Average	1.17 ± 0.08	1.43 ± 0.11
NMR		
Glowinkowski et al. [3]	1.19 ± 0.12	1.54 ± 0.14

[a] τ_a, τ_b, and τ_c indicate correlation times for CH vectors on carbons a, b, and c.

10 solvents. To calculate τ_c from the NMR T_1 measurements, we have assumed a methine carbon–hydrogen bond length of 1.08 Å and a methylene carbon–hydrogen bond length of 1.09 Å.

The semiquantitative agreement in Table 3 indicates that subtle features of the local dynamics of polyisoprene are being successfully captured in the computer simulation. It is worth noting that the breakdown of the nT_1 rule in polyisoprene has sometimes been attributed to dipolar relaxation of the methine carbon with nonbonded protons [15]. The agreement between simulation and experiment without considering this interaction clearly indicates that the origin of this phenomenon is dynamical in nature (i.e., the correlation times associated with different CH vectors in the chain backbone are different). The behavior can be rationalized in physical terms. The vector on carbon c cannot significantly reorient without the entire double bond group reorienting. Thus it is not surprising that this vector has the longest correlation time. Figure 2 shows that the two methylene groups move in response to significantly different torsional potentials. The CH vector neighboring the lower potential (carbon a) has a shorter correlation time.

Monnerie and co-workers have described NMR results on dilute polyisoprene solutions with a model that emphasizes the importance of librational motions [16]. The very fast initial decays of the correlation functions shown in Fig. 4 are likely associated with such motions. Further evidence for librational motions in this system will be discussed in a future publication [10].

Absolute Correlation Times

It is possible to compare the absolute correlation times from the simulations with experiments if a connection can be made between the friction constant and the viscosity. Paul and Mazo have found that translational diffusion data on alkanes can be explained assuming that Stokes' law is valid and the hydrodynamic radius of a methylene group is taken to 0.77 Å [17]. This implies that the methylene friction constant used in the simulations is appropriate for a viscosity of 1.6 cP. Experimentally, it has been found that τ_c for the vector connecting carbon c to its bonded proton is 90 ± 10 ps in toluene at 273

K ($\eta = 0.77$ cP) [3]. In the high-friction limit, the simulation results scale with the friction coefficient. Scaling τ_c from the simulations by the viscosity yields 53 ± 5 ps at $\eta = 0.77$ cP and $T = 273$ K, within a factor of 2 of the experimental results. Similar agreement between NMR correlation times and Brownian dynamics simulations have been reported for a 9-carbon alkane by Brown, et al. [18].

Unfortunately, the comparison of correlation times is more ambiguous than indicated above. The experimental correlation times do not scale linearly with the viscosity and there is evidence that the frequency dependence of the friction coefficient must be considered [3]. Bagchi and Oxtoby have shown in a model calculation that effects of frequency-dependent friction will be minimal in the low-viscosity region ($\eta < 0.5$ cP) [19]. Thus the comparison we have made in the previous paragraph is probably the best that can be made. Given experimental results indicating the importance of frequency-dependent friction, it might be necessary to include time correlations of the random forces in future Brownian dynamics simulations. Alternatively, Brownian dynamics in the high-friction limit may still be appropriate, but a more complex relationship than Stokes' law may be required to connect viscosity to friction coefficients.

Activation Energies

From simulation results at two temperatures we have calculated activation energies for the three NMR correlation times. Within statistical error, they are all equal to 7.6 ± 0.4 kJ/mol. This is quite similar to the energy barrier for torsion of the $\gamma\beta$ bond (9 kJ/mol), although this similarity does not prove that rotation about this bond is the controlling factor in conformational dynamics. The activation energy obtained from the NMR experiments (after accounting for the solvent contribution) is 13 ± 2 kJ/mol. It is possible that the difference between the experimental and simulated activation energies is due to errors in the potentials used for torsions.

SUMMARY

The major focus of this chapter has been a comparison of Brownian dynamics simulations with experimental results on the local segmental dynamics of polyisoprene. It is worthwhile to summarize some of the approximations made in these simulations: (1) independent torsional potentials; (2) use of united atoms, particularly the replacement of the $CHCCH_3$ group by a single particle; and (3) unrealistically loose stretching and bending potentials. In our judgment, the first two are probably the most significant. In addition, it is still an open question whether high-frequency barrier crossings can be described successfully by a Brownian dynamics simulation with random forces that are not time correlated.

Despite the approximations and potential problems discussed above, the simulations were in semiquantitative agreement with experimental measurements of the local dynamics of polyisoprene. For the present we assume that this agreement is not fortuitous. A detailed conformational analysis of the simulation trajectories should yield important insights into the fundamental mechanisms for local dynamics in polyisoprene.

Brownian dynamics simulations may prove to be a powerful tool for understanding local segmental dynamics in nonpolar systems such as polyisoprene. It is expected that this type of simulation would not be appropriate when specific interactions such as hydrogen bonding or ionic interactions are present.

ACKNOWLEDGMENTS

This work was supported by the National Science Foundation (DMR-8822076). We gratefully acknowledge helpful discussions with Eugene Helfand and Radley Olson. We also acknowledge a generous grant of computing time from IBM.

REFERENCES

1. A. A. Jones and W. H. Stockmayer, *J. Polym. Sci., Polym. Phys. Ed.* **15**, 847 (1977).
2. D. A. Waldow, B. S. Johnson, P. D. Hyde, M. D. Ediger, T. Kitano, and K. Ito, *Macromolecules* **22**, 1345 (1989).
3. S. Glowinkowski, D. J. Gisser, and M. D. Ediger, *Macromolecules* **23** (in press).
4. N. G. van Kampen, *Stochastic Processes in Physics and Chemistry,* North-Holland, Amsterdam, 1981.
5. W. F. van Gunsteren, H. J. C. Berendsen, and J. A. C. Rullman, *Mol. Phys.* **44**, 69 (1981).
6. R. M. Levy, M. Karplus, and J. A. McCammon, *Chem. Phys. Lett.* **65**, 4 (1979).
7. J. H. Weiner and M. R. Pear, *Macromolecules* **10**, 317 (1977).
8. E. Helfand, Z. R. Wasserman, and T. A. Weber, *Macromolecules* **13**, 526 (1980).
9. T. A. Weber and E. Helfand, *J. Phys. Chem.* **87**, 2881 (1983).
10. D. B. Adolf and M. D. Ediger, *Macromolecules* (to be published).
11. N. L. Allinger, *J. Am. Chem. Soc.* **99**, 8127 (1977).
12. (a) E. Helfand, *Bell Syst. Tech. J.* **58**, 2289 (1979); (b) H. S. Greenside and E. Helfand, *Bell Syst. Tech. J.* **60**, 1927 (1981).
13. S. W. Provencher and R. H. Vogel, in *Progress in Scientific Computing,* Vol. 2, ed. P. Deuflhard and E. Hairer, Birkhäuser, Boston, 1983, pp. 304–319.

14. K. Hatada, T. Kitayama, Y. Terawaki, Y. Tanaka, and H. Sato, *Polym. Bull.* **2**, 791 (1980).
15. J. Denault and J. Prud'homme, *Macromolecules* **22**, 1307 (1989).
16. R. Dejean de la Batie, F. Lauprêtre, and L. Monnerie, *Macromolecules* **22**, 122 (1989).
17. E. Paul and R. M. Mazo, *J. Chem. Phys.* **48**, 1405 (1968).
18. M. S.Brown, D. M. Grant, W. J. Horton, C. L. Mayne, and G. T. Evans, J. Am. Chem. Soc. **107**, 6698 (1985).
19. B. Bagchi and D. W. Oxtoby, *J. Chem. Phys.* **78**, 2740 (1983).

Molecular Dynamics Simulations of the Dynamic Properties of Polymeric Systems

KURT KREMER
Institut für Festkörperforschung
Forschungszentrum Jülich
D-5170 Jülich
West Germany

GARY S. GREST
Corporate Research Science Laboratory
Exxon Research and Engineering Company
Annandale, New Jersey 08801

ABSTRACT

We present the results of an extensive molecular dynamics simulation of an entangled polymer melt that covers the crossover from Rouse dynamics to reptation. We find that the reptation theory describes the chain dynamics very well. We show that on present-day supercomputers, in order to study this crossover to reptation, it is essential to use a highly idealized, coarse-grained polymer model. We show that a similar calculation incorporating a detailed, microscopic description for even a simple, linear polymer is not possible even on the next generation of supercomputers. We then show how to map our polymer model onto several different polymers and predict the time and length scales for the onset of the expected slowing down beyond the Rouse regime.

INTRODUCTION

Computer simulations have proven to be a powerful tool in the investigation of physical properties of polymers [1–8]. The present volume gives an overview of the various methods that have been employed in order to simulate polymers. Work in this area ranges from rather microscopic models to simple lattice models. A large variety of methods is needed to account for the many different questions that arise. The choice of the appropriate simulation ansatz is an important part of obtaining useful results. This is especially true for the problem of long-time dynamics.

In the present investigation we are interested in studying the dynamical behavior of dense polymeric melts. The main question that we would like to address is how well the reptation concept [5–8] describes the monomer motion on a microscopic length scale (microscopic means on the order of several Kuhn lengths!). We would also like to use our results to understand the typical length and time scales appropriate for different experimental polymeric systems. To do this, one can either simulate a rather simple coarse-grained polymer model and then map it onto real polymers or simulate a detailed, microscopic model. In this chapter we demonstrate how the former is an effective technique considering the computer time available on present-day supercomputers, while the latter is not feasible.

This chapter is organized as follows. In the next section we give a discussion for our choice of simulation model. The following section contains our results, and in the fourth section we discuss the mapping of our model onto several real polymers. The final section contains our conclusions and a brief outlook for the future.

CHOICE OF MODELS

As mentioned in the introduction, we not only want to investigate the general qualitative aspects of the dynamics of dense polymer systems, but would also like to obtain some quantitative insight into the behavior for a number of experimental systems. It is thus tempting to investigate a model that contains as many microscopic parameters as possible. The natural procedure then would be to perform a molecular dynamics simulation in which one solves Newtonian equations of motion for each monomer. Although this is a tempting procedure to follow, it turns out that it would require a prohibitive amount of computing time [4]. Here we will present some detailed predictions for the amount of central processing unit (CPU) hours one would need on a present-day supercomputer for such a simulation.

To carry out the estimate, let's consider the simplest polymer—polyethylene (PE). Rigby and Roe [9] investigated a melt of PE by the means of a molecular dynamics simulation on a Cray X-MP. Their model does not include the hydrogen atoms but incorporates them into a larger effective carbon atom. For the interaction they took a spring potential between the carbon atoms, a bond-bending potential, and a torsional potential in order to account for the backbone structure of the chain. Monomers (in this case carbons) that are not neighbors along the chain sequence of the polymer interact via a standard Lennard-Jones potential. The parameters they took are the generally accepted ones for PE [9]. Since these potentials are nonlinear, the stability of the simulation requires fairly small time steps. Typically, one needs on the order of 50 to 100 integration time steps per oscillation time of the highest-frequency mode of the system. This is the first problem for a realistic simulation of PE. The spring constant k for the C—C bond is very high, allowing only for a very small time stop. Rigby and Roe decreased k by a factor of 7 in order to achieve a rough equality of the typical high-frequency modes in the system. We note here that it is not at all clear whether such a modification is easily possible for complicated structures.

Rigby and Roe [9] studied a system of 10 chains containing 50 carbon atoms each, with a time step of 10^{-14} s. From their data the relaxation time τ_N is estimated to be on the order of 2.5×10^{-8} s. Their optimized program gave a performance of about 50 time steps per CPU second. Thus to follow their 500-particle system on relaxation time required about 14 CPU hours on a Cray X-MP.

Obviously, this is certainly not prohibitively long. In fact, it would not be difficult to make a number of runs for systems containing on the order of 500 carbon atoms. However, a serious problem arises if one wants to study the experimentally interesting regime around the crossover from Rouse dynamics to the reptation regime. This is important not only for the question of the validity of the reptation model, but is also relevant for understanding many other properties: for example, the chain-length dependency of the glass transition. The entanglement molecular weight for PE is $M_e = 1400$, or 100 monomers. To investigate this crossover, one would typically need on the order of $5M_e$ molecular weight or $N = 500$ monomers per chain. Let us assume that one wants to run a system of only 10 chains of length $5M_e$ for one relaxation time of the system. To estimate this time, we assume that the relaxation time can be estimated as

$$\tau(N = 500) = \tau(N = 50)\left(\frac{500}{50}\right)^2 \frac{M(N = 500)}{M_e}$$

$$= 1.25 \times 10^{-5}\,\text{s}$$

Transformed into CPU time, this would require 50,000 hours of CPU time for 1.25×10^9 integration time steps for each such run.

Today this certainly is not possible. On future computers with, let us say, 10^2 times faster CPUs, it might be possible to make a run. However, many such systems with many more chains would be needed for a conclusive investigation, requiring several hundreds of thousands of CPU hours on present-day computers. Thus the perspectives of such an investigation are not very good. The situation becomes completely hopeless if one wants to study the more relevant polymers such as PS or even more complicated ones like PDMS or polycarbonates. In this case the potentials are much more complicated and the entanglement lengths are often significantly larger. A very optimistic estimate would be to expect to need an increase of computer time by a factor on the order of at least a few hundred. Consequently, a complete, detailed simulation of a chemically and industrially interesting polymer on today's fastest supercomputers would easily require at least 10^8 hours or 10^4 years.

To obtain some insight into the physical properties of polymer melts, another approach is needed, since a microscopic approach for the long-time properties of large polymers is a hopeless undertaking. In this context the investigation of Roe and Rigby [9] has turned out to be very important in that it has set realistic time scales for detailed microscopic model simulations. Consequently, the classical approach of using highly simplified models [4,10] is still the only feasible way to gain insight into the dynamical properties of complicated polymer systems. After reviewing our results in the next section, we show that by making a proper mapping of the simple model systems onto different chemical species, it is possible to give quantitative estimates of time and length scales for a variety of different polymeric systems. Since this can be done from a single set of runs for one model, such an approach naturally gives more insight into the universal properties [11] of polymeric systems.

For the present investigations, we employ a dense melt of linear chains of up to $N = 400$ monomers per chain. The monomers interact with each other via a purely repulsive Lennard-Jones potential (in Lennard-Jones units)

$$U_{\text{LJ}}(r) = \begin{cases} 4\epsilon\left[\left(\dfrac{\sigma}{r}\right)^{12} - \left(\dfrac{\sigma}{r}\right)^6\right] + \dfrac{1}{4} & r \leq 2^{1/6}\sigma \\ 0 & r \geq 2^{1/6}\sigma \end{cases} \qquad (1)$$

For nearest neighbors, along the chain we add an anharmonic spring,

$$U_{\text{FENE}}(r) = \begin{cases} -0.5kR_0^2 \ln\left[1 - \left(\dfrac{r}{R_0}\right)^2\right] & r \leq R_0 \\ \infty & r \geq R_0 \end{cases} \tag{2}$$

The parameters are chosen to be $R_0 = 1.5\sigma$ and $k = 30\epsilon/\sigma^2$. This choice was made to prevent bond crossings but also to allow for a large time step $\Delta t = 0.006\tau$, where $\tau = (\sigma^2 m/\epsilon)^{1/2}$, m being the mass of the monomers. The simulations were carried out at a reduced density of $\rho = 0.85\sigma^{-3}$ and temperature $k_B T = 1.0\epsilon$. For standard MD one usually solves the Newtonian equation of motion. However, such a microcanonical simulation is unstable over the many million time steps needed. To prevent stabilization problems, we coupled each monomer very weakly to a frictional background and a heat bath [12]. Thus we solve the equation of motion for each monomer i,

$$m\ddot{\mathbf{r}}_i = -\nabla U_i - m\Gamma\dot{\mathbf{r}}_i + \mathbf{W}_i(t) \tag{3}$$

where $U = U_{\text{LJ}} + U_{\text{FENE}}$. Here Γ is the friction constant and W is a random force and both are coupled via the fluctuation dissipation theorem, resulting in a simulation of canonical ensemble [13]. In the present simulation $\Gamma = 0.5\tau^{-1}$. This value was large enough to stabilize the run but small enough not to produce Rouse-like behavior on length and time scales of the order of a bond length. For the present investigation we used a third- or fifth-order predictor–corrector algorithm. There was no difference in the results; however, the former is about 10% faster than the latter. We ran systems typically of $M = 20$ chains of length $N = 10$ to 200. For $N = 400$ the sample contained only $M = 10$ chains. Finite-size tests with a large sample of $M = 100$ chains of $N = 200$ showed within the error no deviations from the smaller samples. After equilibration† we ran the systems for up to 20×10^6 time steps. For details of the structures of the polymer liquid, see Ref. 14. The overall corresponding time for the whole investigation took around 1500 hours of CPU time of one processor on a Cray X-MP 416. The algorithm [16] uses a linked cluster structure to avoid the $(MN)^2/2$ loop to set up the Verlet neighbor table. This assures a very fast algorithm with a computing time that increases linearly with the number of monomers. Our program presently runs at a speed of $100{,}000/MN$ time steps per second for the foregoing choice of parameters on the Cray X-MP. Note that because of the large time step the Verlet neighbor table had to be set up quite frequently (about once for 20 steps). We should note that a direct comparison of time per integration step for qualitatively very different potentials such as ours and the one used by Roe and Rigby is not useful. What is important is to compare the relaxation times for an equivalent chain length.

---

† For details of the equilibration procedure, see Ref. 10.

RESULTS

It is known that the dynamics of polymer chains in melt changes from Rouse-like for chains that are shorter than the characteristic entanglement length N_e toward a slower dynamics for longer chains. For this slower dynamics several different concepts are presently under discussion, of which the reptation model is the most prominent [7,11]. This crossover from the short-chain Rouse to the long-chain reptation regime is characterized by a transition in the diffusion constant D of the chains [7],

$$D(N) = \frac{k_B T}{N \zeta} \qquad N > N_e$$
$$D(N) \propto N^{-2} \qquad N \gg N_e \tag{4}$$

Here $k_B T$ is the temperature of the system and ζ the monomeric friction coefficient. For the second expression several prefactors involving N_e have been discussed in the literature [16,17]. To cover the time and length regime of interest, the data must show the crossover in the diffusion constant $D(N)$. Figure 1 gives the results for the present investigation. Following Eq. (4), $D(N)N$ should define a plateau giving the monomeric friction coefficient. The data display a clear deviation from the Rouse behavior starting at a chain length somewhere between $N = 25$ and 50. For the longer chains we find reasonable agreement with the expected $D - N^{-2}$ power law. The onset of the deviation from Rouse suggests an N_e on the order of 30 to 40. However, a check with the standard equations [7] yields $N_e \simeq 100$. We should note that the latter typically works well for $N/N_e \gg 1$, which we certainly cannot reach here.

Using the result above we should be in a situation that allows for a

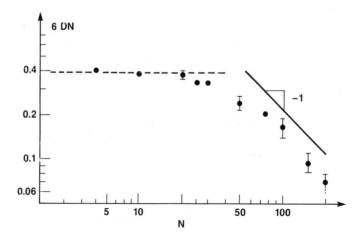

Figure 1 Diffusion constant $6DN$ versus N. D was obtained from the long-time behavior of motion of the center of mass. For $N = 200$ only an upper limit is given.

detailed investigation of the crossovers. One of the key signatures of the reptation model is the mean-square displacement $g_1(t)$ for the monomers. One expects for reptation

$$g_1(t) = \begin{cases} t^{1/2} & t < \tau_0 N_e^2 \\ t^{1/4} & t < \tau_0 N^2 \\ t^{1/2} & t < \dfrac{\tau_0 N^3}{N_e} \\ t^1 & t > \dfrac{\tau_0 N^3}{N_e} \end{cases} \tag{5}$$

while for the Rouse model only the intermediate $t^{1/2}$ regime occurs. Since we can only investigate systems with up to a few entanglement lengths, we confine ourselves to the innermost monomers, which should feel the constraints the most. For a detailed investigation of the outer monomers, see Ref. 10. We have defined $g_1(t)$ as (N even)

$$g_1(t) = \frac{1}{5} \sum_{i=N/2-2}^{N/2+2} \langle (\mathbf{r}_i(t) - \mathbf{r}_i(0))^2 \rangle \tag{6}$$

where we average over the five innermost monomers. Figure 2 show the results in a log-log plot for $g_1(t)$. For short times all the data fall on top of each other, indicating that the inner monomers do not fill the size of the polymer they are in. Only for longer times do we find a typical crossover to a slower motion, which is in good agreement with a $t^{1/4}$ power law. For $N = 30$, the overall diffusion takes over before we reach the reptation regime. Again, we note that the crossover to a slower motion is independent of chain length! Using the slopes of the $t^{1/2}$ and $t^{1/4}$ regimes, we find a crossover time $\tau_e \sim 1800$ and $g_1(\tau_e) = 21$. To estimate N_e we assume that a subchain of length N_e at τ_e is completely related with

$$g_1(\tau_e) \sim 2\langle R_G^2(N_e) \rangle = 21\sigma \tag{7}$$

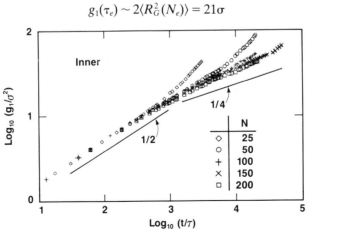

Figure 2 Mean-square displacement $g_1(t)$ versus t/τ averaged over the inner five monomers for five values of N.

and $R_G^2(N_e) = \frac{1}{6}(ll_p)^2 N_e$. With $(ll_p)^2 = 1.7$, we find $N_e \sim 35$. This certainly is also only an estimate, but it agrees very nicely with the onset of the diffusion crossover. Since we here analyzed only middle monomers finite-chain-length effects should be smaller than for overall averages!

From these data we clearly see a crossover from a Rouse regime into a reptation regime. However, this is not sufficient to distinguish between the reptation model [5–7,19,20] and other approaches [18,21]. All reasonable models require a slowed-down monomer motion to account for the increase in the relaxation time. To distinguish the models [9,14,19,20] in the reptation class† from the generalized Rouse models [21], we analyze the Rouse modes $\mathbf{X}_p(t)$ of the chains. For our systems, the Rouse model can be defined as

$$\mathbf{X}_p(t) = \frac{1}{N} \sum_i \mathbf{r}_i(t) \cos \frac{\pi(i-1)}{N-1} - \frac{1}{2N}[\mathbf{r}_1(t) + \mathbf{r}_N(t)] \tag{8}$$

As expected for chains in a melt, the Rouse modes are eigenmodes of the chains. For $N/p < N_e$ we expect Rouse behavior, namely,

$$\frac{\langle \mathbf{X}_p(t) \cdot \mathbf{X}_p(0) \rangle}{\langle X_p^2(0) \rangle} = \exp\left(-\frac{t}{\tau_p}\right) \qquad \tau_p = \frac{\zeta(ll_p)^2 N}{3\pi k_B T} \frac{N}{p^2} \tag{9}$$

This was observed for short chains. The resultant monomeric function coefficient $\zeta = (17 \pm 2)\tau^{-1}$ is in excellent agreement with the value $\zeta = 16\tau^{-1}$ obtained from the diffusion constant (see Fig. 1). For $N/p > N_e$ the relation time is enlarged by a factor of N/N_e in the reptation picture, while the generalized Rouse ansatz of Kavassalis and Noolandi [23] yields an enlargement proportional to N/p^2. Thus we have

$$\tau_p \propto \begin{cases} N^3/p^2 & \text{(reptation, } N/p > N_e) \\ N^3/p^4 & \text{(generalized Rouse, } N/p > N_e) \end{cases} \tag{10}$$

To analyze this we need precise information for times $t > \tau_e$, which is available from the data. Figure 3 shows a comparison between the two power laws for the $N = 200$ systems. The data clearly show that the generalized Rouse scaling strongly disagrees with the data, while for the Rouse scaling all data seem to have the same slope (note that they not necessarily have to be on top of each other). From these data we can say that the reptation model seems to provide a good description of the chain dynamics.

The most intriguing idea of the reptation concept, however, is the picture of the chain moving along its own coarse-grained path. Since we are covering the crossover toward the reptation regime, the question is whether we can visualize this confinement.

The previous results show that $N_e \sim 35$, with $\langle R_G^2(N_e) \rangle = 11$ and thus $\langle R^2(N_e) \rangle = 66$. Following the reptation theory the tube diameter for this confinement should be on the order of $\langle R^2(N_e) \rangle$. Therefore, even for the largest

---

† Hess [19] and Schweizer [20] do not exclude reptation within certain limits. Hess explicitly finds the reptation mode to be dominant. The onset of this is an adjustable parameter.

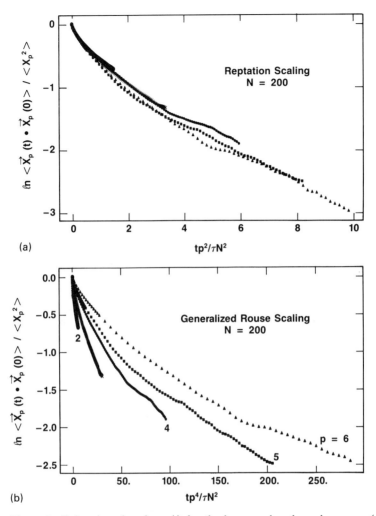

Figure 3 Relaxation plots for $g_p(t)$ for the long-wavelength modes versus (a) $t(p/N)^2$ (reptation scaling); and (b) $t(p^4/N^3)$ (generalized Rouse model scaling) for $N = 200$. The values of p are indicated in (b).

system ($N = 400$) it should be extremely difficult to observe the reptation directly, while for the others ($N \leq 200$), it would be impossible. A way out is to construct a primitive chain (PC). One way to do this is build a chain out of subchains of length N_e: namely,

$$\mathbf{R}_1 = \frac{1}{N_e} \sum_{i=1}^{N_e} \mathbf{r}_i \qquad \mathbf{R}_2 = \frac{1}{N_e} \sum_{i=3}^{N_e+2} \mathbf{r}_i, \ldots \qquad (11)$$

This PC should feel the confinement much stronger. To interpret the motion of the PC, we note that $g_1(\tau_e) = 9$ for the primitive chain. This leads to a tube diameter $d_T = [3g_1(\tau_e)]^{1/2} \simeq 5$. If we now simply plot the configurations at different times, they should only show a tube of diameter d_T. If the chains do not reptate, we expect a region of diameter $[3g_1(\tau_{max})]^{1/2} = 10$, for the longest time $\tau_{max} = 12{,}000\tau$ shown in Fig. 4. The figure clearly visualizes the confinement into a tube, showing that the reptation concept also qualitatively gives a profound picture of the underlying physics.

MAPPING ONTO OTHER SYSTEMS

Up to now we demonstrated that the reptation model describes rather well the motion of a polymer chain in a melt. However, there are two important open questions with respect to these data. First, what is the precise estimate of the entanglement length? In experiments N_e usually is determined by the plateau modulus for very large ($N \gg N_e$) chains. This is not possible with the present simulations data. The second, which will turn out to be directly related to the first, is what we can learn about experimentally interesting linear polymer melts besides the obvious statement that reptation there should work as well. To be able to make quantitative predictions for experiments, we need to map our model chains onto chemical species. Once we have done so, the first question will automatically be answered.

When one discusses static properties of polymer, the persistence lengths are usually mapped onto each other. However, this mapping is not unique and does not take into account any dynamics. The persistence length mapping only gives the limit of the smallest statistical unit we can compare. To make this unique, we must use some additional information. What is needed is to fix the mapping in a way such that both static and dynamic properties can be interpreted. The quantity that governs the behavior of long entangled polymers is the entanglement length N_e or molecular mass M_e. Thus equivalent chains should be subchains of length N_e or M_e. This sets the time and distance scales for comparison between a model system and experimental polymer, but also between different polymers. In so doing, we can also define N_e in the same way that it is done experimentally. Since we cannot directly measure the plateau modulus, we compared the diffusion constants. If we plot N/N_e or M/M_e, respectively, versus the chain diffusion constant $D(N)$ divided by the extrapolated Rouse diffusion constant D_{Rouse}, which is known for short chains, the data should collapse onto a single curve. A fit of N_e to the experimental curve gives the entanglement length, based on the same physics. Figure 5 gives this plot for $N_e = 35$ together with data [22,23] for polystyrene (PS) and polyethylene (PE). For PE our choice of N_e seems to work very well. PS would suggest a slightly larger N_e. However, the two curves of the experiments should coincide as well. It is not completely clear to us whether the deviation is due to errors in the data or whether there are some limitations to universality within this crossover regime. Since the PE data show less scatter and PE does

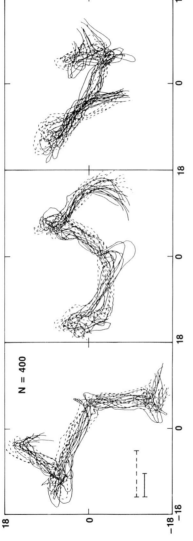

Figure 4 Configurations of the primitive path of three different chains for $N = 400$. The contours are plotted every 600τ and the total elapsed time is $12,000\tau$. The center of mass of the first configuration for each chain is shifted to the center of the box for purposes of illustration. The horizontal dashed bar is the amount the chain moves during the elapsed $24,000\tau$ as determined assuming isotropic motion, while the solid line is the tube diameter from the onset of the $t^{1/4}$ regime in $g_1(t)$ (Fig. 2) for large N.

177

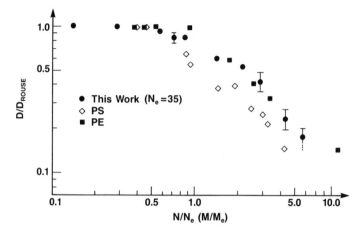

Figure 5 $D(N)/D_{\text{Rouse}}(N)$ versus N/N_e or M/M_e, respectively, for our simulation data ($N_e = 35$), PE with $M_e = 1400$ (Ref. 22), and PS with $M_e = 1800$ (Ref. 23).

not contain any side groups as our model, we follow the PE mapping. Thus our initial choice of $N_e = 35 \pm 5$ is in agreement with the standard viscoelastic definitions.

To proceed, the length scales are given simply by setting $\langle R^2(M_e) \rangle$ of the chemical species of interest equal to $\langle R^2(N_e) \rangle$ of our model. Table 1 gives examples for a variety of different systems. For the time scale a similar trivial mapping follows if the diffusion constants are available. One has only to equate the Rouse diffusion constants of the systems for N_e and M_e, respectively. Together with the length scaling mentioned above, this gives the time scaling directly. However, the diffusion constant is not available in all cases. Then we use the monomeric friction constant as obtained from viscoelastic measurements. Here, however, we [10] cannot use the small N bead friction. The bead friction is taken from the plateau modulus extrapolated down to $N = 2N_e$ by using the Rouse equation. It is known that the monomeric friction coefficient increases with increasing chain length and tends toward its asymptotic value for $N \geq 2N_e$. Therefore, we use $\zeta = 25\tau^{-1}$ as we find for the long chains ($N \geq 75$) instead of our short-chains value. We can then substitute the friction coefficients into the Rouse diffusion equation and obtain a time scaling. Doing this provides us with a prediction of the time for the crossover from Rouse to reptation dynamics. The results are given in Table 1 (see also Ref. 11).

As one can see from Table 1, only PTHF [24] has a time that is short enough to be observed by the old spin-echo machines, which had a resolution of $t_{\text{max}} = 10^{-8}$ s. Higgins and Root [24] observed a slowing down of the motion of the monomers by neutron spin echo and attributed it to the occurrence of reptation. The time τ_e we find for the crossover of PTHF is almost exactly reproduced by their experiment. These experiments were strongly questioned

TABLE 1 Mapping of the Bead-Spring Polymer Model Described Herein onto Experimental Polymers, Assuming That $N_e = 35$[a]

System	T	Monomer Mass	M_e	Equiv. Number of Beads	Equiv. Mol. Mass	ll_p	$1\sigma =$	d_T	$\sqrt{g_1(1\tau_e)}$	$1\tau =$	τ_e
MD system	$1\epsilon/K_B$	1	35	1	1	1.3σ	—	7.7σ	4.5σ	—	$1800\,\tau$
PS[b]	485 K	104	18,000	4.95	515	7.4 Å	12.3 Å	97 Å	56 Å	3.1×10^{-8} s	5.5×10^{-5} s
PE[c]	448 K	14	1,350	2.76	38.6	4 Å	5.1 Å	39 Å	23 Å	6.6×10^{-11} s	1.2×10^{-7} s
PDMS[d]	300 K	74	9,000	3.47	257	6.2 Å	8.7 Å	68 Å	39 Å	2.3×10^{-10} s	4.1×10^{-7} s
PTHF[e]	418 K	72	1,440	0.57	41	8 Å	4.6 Å	35 Å	21 Å	1.8×10^{-12} s	3.2×10^{-9} s
PEP[f]	500 K	70	2,950	1.20	84	7.74 Å	6.5 Å	50 Å	26 Å	0.55×10^{-11} s	1.0×10^{-8} s
PI[f]	307 K	68	4,100	1.73	117	6.6 Å	6.7 Å	51 Å	29.7 Å	1.0×10^{-10} s	1.8×10^{-7} s

[a] The mapping procedure is described in the text. Taking $N_e = 35 \pm 5$ affects the value predicted for $g_1(\tau_e)$ by approximately 15% and 30%, respectively, the uncertainty in the mapping for PS causing a doubling of the effective uncertainty.

[b] For PS ll_p are taken from Ferry [27] for PS in θ-solvent. For mapping the bead friction due to the diffusion constant of Antonietti et al. [23] was used. A consistency check with ζ_0 from viscoelasticity gave only about 10% deviation! Note that $N_e = 50$ would change the time mapping, giving $\tau_e = 2.6 \times 10^{-5}$ s.

[c] Persistence length as given by Flory [28]. M_e and the diffusion constant taken from Pearson et al. [22].

[d] Persistence length and bead friction due to Ferry [27]. Note that the viscoelastic bead friction is obtained from entangled polymers. The actual scattering experiment was performed at $T = 373$ K. Assuming everything besides ζ fixed, τ_e changes to 1×10^{-7} s. However, the more recent experiment suggests a strong increase in M_e with T. Thus τ_e(PDMS, $T = 373$ K) $= 1 \times 10^{-7}$ s seems to be an absolute lower limit! For the Rouse diffusion equation used for comparison, we use the long-chain bead friction.

[e] Persistence length M_e and bead friction due to Pearson (private communication). Here also the long-chain bead friction has been taken.

[f] ζ, ll_p, M_e from Fetters (private communication). For ζ(PI), N_e was taken to be $2.5 N_e$. $2N_e$ would reduce τ_e by about 30%.

179

by the data of Richter et al. [25] obtained for PDMS. They found only Rouse behavior. However, our mapping show that PDMS at the conditions employed has a $\tau_e = 4.1 \times 10^{-7}$ s, which was about a factor of 40 beyond the capabilities of the instrument. Therefore, the two experiments do not contradict each other at all. The PDMS system was just completely out of range. In addition, the PDMS chains were probably too short. End fluctuations and the onset to the creep term would have destroyed the reptation signal even if the times of interest had been reached. We should mention that recent experiments of Richter et al. [26] on PEP show the slower decay of the scattering function $S(q, t)$ using a newer spectrometer. Our mapping works for PEP as well as for PTHF.

CONCLUSIONS

In this chapter we have shown that a simulation of the dynamics of a polymeric melt is possible only for highly simplified and idealized models. Even the next generations of computers will not provide the computing power to go much beyond the present investigation with respect to molecular details. However, we believe that one can learn more about the general polymeric phenomena by sticking to an idealized system.

Our data clearly show that the reptation model provides a very good description of monomeric motion beyond the scale of the entanglement length. Analyzing the Rouse modes, we were able to distinguish between the reptation models and the generalized Rouse models. We were also able to resolve a longstanding controversy between different neutron scattering studies [24,25] by demonstrating how to map our model polymer on experimental polymeric systems as well as different experimental systems among themselves.

The next obvious step in the analysis of dense polymeric systems will be to cross-link the melts. Since the configurations were generated in the course of the present study, this is possible with a relatively reasonable amount of computer time.

ACKNOWLEDGMENTS

We acknowledge valuable discussions with many colleagues, especially K. Binder, I. Carmesin, B. Dünweg, L. J. Fetters, W. W. Graessley, D. S. Pearson, P. A. Pincus, N. Pistoor, H. Sillescu, T. A. Witten, and D. Richter. We acknowledge support from NATO Travel Grant 86/680. K. K. acknowledges the hospitality of Exxon Research and G.S.G., Universitat Mainz, and IFF-KFA Jülich. These calculations were made possible by a generous grant of CPU time from the German Supercomputer HLRZ, Jülich.

REFERENCES

1. J. N. Hammersley and D. Handscomb, *Monte Carlo Methods,* Chapman & Hall, London, 1983.
2. A. Baumgärtner, *Annu. Rev. Phys. Chem.* **35,** 419 (1985).
3. K. Kremer and K. Binder, *Comput. Phys. Rep.* **7,** 259 (1988).
4. K. Binder in *Molecular Level Calculations of the Structure and Properties of Non-crystalline Polymers,* ed. J. Biscerano, Dekker, New York, 1989.
5. S. F. Edwards, *Proc. Phys. Soc.* **92,** 9 (1967).
6. P.-G. deGennes, *J. Chem. Phys.* **55,** 572 (1971).
7. N. Doi and S. F. Edwards, *J. Chem. Soc., Faraday Trans. 2* **71,** 1789, 1802, 1818 (1978).
8. M. Doi and S. F. Edwards, *The Theory of Polymer Dynamics,* Clarendon Press, Oxford, 1986.
9. D. Rigby and R. J. Roe, *J. Chem. Phys.* **87,** 7285 (1987); **89,** 5280 (1988).
10. K. Kremer, G. S. Grest, and I. Carmesin, *Phys. Rev. Lett.* **61,** 566 (1988); K. Kremer and G. S. Grest, *J. Chem. Phys.* **92,** 5057 (1990). The latter paper contains an extensive list of other publications on the problems of dynamics of polymer melt.
11. P.-G. de Gennes, *Scaling Concepts in Polymer Physics,* Cornell University Press, Ithaca, N.Y., 1979.
12. G. S.Grest and K. Kremer, *Phys. Rev.* **A33,** 3628 (1988).
13. T. Schneider and E. Stoll, *Phys. Rev.* **B17,** 1302 (1978).
14. J. G. Curro, K. G. Schweizer, G. S. Grest, and K. Kremer, *J. Chem. Phys.* **91,** 1357 (1989).
15. G. S. Grest, B. Dünweg, and K. Kremer, *Comput. Phys. Commun.* **55,** 269 (1989).
16. W. W. Graessley, *Stud. Polym. Sci.* **2,** 163 (1987).
17. D. S. Pearson, *Rubber Chem. Technol.* **60,** 439 (1987).
18. A. Kolinski, J. Skolnick, and R. Yaris, *J. Chem. Phys.* **84,** 1992 (1987); **86,** 1567 (1986).
19. W. Hess, *Macromolecules* **19,** 1395 (1986); **20,** 2589 (1987); **21,** 2670 (1988).
20. K. Schweizer, *J. Chem. Phys.* **91,** 5822 (1989).
21. T. A. Kavassalis and J. Noolandi, *Macromolecules* **21,** 2869 (1988).
22. D. S. Pearson, G. Verstrate, E. van Meerwall, and F. C. Schilling, *Macromolecules* **20,** 1133 (1987).
23. M. Antonietti, H. K. Foelsch, and H. Sillescu, *Makromol. Chem.* **188,** 2317 (1987).
24. J. S. Higgins, *Physics* **B136,** 201 (1986); J. S. Higgins and J. E. Root, *J. Chem. Soc., Faraday Trans. 2* **81,** 757 (1985).
25. D. Richter, A. Baumgärtner, K. Binder, B. Ewen, and J. B. Hayter, *Phys. Rev. Letter* **47,** 109 (1981).
26. D. Richter, B. Farago, L. J. Fetters, J. S. Huang, B. Ewen, and C. Lartigue, *Phys. Rev. Left.* **64,** 1389 (1990).
27. J. D. Ferry, *Viscoelastic Properties of Polymers,* Wiley, New York, 1980.
28. P. Flory, *Statistical Mechanics of Chain Molecules,* Wiley-Interscience, New York, 1969.

13

Simulation of Polymer Chain Dynamics with a Nonlattice Model

DAVID EICHINGER and *DAVID E. KRANBUEHL*
Chemistry Department
College of William & Mary
Williamsburg, Virginia 23185

PETER H. VERDIER
Polymers Division
National Institute of Standards and Technology
Gaithersburg, Maryland 20899

ABSTRACT

The relaxation behavior of bead-stick models of polymer chains not constrained to move on lattices has been studied by Monte Carlo simulation. For the nonlattice model reported here, as for lattice models reported in earlier work, the effects of excluded volume constraints on the long internal relaxation times of the chains are found to be much stronger than the effects of excluded volume constraints on equilibrium dimensions.

KEYWORDS: Bead-stick; Excluded volume; Freely jointed chain; Monte Carlo; Nonlattice model; Polymer chain dynamics; Relaxation time; Simulation

INTRODUCTION

The use of lattice-model chains and Monte Carlo simulation techniques has been the basis of numerous studies of the dynamics of random-coil polymer chains in condensed phases. In this model, which has been developed over the past 30 years by Verdier, Stockmayer, Kranbuehl [1–23], and others [24–33], a sufficient number of monomer units to constitute a statistical segment in a random-coil chain is represented by one bead-stick unit. The lattice chain model has been especially useful because it is readily amenable to the introduction of hard-core excluded-volume interactions. These repulsive forces between segments of the polymer chain give rise to the expansion of chain dimensions, lengthening of relaxation times, and ultimately chain entanglement for sufficiently long chains and/or high segment densities. The effect of these excluded volume interactions on polymer chain dynamics is particularly difficult to treat analytically. Although significant progress has been made using expansion techniques about the no-excluded-volume limit, using tube and snake models in the entanglement limit and a variety of other averaging techniques [34–36], no single analytical model exists that incorporates excluded-volume interactions and treats chain length and segment density continuously.

Monte Carlo simulations of lattice models with and without excluded volume, on the other hand, have been particularly effective for studying excluded volume effects and the associated effects of chain connectivity and entanglement on the dynamic properties of polymer chains as functions of chain length and segment density [20,34]. At the same time, increasing use of lattice models and the discussion generated by their predictions have heightened concern about the possibility of anomalous effects due to the lattice constraints and the choices of bead movement rules. This worrisome question, which is important in its own right, has become particularly so in light of the very strong chain-length dependence of the polymer's longest relaxation time, which is found when excluded volume is present. To remove the possibility of lattice effects on chain dynamics, we have carried out simulations in which no lattice constraints are present. We shall refer to these bead-stick models without lattice constraints, sometimes called "freely jointed chains," as "nonlattice models."

NONLATTICE MODEL

A random-coil polymer chain $N - 1$ units long is modeled by a string of N connected beads. The vectors connecting each bead with the next are all the same length. There is no restriction on the angle between successive vectors. Brownian motion of the chain is simulated by sequences of elementary moves, each of which consists of selecting either a single bead or a connected pair of beads at random and attempting a local move of the selected bead or bead pair, leaving the rest of the chain unmoved. The attempted move may be described as a rotation of the selected bead or beads about the axis defined by the two neighboring beads along the chain. For the model reported here, the chance of choosing a single bead, rather than a bead pair, was $\frac{1}{2}$, all beads and all bead pairs were chosen with equal probability, and the angle of rotation was chosen at random in the interval $(-\pi, \pi)$.

All the simulations were carried out both with and without excluded volume constraints. Such constraints are easily incorporated into lattice models by forbidding multiple occupancy of lattice sites. Their inclusion in nonlattice chain dynamics is only slightly less simple. In this case the bead diameter is set equal to the length of the vector between connected beads along the chain, and overlap between beads is forbidden. If a new position is forbidden, no move is made, but a unit of simulated time is counted whether or not a move takes place.

As the simulation proceeds, the complete chain configuration is repeatedly sampled and stored. The stored chain configurations are used to generate sampled values of the end-to-end vector $\mathbf{l}$, the position of the center of mass of the chain, and other quantities of interest, as functions of simulated time t. These sampled values are used to form estimates of the autocorrelation function $\rho(\mathbf{l}, \mathbf{l}, t) = \langle \mathbf{l}(0) \cdot \mathbf{l}(t) \rangle / \langle l^2 \rangle$ in $\mathbf{l}$, translational diffusion constants, and other desired dynamical and equilibrium properties.

RESULTS

Simulations are reported here of chains of from 9 to 99 beads, of and from equilibrium conformations. As in previous simulations of dynamical bead-stick models both with and without lattice constraints, semilogarithmic plots of $\rho(\mathbf{l}, \mathbf{l}, t)$ versus t appear to become linear at sufficiently long times. As in our earlier work, we have extracted long relaxation times τ_1 by fitting ρ as a function of t by unweighted least squares to the form $\rho = a \, \exp(-t/\tau_1)$ at times longer than that required for ρ to fall below 0.6. Values of a and τ_1 so obtained are shown in Table 1.

The effects of excluded volume on chain dynamics may be exhibited directly by forming the ratios $R_\tau = \tau_1/{}^0\tau_1$ of values of τ_1 for chains with excluded volume to the corresponding values ${}^0\tau_1$ for chains of the same length and the same move rules except for the absence of excluded volume constraints. Values of R_τ versus N are given in Table 1, and are shown in Fig. 1 in a log-log plot. The slopes of such plots may be used to estimate the power-law dependence γ of R_τ on N: $R_\tau \propto N^\gamma$. For the model reported here, γ is seen to be about $\frac{1}{2}$, in approximate agreement with earlier work on simple cubic, face-centered cubic, and body-centered cubic lattices.

CONCLUSIONS

The relaxation properties of nonlattice bead-stick chains with and without excluded volume constraints are remarkably similar to those found in earlier studies of lattice-model chains. For an elementary move consisting of random rotation of a region randomly chosen to consist of either one or two beads, excluded volume constraints increase the chain-length dependence of the longest relaxation time by about the one-half power of chain length.

TABLE 1 Coefficients a and Relaxation Times τ_1 for Nonlattice Bead-Stick Chains of N Beads with Excluded Volume Constraints[a]

N	a		τ_1		${}^0a$		${}^0\tau_1$		R_τ	
9	0.969	(0.082)	0.391	(0.054)	0.890	(0.237)	0.137	(0.027)	2.85	(0.69)
15	0.942	(0.035)	0.467	(0.012)	0.897	(0.154)	0.116	(0.025)	4.03	(0.87)
33	0.937	(0.037)	0.588	(0.131)	0.921	(0.074)	0.098	(0.009)	6.0	(1.4)
45	0.832	(0.023)	1.001	(0.048)	0.721	(0.129)	0.131	(0.025)	7.6	(1.5)
63	0.910	(0.031)	0.813	(0.098)	0.878	(0.078)	0.101	(0.013)	8.0	(1.4)
99	0.832	(0.075)	1.044	(0.075)	0.831	(0.022)	0.103	(0.024)	10.1	(2.5)

[a] Values were obtained by fitting limiting long-time behavior of autocorrelation functions $\rho(\mathbf{l}, \mathbf{l}, t)$ for the end-to-end vector $\mathbf{l}$ as a function of time t to the form $a \, \exp(-t/\tau_1)$. Also shown are corresponding quantities ${}^0a$ and ${}^0\tau_1$ obtained for chains without excluded volume constraints, and ratios $R_\tau = \tau_1/{}^0\tau_1$. Values of τ_1 are in units of N^3 elementary moves. Numbers in parentheses are sample standard deviations of the mean.

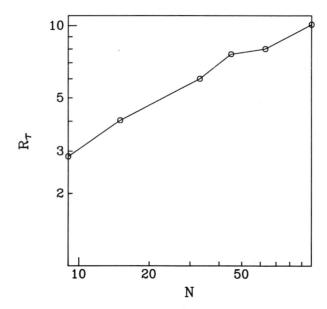

Figure 1   Log-log plot of ratios R_τ of long relaxation times τ_1 for nonlattice bead-stick chains of N beads with excluded volume constraints to the corresponding quantities $^0\tau_1$ for chains without excluded volume constraints.

Overall, these nonlattice results, taken together with the results of recent simulations on several cubic lattices, show with remarkable consistency that the effects of excluded volume constraints on the long relaxation times of chains are much stronger than the effects of those constraints on equilibrium chain dimensions.

REFERENCES

1. P. H. Verdier and W. H. Stockmayer, *J. Chem. Phys.* **36,** 227 (1962).
2. P. H. Verdier, *J. Chem. Phys.* **43,** 2546 (1965).
3. P. H. Verdier, *J. Chem. Phys.* **45,** 2118 (1966).
4. P. H. Verdier, *J. Chem. Phys.* **45,** 2122 (1966).
5. R. A. Orwoll and W. H. Stockmayer, *Adv. Chem. Phys.* **15,** 305 (1969).
6. P. H. Verdier, *J. Comput. Phys.* **4,** 204 (1969).
7. P. H. Verdier, *J. Chem. Phys.* **52,** 5512 (1970).
8. D. E. Kranbuehl and P. H. Verdier, *Polym. Prepr., Am. Chem. Soc., Div. Polym. Chem.* **12,** 625 (1971).
9. W. H. Stockmayer, W. Gobush, and R. Norvich, *Pure Appl. Chem.* **26,** 537 (1971).
10. D. E. Kranbuehl and P. H. Verdier, *J. Chem. Phys.* **56,** 3145 (1972).
11. D. E. Kranbuehl, P. H. Verdier, and J. M. Spencer, *J. Chem. Phys.* **59,** 3861 (1973).

12. P. H. Verdier, *J. Chem. Phys.* **59**, 6119 (1973).
13. D. E. Kranbuehl and P. H. Verdier, *Polymer Prepr., Am. Chem. Soc., Div. Polym. Chem.* **17**, 144 (1976).
14. P. H. Verdier and D. E. Kranbuehl, *Polymer Prepr., Am. Chem. Soc., Div. Polym. Chem.* **17**, 148 (1976).
15. D.E. Kranbuehl and P. H. Verdier, *J. Chem. Phys.* **67**, 361 (1977).
16. P. H. Verdier, *J. Chem. Phys.* **70**, 5708 (1979).
17. D. E. Kranbuehl and P. H. Verdier, *J. Chem. Phys.* **71**, 2662 (1979).
18. D. E. Kranbuehl and P. H. Verdier, *Polymer Prepr., Am. Chem. Soc., Div. Polym. Chem.* **21**, 195 (1980).
19. D. E. Kranbuehl and P. H. Verdier, *Polymer* **24**, 383 (1983).
20. P. Romiszowski and W. H. Stockmayer, *J. Chem. Phys.* **80**, 485 (1984).
21. D. E. Kranbuehl and P. H. Verdier, *Macromolecules* **17**, 749 (1984).
22. D. E. Kranbuehl and P. H. Verdier, *Macromolecules* **18**, 1638 (1985).
23. P. H. Verdier and D. E. Kranbuehl, *Macromolecules* **20**, 1362 (1987).
24. O. J. Heilmann, *Mater. Fys. Medd. Dan. Vid. Selsk.* **37**-2 (1968).
25. J. Rotne and O. J. Heilmann, *Proceedings of the 7th International Congress on Rheology,* Swedish Society of Rheology, 1976, p. 510.
26. M. Lax and C. Brender, *J. Chem. Phys.* **67**, 1785 (1977).
27. T. M. Birshtein, V. N. Gridnev, Y. Y. Gotlib, and A. M. Skvortsov, *Vysokomol. Soyed.* **A19**, 1398 (1977).
28. J. G. Curro, *Macromolecules* **12**, 463 (1979).
29. A. Baumgärtner, *Annu. Rev. Phys. Chem.* **35**, 419 (1984).
30. M. Dial, K. S. Crabb, C. C. Crabb, and J. Kovac, *Macromolecules* **18**, 2215 (1985).
31. A. Kolinski, J. Skolnick, and R. Yaris, *J. Chem. Phys.* **84**, 1922 (1986); **86**, 1567, 7164, 7174 (1987).
32. J. Naghizadeh and J. Kovac, *Phys. Rev. Lett.* **59**, 1710 (1987).
33. C. C. Crabb, D. F. Hoffman, Jr., M. Dial, and J. Kovac, *Macromolecules* **21**, 2230 (1988), and references cited therein.
34. M. Doi and S. F. Edwards, *J. Chem. Soc., Faraday Trans. 2* **74**, 1789, 1818 (1978).
35. P.-G. de Gennes, *Scaling Concepts in Polymer Physics,* Cornell University Press, Ithaca, N.Y., 1979.
36. K. F. Freed, *Renormalization Group Theory of Macromolecules,* Wiley, New York, 1987.

Simulation Algorithm for Reptation Theories Yielding Viscometric Functions without Errors Due to Time Discretization

HANS CHRISTIAN ÖTTINGER
Institut für Polymere
ETH-Zentrum
CH-8092 Zürich
Switzerland

ABSTRACT

A recently developed algorithm for simulating the Doi–Edwards and Curtiss–Bird models for concentrated polymer solutions and melts in homogeneous flows is improved. For the improved algorithm, time-discretization errors are exponentially small and hence completely negligible. For steady shear flow it is demonstrated that this simulation algorithm is capable of supplying high-precision results for the viscometric functions over a wide range of shear rates, in particular, even at high shear rates. Furthermore, the new simulation algorithm constitutes a handy and powerful tool for investigating complex flow situations and various generalizations of the original Doi–Edwards and Curtiss–Bird models.

INTRODUCTION

In a recent paper it was shown how the dynamics of polymers assumed in reptation theories for the rheology of undiluted polymer systems can be described by a stochastic process that satisfies a simple stochastic differential equation [1]. This stochastic process, which is referred to herein as the *reptation process*, involves only a single unit vector $U(t)$ and a single number $S(t)$ from the interval $[0, 1]$ characterizing the tangential direction (U) of a chain segment at a certain position (S) within a given chain. Both $U(t)$ and $S(t)$ are random variables. The deterministic differential equation for the time evolution of $U(t)$ due to the externally applied flow field and the stochastic differential equation for $S(t)$ describing reptational motion are coupled only through the boundary conditions imposed on U when S is reflected at the upper or lower limits of its range $[0, 1]$ (furthermore, the boundary conditions, together with the initial conditions, lead to the stochastic nature of the unit vector U).

The polymer dynamics in the two most commonly used reptation theories, the Doi–Edwards (DE) [2–5] and Curtiss–Bird (CB) [6–11] models, is described by the same reptation process $(U(t), S(t))$ (see also the comprehensive textbooks by Bird et al. [12] and by Doi and Edwards [13]). However, the DE and CB models differ in the way in which the stress tensor and hence the rheological properties are calculated as ensemble averages over the trajectories of the reptation process. Formally, the DE model may be regarded as a special case of the CB model.

An important advantage of reformulating the diffusion equation usually employed to describe the polymer dynamics in reptation theories in terms of a stochastic differential equation for the reptation process $(U(t), S(t))$ lies in the following fact. By numerical integration of the resulting stochastic differential equation for $(U(t), S(t))$ one immediately obtains a simulation algorithm for reptation theories [1]. In doing so, the time discretization which is unavoidable in a numerical integration causes a deviation of the simulation results from the desired properties of the reptation process with continuous time evolution. For the algorithm obtained by naive discretization of the underlying stochastic differential equation, the error due to time discretization is proportional to the square root of the (small) time-step width Δt. With a modified algorithm, the time-discretization errors could be reduced to be linear in Δt, thus facilitating extrapolation to the continuous-time limit [1]. *It is our purpose in this chapter to develop a further improved simulation algorithm for the reptation process in which the error due to time discretization can be neglected.* For the algorithm suggested below, it is unnecessary to perform simulations for various time-step widths Δt and to extrapolate the results to zero time-step width.

In the following section we develop the new simulation algorithm with exponentially small time-discretization errors. To demonstrate the power of the simulation method, we then compare the results obtained from simulating the DE and CB models in steady shear flow over a wide range of shear rates to the results obtained by other numerical methods. The chapter is concluded with a brief outlook.

ALGORITHM

By naive discretization of the stochastic differential equation for the reptation process $(U(t), S(t))$ one arrives at a recursive algorithm for calculating $U_j = U(j \, \Delta t)$ and $S_j = S(j \, \Delta t)$ for $j = 0, 1, 2, \ldots$. We briefly summarize the resulting algorithm, which was derived and described in more detail in Ref. 1:

1. For given U_j, one calculates the (preliminary) unit vector

$$U_{j+1} = \frac{U_j + \kappa \cdot U_j \, \Delta t}{|U_j + \kappa \cdot U_j \, \Delta t|} \tag{1}$$

where κ is the transposed velocity-gradient tensor, assumed to be independent of position (homogeneous flow field). For a time-dependent flow field, the tensor κ in Eq. (1) is to be taken at the time $t = j \, \Delta t$. Equation (1), which is a discretized version of the deterministic differential equation for $U(t)$, describes the effect of the flow field, which orients the segments of the polymers in a deterministic way. The normalization in Eq. (1) makes sure that in the course of the simulation the length of the vector U_j always remains exactly equal to 1.

2. For a given S_j, one calculates the (preliminary) position label

$$S_{j+1} = S_j + \sqrt{(2/\lambda)} \, \Delta t \, W_j \tag{2}$$

Algorithm   191

where λ is the fundamental time constant occurring in reptation theories ("disengagement time") and the quantities W_j are independent Gaussian random numbers with zero mean and variance 1 (i.e., $\langle W_j \rangle = 0$ and $\langle W_j W_k \rangle = \delta_{jk}$). Equation (2) expresses the fact that the quantity S_j describing the position within a given chain at which the polymer has the direction $\mathbf{U}_j$ can change only through reptational motion, that is, via diffusion of the chain along its backbone. In the continuous-time limit, Eq. (2) characterizes the Wiener process invented to model Brownian motion.

    3. The preliminary configuration $(\mathbf{U}_{j+1}, S_{j+1})$ needs to be further modified when the value of S_{j+1} calculated according to Eq. (2) lies outside the interval $[0, 1]$. In this case S_{j+1} is replaced by the value obtained via reflection at the boundary passed in leaving the interval $[0, 1]$, that is, $S_{j+1} \rightarrow -S_{j+1}$ for $S_{j+1} < 0$ and $S_{j+1} \rightarrow 2 - S_{j+1}$ for $S_{j+1} > 1$. Furthermore, when a reflection occurs, $\mathbf{U}_{j+1}$ is replaced by a random unit vector chosen from a uniform distribution over the unit sphere. The choice of a random unit vector when S reaches a boundary is usually interpreted as a consequence of the exploration of a new environment by one of the two segments at the chain ends.

    To start a simulation, one needs to specify the initial conditions. For equilibrium initial conditions at $t_0 = 0$, S_0 is to be chosen as a random number from the interval $[0, 1]$, and the unit vector $\mathbf{U}_0$ is also to be chosen at random. For steady flows, one can start with an equilibrium initial condition and evolve it in time until relaxation to the steady state takes place, before estimating the stress tensor.

    The simulation algorithm above may be interpreted as follows. In concentrated solutions and melts, the motion of each polymer is restricted by the surrounding polymers, which may be taken into account through a confining tube (or, alternatively, through anisotropic frictional properties). The unit vector $\mathbf{U}_j$ may be interpreted as the direction of a particular tube segment (or as the direction of lowest friction), and the number S_j describes which segment of a particular polymer chain is at time $j \, \Delta t$ in the tube segment with the orientation $\mathbf{U}_j$. The tube, and hence also the polymer in the tube, are deformed by the flow field [cf. Eq. (1)], and as a consequence of the reptational motion of the polymer in the tube, the orientation of the tube segment is forced on a permanently changing segment within the particular reptating chain [cf. Eq. (2)]. When one of the chain ends leaves the particular tube segment, a new orientation is created at random [cf. . . . the boundary condition]. The fact that the polymer dynamics as described in the DE and CB models can be represented by the stochastic time evolution of a single unit vector $\mathbf{U}_j$ and a number $S_j \in [0, 1]$ makes clear that these models may be regarded as "one-segment theories." The properties of an entire chain or of an entangled many-chain system are taken into account only in a mean-field-type manner.

    For steady shear flow, which is characterized by a velocity field of the form $v_x = \dot{\gamma} y$, $v_y = 0$, $v_z = 0$ with constant shear rate $\dot{\gamma}$ (i.e., $\kappa_{xy} = \dot{\gamma}$ is the only nonzero component of the velocity gradient tensor), the deviations of the simulation results for the viscometric functions from the exact results for small time-step widths Δt were found to be of order $\sqrt{\Delta t}$ [1]. Since the naive simulation algorithm guarantees that S_j has the exact distribution of the random variables $S(j \Delta t)$ for $j = 0, 1, 2, \ldots$, and for steady shear flow, Eq. (1) reproduces the exact deterministic time evolution caused by the flow field, errors of order $\sqrt{\Delta t}$ must be a consequence of the implementation of the boundary conditions. Even if both S_j and S_{j+1} as calculated according to Eq. (2) lie in the interval $[0, 1]$, there is a certain, nonzero probability that the continuous-time process $S(t)$ has left the interval $[0, 1]$ between $j \Delta t$ and $(j + 1) \Delta t$, especially when the values of S are close to the boundaries 0 or 1. Hence there can occur reflections that should trigger the process $U(t)$ to start with a new random unit vector which, however, go unnoticed in a simulation with discrete time step. It has been recognized in Ref. 14 that such unobserved reflections at the boundaries are responsible for errors of order $\sqrt{\Delta t}$. Notice that in Ref. 14, (first) passages of the boundaries are considered rather than reflections at the boundaries. These properties are closely related through the reflection principle for the Wiener process, which states that an original trajectory and the trajectory obtained by reflection at the boundary at the time of reaching the boundary have equal probabilities [15]. The reflection principle expresses the fact that the time evolution of the Wiener process after reaching the boundary is symmetric. This close relationship between original and reflected trajectories is exploited several times in the subsequent discussion.

In Ref. 14, the conditional probability, P, for a reflection at the boundary $B = 0$ or $B = 1$ for given values of S_j and S_{j+1} and *no observed reflection* has been calculated to be

$$P = \exp\left[-\frac{\lambda}{\Delta t}(B - S_j)(B - S_{j+1})\right] \tag{3}$$

Obviously, this probability is exponentially small and hence completely negligible except when the distance between the boundary B and S_j is at most of order $\sqrt{\Delta t}$ (like $B - S_j$, the difference $B - S_{j+1}$ is of the order $\sqrt{\Delta t}$ because, according to Eq. (2), $|S_j - S_{j+1}|$ is also of order $\sqrt{\Delta t}$); for this reason, one need consider only one (the closer) boundary in each small time step.

The improved algorithms of Refs. 1 and 14 were obtained by choosing a new random vector U_{j+1} with probability P even when no reflection at a boundary has been observed in the step from S_j to S_{j+1} in order to compensate for the unobserved reflections. Then, the leading-order corrections are of the order Δt because the only remaining error is due to the fact that the precise time at which the last (observed or unobserved) reflection in the interval $[j \Delta t, (j + 1)\Delta t]$ occurred is unknown. The algorithm proposed in this chapter is based on the precise determination of the random time at which the last reflection at a boundary (or equivalently, passage of a boundary) occurred. This random time may be found to arbitrarily high precision by repeated bisection of each time interval in which an observed or unobserved reflection occurs.

Suppose that $S = S_j \in [0, 1]$ and $S' = S_j + \sqrt{(2/\lambda)\,\Delta t}\,W_j$. We consider the possibility of a reflection at the boundary $B = 0$ ($B = 1$) for $S + S' \le 1$ ($S + S' > 1$). For $S' \notin [0, 1]$, a reflection was observed. If S' lies in the interval $[0, 1]$ but an unobserved reflection at the boundary B occurred in the interval $[j\,\Delta t, (j + 1)\,\Delta t]$ according to the criterion above, S' is replaced by $2B - S'$ (according to the reflection principle for the Wiener process, S' and $2B - S'$ occur with the same probability). For observed or unobserved reflections, we determine the precise time of the last reflection at B (the last passage of B in going from S to S') in the following manner. We first calculate the random position S'' at the intermediate time $(j + \frac{1}{2})\,\Delta t$, which is $S'' = \frac{1}{2}(S + S') + \frac{1}{2}\sqrt{(2/\lambda)\,\Delta t}\,W$, where W is a Gaussian random number with zero mean and variance 1 which is independent of all the other random numbers used in the simulation. [Notice that for a time step of half the original width the amplitude of the random term has to be reduced by a factor of 2 compared to the full time step in Eq. (2).] If $S'' \in [0, 1]$, the boundary must be passed in the second half of the interval $[j\,\Delta t, (j + 1)\,\Delta t]$; in other words, the last reflection for the Wiener process in the interval $[0, 1]$ with reflecting boundaries must have occurred in the interval $[(j + \frac{1}{2})\,\Delta t, (j + 1)\,\Delta t]$, and we use S'' as the new initial value at time $t = (j + \frac{1}{2})\,\Delta t$. If $S'' \notin [0, 1]$, it is not clear whether the final value $S' \notin [0, 1]$ is reached from the intermediate value S'' with or without further passage of the boundary by the underlying continuous-time process. The probability for an unobserved passage in the interval $[(j + \frac{1}{2})\,\Delta t, (j + 1)\,\Delta t]$ is given by Eq. (3) with S_j and S_{j+1} replaced by S'' and S', and Δt replaced by $\Delta t/2$ for the smaller time step. If there is no further passage of the boundary, the last reflection must have occurred in the first half of the interval $[j\,\Delta t, (j + 1)\,\Delta t]$; otherwise, the reflection occurred in the second half of that interval. The same procedure of bisection is then repeated for the resulting, shorter time intervals with the corresponding initial and final values S and S'' (or, S'' and S') until the time of the last reflection is determined with the desired precision. For this algorithm, the only error due to the finite time-step width Δt is caused by reflections at both boundaries in a single time step. The probability for such double reflections is exponentially small and hence completely negligible for sufficiently small time steps (see the following section for further details).

RESULTS

For steady shear flow, the algorithm described in the preceding section was used to perform simulations for several values of the dimensionless shear rate $\lambda\dot{\gamma}$. The various components of the stress tensor for the CB model [7] were evaluated as described in Ref. 1. The resulting viscosity, η, and normal-stress coefficients, Ψ_1 and Ψ_2, have been decomposed in the usual way [9]:

$$\eta(\dot{\gamma}) = \eta_\Omega(\dot{\gamma}) + \epsilon\eta_Y(\dot{\gamma}) \tag{4}$$

$$\Psi_1(\dot{\gamma}) = \Psi_{1,\Omega}(\dot{\gamma}) + \epsilon\Psi_{1,Y}(\dot{\gamma}) \tag{5}$$

$$\Psi_2(\dot{\gamma}) = \Psi_{2,\Omega}(\dot{\gamma}) + \epsilon\Psi_{2,Y}(\dot{\gamma}) \tag{6}$$

where the parameter ϵ is the link-tension coefficient appearing in the CB model ($0 \le \epsilon \le 1$). The functions with subscript Ω are the respective viscometric functions of the DE model ($\epsilon = 0$). In Ref. 9, the identity $\Psi_{2,\Omega}(\dot{\gamma}) = -\Psi_{2,Y}(\dot{\gamma})$ was derived, which reduces the number of unknown functions in Eqs. (4) to (6) to five.

The simulations for all values of $\lambda\dot{\gamma}$ were performed with time-step width $\Delta t = 0.01\lambda$. For this time-step width, the probability for $|S_{j+1} - S_j|$ calculated according to Eq. (2) to be larger than 1 is about 1.5×10^{-12}. Therefore, reflections at both boundaries of the interval $[0,1]$ will practically not occur in a simulation of at most 2×10^9 time steps, and a simulation with $\Delta t = 0.01\lambda$ produces no error due to time discretization. For low and intermediate shear rates, the simulation for each value of the shear rate ($\lambda\dot{\gamma} = 0.2, 0.5, 1, 2, 5, 10, 20, 50, 100$) consisted of a total number of 2×10^9 time steps, and the times at which reflections occurred were calculated with an error of at most $1.9 \times 10^{-8}\lambda$. For high shear rates, the simulation for each value of the shear rate ($\lambda\dot{\gamma} = 200, 500, 1000, 2000, 5000$) consisted of 1.5×10^9 time steps, and the times at which reflections occurred were calculated with an error of at most $1.9 \times 10^{-11}\lambda$. By simulating an ensemble of 10,000 trajectories in the innermost loops, the most time-consuming parts of the simulation program could be vectorized. Each run for a single value of the shear rate required between 3.5 and 4 hours of central processing unit (CPU) time on a Cray-2 computer.

In Table 1 we give a detailed comparison between our simulation results for η_Ω and η_Y and the results obtained by numerical integration of the CB expression for the stress tensor. (In Table 1 the numbers in parentheses represent the statistical error in the last figure displayed as estimated from the fluctuations in the results for the 10,000 trajectories simulated.) The values referred to as "exact" in Table 1 were essentially taken over from Ref. 9, where four significant figures have been displayed. An independent calculation of the functions η_Ω and η_Y by means of a numerical integration procedure was performed by B. J. Geurts, who kindly provided detailed tables of the results published in Fig. 4 of Ref. 16. For low shear rates, Geurts' results allowed a more detailed comparison for η_Y. At higher shear rates, discrepancies in the numerical results obtained by Bird et al. and by Geurts were taken into account by reducing the number of significant figures displayed such that the respective data are roughly consistent with the column labeled "exact" in Table 1.

The agreement between the simulation and exact results for η_Ω and η_Y in

TABLE 1 The Two Contributions to the Dimensionless Viscosity Predicted
by the Curtiss–Bird Model

	$\dfrac{\eta_\Omega}{NnkT\lambda}$		$\dfrac{\eta_Y}{NnkT\lambda}$	
$\lambda\dot\gamma$	Exact	Simulation	Exact	Simulation
0.2	1.666×10^{-2}	$1.66(1) \times 10^{-2}$	1.1115×10^{-2}	$1.1116(2) \times 10^{-2}$
0.5	1.661×10^{-2}	$1.664(5) \times 10^{-2}$	1.1135×10^{-2}	$1.1133(2) \times 10^{-2}$
1	1.644×10^{-2}	$1.643(2) \times 10^{-2}$	1.1201×10^{-2}	$1.1199(2) \times 10^{-2}$
2	1.585×10^{-2}	$1.585(1) \times 10^{-2}$	1.1411×10^{-2}	$1.1413(2) \times 10^{-2}$
5	1.337×10^{-2}	$1.337(1) \times 10^{-2}$	1.2027×10^{-2}	$1.2030(2) \times 10^{-2}$
10	9.86×10^{-3}	$9.863(2) \times 10^{-3}$	1.218×10^{-2}	$1.2185(2) \times 10^{-2}$
20	6.02×10^{-3}	$6.0188(7) \times 10^{-3}$	1.111×10^{-2}	$1.1108(1) \times 10^{-2}$
50	2.42×10^{-3}	$2.4169(2) \times 10^{-3}$	7.90×10^{-3}	$7.9016(9) \times 10^{-3}$
100	1.05×10^{-3}	$1.0537(1) \times 10^{-3}$	5.27×10^{-3}	$5.2742(6) \times 10^{-3}$
200	4.26×10^{-4}	$4.2636(3) \times 10^{-4}$	3.21×10^{-3}	$3.2140(4) \times 10^{-3}$
500	1.20×10^{-4}	$1.2038(1) \times 10^{-4}$	1.52×10^{-3}	$1.5218(2) \times 10^{-3}$
1000	4.48×10^{-5}	$4.4840(4) \times 10^{-5}$	8.27×10^{-4}	$8.267(1) \times 10^{-4}$
2000	1.6×10^{-5}	$1.6423(2) \times 10^{-5}$	4.38×10^{-4}	$4.3751(8) \times 10^{-4}$
5000	4.3×10^{-6}	$4.2834(5) \times 10^{-6}$	1.84×10^{-4}	$1.8394(4) \times 10^{-4}$

Table 1 is very satisfactory. The simulations yield remarkably precise results. For practical purposes, one could certainly sacrifice the last significant figure; when error bars 10 times larger are acceptable, the required computer time is reduced by a factor of 100 and is hence of the order of 2 or 3 minutes for each shear rate. The computer simulation is problematic at low shear rates, where a small off-diagonal component of the stress tensor has to be estimated and divided by the small shear rate in order to obtain η_Ω. On the other hand, the results obtained by numerical integration are very reliable at low shear rates, whereas the evaluation of the required integrals at high shear rates is rather difficult (this can also be concluded from the inconsistencies in the numerical results by Bird et al. and by Geurts).

The agreement between the simulation and exact results for the functions characterizing the normal-stress coefficients in the definitions (4) to (6) is similarly satisfactory. Even for very high shear rates, one finds at least a four-significant-figure agreement between the respective results for $\Psi_{1,\Omega}$ and $\Psi_{1,Y}$. For all shear rates, the relationship $\Psi_{2,\Omega}(\dot\gamma) = -\Psi_{2,Y}(\dot\gamma)$ is fulfilled by the simulation data.

OUTLOOK

Simulation of the reptation process $(\mathbf{U}(t), S(t))$ describing the polymer dynamics in reptation theories requires (1) calculation of the deterministic time evolution of the unit vector $\mathbf{U}(t)$, (2) simulation of the Wiener process $S(t)$ on the interval $[0, 1]$ with reflecting boundary conditions, and (3) implementation of the boundary conditions for the process $\mathbf{U}(t)$. In this chapter we suggested a simulation method in which the third problem is solved exactly, that is, in which the times at which the deterministic time evolution of $U(t)$ has to be started with a new random unit vector [when $S(t)$ is reflected at a boundary] can be calculated with arbitrarily high precision. Since the Wiener process on the interval $[0, 1]$ can be simulated in a rigorous manner (problem 2), *the stochastic part of the simulation is fully solved,* and the remaining problem is calculation of the deterministic time evolution of the unit vector $U(t)$ caused by the externally applied flow field. For steady shear flow, the recursive equation (1) produces the exact solution $U(t)$ (at the discrete times $t = j \Delta t$), such that the entire simulation procedure is rigorous. For more complex flow situations, one needs to employ efficient methods for solving the deterministic time-evolution equation for $U(t)$ in order to obtain a powerful simulation algorithm; various standard algorithms for solving deterministic differential equations are available in most software libraries.

The most important advantage of computer simulations, aside from their flexibility in treating complex flow situations, is the fact that similar simulations techniques can also be used to investigate generalizations of the DE and CB models. A very interesting generalized reptation theory is the "reptating-rope model" developed by Jongschaap and Geurts [16–20]. In this model an attempt is made to introduce effects of correlations between the chain orientations at different positions within a single chain. It is therefore not surprising that computer simulation of the reptating-rope model requires the simulation of two copies of the reptation process. Details on the simulation algorithm, which also illuminates the physical content of the final model equations for the reptating-rope model, and detailed results will be given in a forthcoming paper [21].

ACKNOWLEDGMENTS

I wish to thank B. J. Geurts for providing tables of his numerical data for the viscosity predicted by the Curtiss–Bird model. The simulations were performed on the University of Stuttgart's Cray-2 computer.

REFERENCES

1. H. C. Öttinger, *J. Chem. Phys.* **91,** 6455 (1989).
2. M. Doi and S. F. Edwards, *J. Chem. Soc., Faraday Trans. 2* **74,** 1789 (1978).
3. M. Doi and S. F. Edwards, *J. Chem. Soc., Faraday Trans. 2* **74,** 1802 (1978).
4. M. Doi and S. F. Edwards, *J. Chem. Soc., Faraday Trans. 2* **74,** 1818 (1978).
5. M. Doi and S. F. Edwards, *J. Chem. Soc., Faraday Trans. 2* **75,** 38 (1979).
6. C. F. Curtiss and R. B. Bird, *J. Chem. Phys.* **74,** 2016 (1981).
7. C. F. Curtiss and R. B. Bird, *J. Chem. Phys.* **74,** 2026 (1981).
8. R. B. Bird, H. H. Saab, and C. F. Curtiss, *J. Chem. Phys.* **86,** 1102 (1982).

9. R. B. Bird, H. H. Saab, and C. F. Curtiss, *J. Chem. Phys.* **77,** 4747 (1982).

10. H. H. Saab, R. B. Bird, and C. F. Curtiss, *J. Chem. Phys.* **77,** 4758 (1982).

11. X. Fan and R. B. Bird, *J. Non-Newtonian Fluid Mech.* **15,** 341 (1984).

12. R. B. Bird, C. F. Curtiss, R. C. Armstrong, and O. Hassager, *Dynamics of Polymeric Liquids,* Vol. 2, 2nd ed. Wiley-Interscience, New York, 1987.

13. M. Doi and S. F. Edwards, *The Theory of Polymer Dynamics,* Clarendon Press, Oxford, 1986.

14. W. Strittmatter, *J. Stat. Phys.* (submitted) (1989).

15. M. Loève, *Probability Theory,* Vol. 2, 4th ed. Springer-Verlag, New York, 1977.

16. B. J. Geurts, *J. Non-Newtonian Fluid Mech.* **31,** 27 (1989).

17. R. J. J. Jongschaap, in *Progress and Trends in Rheology II,* Proceedings of the Second Conference of European Rheologists, ed. H. Giesekus and M. F. Hibberd, Supplement to *Rheol. Acta,* 1988, pp. 99–102.

18. R. J. J. Jongschaap and B. J. Geurts, *Rolduc Polymer Meeting,* Vol. 2, 1987.

19. B. J. Geurts and R. J. J. Jongschaap, *J. Rheol.* **32,** 353 (1988).

20. B. J. Geurts, *J. Non-Newtonian Fluid Mech.* **28,** 319 (1988).

21. H. C. Öttinger, *J. Chem. Phys.* **92,** 4540 (1990).

Dynamics
of Gel Electrophoresis

D. GERSAPPE and M. OLVERA DE LA CRUZ
Department of Materials Science and Engineering
Northwestern University
Evanston, Illinois 60208

ABSTRACT

An off-lattice computer simulation based on the Langevin equation of motion is used to study gel electrophoresis in a periodic gel. The simulated chain dynamics and the dynamics predicted by tube reptation theories differ quite dramatically. The loss of length-dependent mobility is shown as a consequence of increasing the field strength. The dynamics of the chain during orthogonal pulsed field gel electrophoresis is discussed and the effect of gel inhomogeneity on the statistics of the chain is studied.

INTRODUCTION

Gel electrophoresis is a technique that is widely used to separate proteins and nucleic acids, in particular DNA. In its simplest form the technique consists of applying an electric field to a gel that contains the molecules of interest. Since proteins carry a net charge at any pH value other than their isoelectric point, they migrate under the influence of an electric field. In the limit of small electric fields, the mobility of linear proteins is inversely proportional to their length. So, after sufficient time is allowed for migration, proteins of different sizes separate physically in the gel. However, as the electric field is increased, the length dependence of the mobility is sharply reduced. Consequently, the technique is unable to separate very long chains [1]. To optimize the separation process, more sophisticated techniques have been developed [2,3]. The dynamics of the gel electrophoresis process is still not fully understood, and experimental developments have relied solely on empirical observations.

To obtain a better understanding of the dynamics of gel electrophoresis, we have developed a computer simulation of the process in a periodic gel. The results of the simulation illustrate a diffusional process that cannot be explained in terms of the existing tube-reptation theories of gel electrophoresis [4,5].

PAST APPROACHES

In tube theories the fluctuations of the internal segments of the chain through the gel junction points are neglected on the assumption that there is a large entropic penalty associated with this. Thus the chain is constrained to move along a tube path, the diameter of which is the average distance between gel

199

junction points. However, as the field is increased, the fluctuations of the internal segments become increasingly important. Only in the limit of zero field can these fluctuations be neglected.

The failure of tube theories to describe the dynamics of gel electrophoresis is further amplified by experimental evidence [6]. Experiments carried out show that the mobility of the chains saturate at much lower fields than predicted by tube theories. Further, while tube theories predict that the mobility of a chain, at saturation, reduces to that of a free chain, the mobility determined experimentally plateaus to a much lower value.

In the past the problem has been tackled via a Monte Carlo simulation, which was carried out on-lattice by Olvera et al. [7]. They found that the Monte Carlo method leads to large metastabilities as the mobility of the chain goes as $\exp(-EL)$, where E is the effective field and L the length of the chain, contrary to what is expected. Also, it should be noted that the lattice model itself is unsuitable for studying the dynamics of the process; in the limit of small fields and tight gels, for example, the motion that contributes most to the mobility of the chain is the continuous twisting of the bond angles, not the discrete "hops" allowed by the lattice model.

Recent simulations by Deutsch [8] and by Shaffer and Olvera [9] illustrate the importance of including all the internal degrees of freedom of the chain. Deutsch simulates the motion of the chain directly by solving numerically the chain's Langevin equation of motion. He finds that at large fields the motion of the chain deviates strongly from that predicted by tube theories. In his simulation the chain does not remain extended in the direction of the field but hooks and unhooks itself from the obstacles. This is also seen in the simulation carried out by Shaffer and Olvera, where the chain was modeled as a Rouse chain (i.e., as a series of beads connected by entropic springs). They then discretized the Langevin equation and used it to describe the dynamics of the chain.

The simulation of Shaffer and Olvera agrees qualitatively with the results of Deutsch, but the fields used were much smaller. The breakdown of the reptation model was observed even at these low fields. Initially, they generated the chain in a random coil configuration. When the electric field was switched on, they found that regions with a higher concentration of monomers tended to leak through the gel junction points, pulling the chain behind them. At smaller fields, the regions that were monomer "rich" tended to diffuse toward the chain ends, resulting in short, open J-shaped chain conformations with a denser region around the end segment. If this region were sufficiently dense, it could pull the end segment, stretching the chain again. As the field was very small, large instabilities that formed far from the ends rarely occurred. Therefore, the J shapes that they observed were short and open, but the chain still underwent contractions. Shaffer and Olvera found that the radius of gyration of the chain underwent oscillations, the periods of which were both length and field dependent (Fig. 1). Loss of the inverse length dependence of mobility as chain length was increased was also observed.

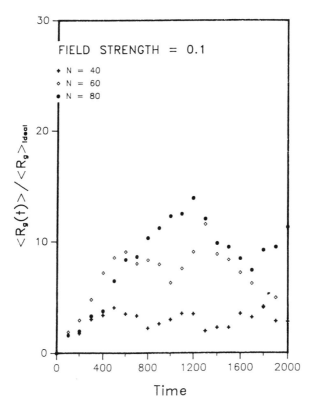

Figure 1   Average radius of gyration $\langle R_g \rangle$ scaled by the unperturbed value $\langle R_g \rangle_{\text{ideal}}$ (in the absence of an electric field) versus time for various chain lengths. E_s, the electric field, is 0.1 and kept constant. Each point in the figure corresponds to an average over 10 different initial conformations of the chain.

RESULTS OF NEW APPROACH

Empirically developed techniques using rotating or inverting pulsed fields have been shown to be more efficient in separating long chains [10]. Here we study the effect on the dynamics of the chain of rotating the field 90°. We found that even at smaller electric fields the mechanism by which the chain migrates cannot be described by tube theories, as it is governed by the configuration of the chain when the field is switched. If the field is rotated when the chain is a configuration where the dense regions are concentrated at the ends, the end that is more dense is the one that pulls the chain (Fig. 2). However, if the field is rotated when the chain is almost kink-free, instabilities form along the chain length and when they are sufficiently close to the ends, they can actually pull the end segments toward them and leak through the entanglements (Fig. 3).

From our simulation we see that the chain dynamics are no longer controlled by the leading segment but rather, by the local regions of high

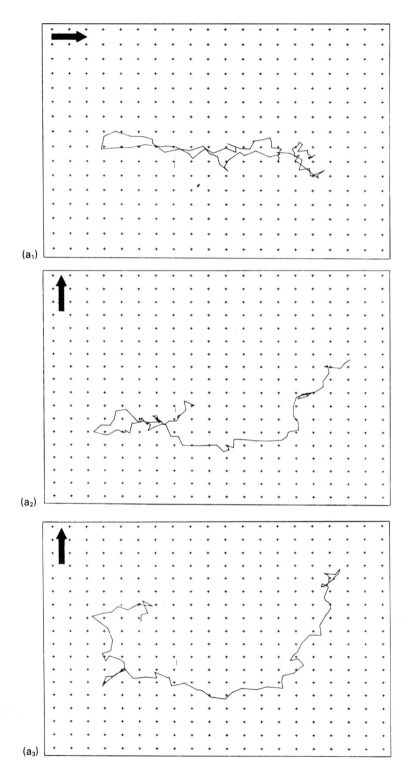

(a₁)

(a₂)

(a₃)

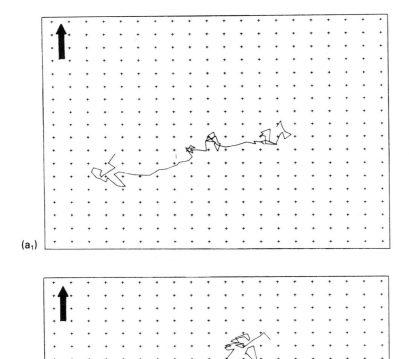

(a₁)

(a₂)

Figure 3 Sequence steps $3a_1$ to $3a_2$ show the conformation of the chain correspond-ing to $E_s = 0.1$, after the field is rotated by 90°. The snapshots were taken at 200 and 400 time steps after switching the field. (The arrows indicate the direction of the field.)

◄ **Figure 2** Sequence of the chain conformation when the electric field is rotated by 90°. Sequence steps $2a_1$ to $2a_3$ show the conformation of the chain for an electric field $E_s = 0.15$. The snapshots were taken at 0, 200, and 400 time steps after switch-ing the field. (The arrows indicate the direction of the field.)

monomer concentration that form along the chain. As these monomer-"rich" regions can leak through the entanglements, it is imperative when studying the dynamics of pulsed field gel electrophoresis to include all internal degrees of freedom of the chain.

Our simulation also reproduces the experimental results of Schwartz and Koval [10]: that is, perpendicular fields prevent the collapse of the chain and give rise to open U-shaped conformations. To optimize the separation technique, however, it is important to know the stretching time of a given chain length as a function of field strength.

The simulation that we carried out used a periodic gel, and to find out the effect of gel inhomogeneity on the dynamics of the process, we are studying the effect of disorder on the chain statistics. We do not expect the dynamics of the process to be affected much by gel inhomogeneity.

Currently, we are studying the statistical properties of a chain embedded in a random medium. As there is no driving force here, we use an on-lattice Monte Carlo method. We generate a SAW by using the dynamical reptation method and move it using a standard Metropolis algorithm for 10 Monte Carlo steps, where each Monte Carlo step involves N^2 movements (N = length of the chain).

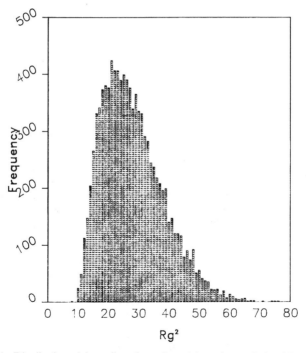

Figure 4   Distribution of the radius of gyration of the polymer chain with excluded volume in an inhomogeneous gel. The concentration of the gel is 0.2, and the chain length is 40.

We found that the exponent which relates the radius of gyration of the chain to its length is the same as it would be were no impurities present. Further, when we calculate the distribution of the radius of gyration (Fig. 4), we find a distribution very similar to the case of no impurities.

REFERENCES

1. M. W. McDonnel, M. N. Simon, and F. W. Studier, *J. Mol. Biol.* **110,** 119 (1977).
2. N. C. Stellagen, *J. Biomol.* **3,** 299 (1985).
3. D. C. Schwartz and D. R. Cantor, *Cell* **37,** 67 (1984).
4. O. J. Lumpkin and B. H. Zimm, *Biopolymers* **21,** 2315 (1982); O. J. Lumpkin, P. Dejardin, and B. H. Zimm, *Biopolymers* **24,** 1575 (1985).
5. G. W. Slater and J. Noolandi, *Biopolymers* **24,** 2181 (1985).
6. H. Hervert and C. P. Bean, *Biopolymers* **26,** 727 (1987).
7. M. Olvera de la Cruz, J. M. Deutsch, and S. F. Edwards, *Phys. Rev.* **A33,** 2047 (1986).
8. J. M. Deutsch, *Science* **240,** 922 (1988).
9. E. O. Shaffer and M. Olvera de la Cruz, *Macromolecules* **22,** 1351 (1989).
10. D. C. Schwartz and M. Koval, *Nature* **338,** 520 (1989).

Computation and Display of the Diffusive Behavior of Polymer Chains

R. F. T. STEPTO
Polymer Science and Technology Group
Manchester Materials Science Centre
University of Manchester and UMIST
Grosvenor Street
Manchester, M1 7HS, United Kingdom

ABSTRACT

Two applications of Monte Carlo computations to the diffusive behavior of polymer chains are described. In the first, the quantitative prediction of experimental diffusion coefficients of short polymethylene chains in various solvents is considered using full Oseen and Kirkwood–Riseman hydrodynamics. In both cases the segmental friction is found to be a function of chain length and solvent. Also, the differences between Oseen and Kirkwood–Riseman hydrodynamics are negligible. The variations in segmental friction found are attributed to the neglect of chain dynamics and solvent structure in the hydrodynamic models and point to the necessity of using molecular dynamics computations of chains in discontinuous solvents.

The second application explores the modeling center-of-mass diffusion in real time using Monte Carlo computations and experimental values of diffusion coefficients. It is shown that the variation of diffusion coefficient with chain length for polymethylene chains can be modeled by assuming that the elementary computer moves used in the Monte Carlo computations occur simultaneously in real time. For the particular computations used and the experimental data considered, the number of simultaneous moves is found to be proportional to chain length. Applications of this method of modeling center-of-mass diffusion are discussed briefly.

INTRODUCTION

In Monte Carlo (MC) computations of the conformations of polymer chains, the coordinates of all skeletal groups (segments) in each chain conformation generated are known. If Metropolis sampling is used [1], randomly selected skeletal-bond conformations are changed to generate new chains. The coordinates of the skeletal groups may be located in an external coordinate system in which chain conformations occur with a frequency proportional to their Boltzmann factors.

Knowledge of the coordinates of all the segments means that many conformation-dependent properties can be evaluated from one simulation: for example, particle scattering functions, molecular shapes (equivalent ellipsoids), end-to-end distance distributions, diffusion coefficients, and so on. Moreover, if realistic chain models are used, calculations relate to particular temperatures and experimental data may be interpreted quantitatively (see Ref. 2 and references quoted therein). The Metropolis sampling of chain conformations of realistic (off-lattice) chains requires an algorithm that reduces rounding errors when sections of chains are moved. Such an algorithm has been described [2].

In this chapter we use the results of MC computations with Metropolis sampling, employing a well-established rotational isomeric state (RIS) model of the polymethylene (PM) chain [3] for the quantitative fitting of experimental diffusion coefficients [4] (D) at 25°C, and a tetrahedral lattice model with excluded volume [1,5] for direct modeling of center-of-mass diffusion in external coordinates. The first part shows clearly the effects of neglecting chain dynamics. The second part shows the possibility of relating MC-based molecular movements to center-of-mass movements in real time.

FITTING OF EXPERIMENTAL DIFFUSION COEFFICIENTS OF POLYMETHYLENE CHAINS IN VARIOUS SOLVENTS

Theory and Computation

It has been shown [4] that the use of linear hydrodynamics with rigid chain conformations in a solvent continuum gives

$$\frac{D}{kT} = \frac{1}{f} = \frac{1}{3\zeta} \langle \mathrm{Tr}(\mathbf{K}^{-1}) \rangle_{\mathrm{conf}} \tag{1}$$

where ζ is the segmental friction coefficient, f the molecular friction coefficient, $\zeta\mathbf{K}$ the molecular friction tensor, and $\langle \cdot \rangle_{\mathrm{conf}}$ denotes the average over chain conformations.

In detail, $\mathbf{K}^{-1}$ for a given chain conformation is evaluated through the equation

$$\mathbf{K} = \sum_{i,j=1}^{x} \mathbf{Q}_{ij} \tag{2}$$

where x is the number of friction centers (chain segments for PM chains) and the $\mathbf{Q}_{ij}$ are evaluated from the simultaneous equations

$$\sum_{k=1}^{x} \mathbf{P}_{ik} \cdot \mathbf{Q}_{kj} = \zeta_{ij} \mathbf{I} \tag{3}$$

Here

$$\mathbf{P}_{ij} = \delta_{ij} \mathbf{I} + \zeta \mathbf{T}_{ij} \tag{4}$$

where $\zeta\mathbf{T}_{ij}$ is the hydrodynamic interaction tensor, $\zeta\mathbf{Q}_{ij}$ the friction tensor describing the hydrodynamic force on segment i due to segment j, and $\mathbf{P}_{ij}/\zeta$ the corresponding mobility tensor.

In the Oseen approximation,

$$\zeta \mathbf{T}_{ij} = \frac{\zeta}{8\pi\eta_0\, l\tilde{r}_{ij}}\, (\mathbf{I} + \mathbf{n}_{ij}\,\mathbf{n}_{ij}) \qquad \mathbf{T}_{ii} = \mathbf{0} \tag{5}$$

where η_0 is the solvent (continuum) viscosity, $\tilde{r}_{ij}$ the distance between segments i and j for unit bond length, l the actual bond length, and $\mathbf{n}_{ij}\,\mathbf{n}_{ij}$ is the self-direct product of the unit vector from i to j. Equation (5) may be rewritten

$$\zeta \mathbf{T}_{ij} = \frac{3}{4}\left(\frac{\kappa}{\tilde{r}_{ij}}\right)(\mathbf{I} + \mathbf{n}_{ij}\,\mathbf{n}_{ij}) = \frac{3\kappa}{4}\, \tilde{\mathbf{T}}_{ij} \tag{6}$$

with

$$\kappa = \frac{\zeta}{6\pi\eta_0\, l} \tag{7}$$

and $\tilde{\mathbf{T}}_{ij}$ the Oseen hydrodynamic interaction tensor for unit bond length. κ is a parameter scaling the strength of intersegmental hydrodynamic interactions.

In the limit of weak interactions ($\zeta \mathbf{T}_{ij}$ small), the inversions required for the evaluation of $\mathbf{Q}_{ij}$ from $\mathbf{P}_{ij}$ according to Eq. (3), and $\mathbf{K}^{-1}$ from $\mathbf{K}$ according to Eq. (1) may be performed by retaining only the first two terms in the binomial expansions of $\mathbf{P}_{ij}^{-1}$ and $\mathbf{K}^{-1}$. These approximations lead [4] to the relationship

$$\frac{1}{3}\,\mathrm{Tr}(\mathbf{K}^{-1}) = \frac{1}{x} + \frac{\kappa}{4x^2} \sum_{i,j=1}^{x} \mathrm{Tr}(\tilde{\mathbf{T}}_{ij}) \tag{8}$$

Further, use of Eq. (1) and the Oseen tensor for $\tilde{\mathbf{T}}_{ij}$ gives immediately the well-known Kirkwood–Riseman (K-R) equation

$$\frac{\zeta}{f} = \frac{1}{x} + \frac{\kappa}{x^2}[\tilde{R}^{-1}] \tag{9}$$

where $[\tilde{R}^{-1}]$ represents $\sum_{i,j=1}^{x}\langle 1/\tilde{r}_{ij}\rangle_{\mathrm{conf}}$. Note that the present derivation [4] of Eq. (9) shows that it does *not* depend on preaveraging of the Oseen tensor. It requires only limitingly weak hydrodynamic interactions. Further, *post*-averaging over chain conformations is required in both the Oseen and K-R approximations, as the derivation of Eq. (1) uses the condition that the instantaneous segmental and center-of-mass velocities are always equal (the rigid-body assumption).

Computations of D or ζ/f for PM have been carried out using Oseen hydrodynamics [4,6] [Eq. (1) to (7)] and K-R hydrodynamics [7,8] [Eq. (9)]. In both cases, Metropolis sampling was used to generate the chain conformations from which the required values of r_{ij} were evaluated. κ was an independently chosen parameter. The computations using Oseen hydrodynamics are more lengthy than those using K-R hydrodynamics and they have to be carried out individually for each value of κ. They have therefore been carried out and matched with experiment only for PM in benzene.

Discussion of Results: Interpretation of Segmental Friction

The results of the computations are summarized in Table 1 and Fig. 1 (cf. Ref. 9). Table 1 shows the values of κ required to *fit* existing experimental values of D for short PM chains in various solvents. From the fitting of the results in benzene, it is clear that the use of Oseen hydrodynamics brings only small changes to the values of κ found using K-R hydrodynamics. Importantly, both give κ (i.e., segmental friction) increasing with chain length.

Figure 1 shows the reciprocal Stokes–Einstein diffusion radius in benzene as a function of x. The Oseen and K-R values were calculated using the values of κ appropriate for $x = 6$ (i.e., 0.35 and 0.37, respectively). (At $x = 5$, the effects of the terminal —CH_3 groups are more noticeable.) It is clear from Fig. 1 that it is impossible to fit the experimental values of D using rigid-body hydrodynamics in a continuum and a constant value of κ. Moreover, only minor improvements are achieved by using Oseen hydrodynamics. The same will be true for the other systems in Table 1. Hence, in the rigid-body approximation, the K-R equation may just as well be used to interpret experimental values of D.

It has been found empirically that over the range of chain lengths studied, the values of κ (or ζ) increase approximately in proportion to the effective bond length (b) of the PM chain ($b = \langle r_0^2 \rangle/n; n = x - 1$), a result also confirmed

TABLE 1 Values of Segmental Hydrodynamic Interaction Parameter (κ) that Agree with Experimental Values of D for PM Chains in Various Solvents According to K-R and Oseen Hydrodynamics[a]

| | κ | | | | | |
x	$CCl_4/$ 25°C/K-R	Benzene/ 25°C/K-R	Benzene/ 25°C/Oseen	Tetralin/ 22.2°C/K-R	Quinoline/ 25°C/K-R	Decalin/ 25°C/K-R
5	0.468	0.398	0.335	0.238	0.220	0.201
6	0.407	0.369	0.347	0.224	0.195	0.189
7	0.433	0.381	0.359	0.217	0.198	0.179
8	0.409	0.394	0.357	—	0.211	0.178
9	—	0.411	0.376	0.242	0.221	0.188
10	0.422	0.422	0.383	—	0.233	0.182
12	0.429	0.422	0.392	0.220	0.251	0.192
14	—	—	—	0.247	—	0.196
16	0.439	0.455	0.405	—	0.280	—
18	0.587	—	—	0.251	—	0.218
19	—	0.468	0.422	—	0.315	—
20	0.512	—	—	—	—	—
24	—	0.526	0.454	—	—	—
28	0.541	0.551	0.496	0.322	—	0.269

[a] From Refs. 4 and 9.

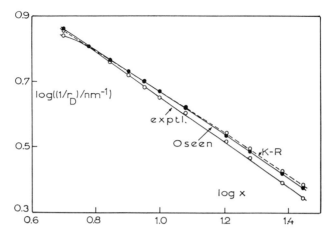

Figure 1 Polymethylene chains in benzene at 25°C. Reciprocal Stokes–Einstein diffusion radius $(1/r_D)$ as a function of chain length (x). Comparison of experimental values and values calculated according to K-R and Oseen hydrodynamics with segmental friction (κ) left constant at the values appropriate for $x = 6$ (K-R: $\kappa_6 = 0.37$; Oseen: $\kappa_6 = 0.35$).

for other systems with weak hydrodynamic interactions [6,8,10]. Thus one may write

$$\zeta = c_b(6\pi\eta_0 b) \tag{10}$$

with $c_b b$ the effective Stokes radius of a segment. Table 2 shows that c_b is constant to within experimental error. However, it does depend on solvent. The average value of c_b, $\langle c_b \rangle$, shows a tendency to decrease with increase of molar volume (V_0) and viscosity of solvent. Such a change is consistent with increased slippage between segments and the supposed solvent continuum [7,9]. Some dependence on molecular shape probably also exists.

From Eqs. (9) and (10), κ is seen to be the ratio of the Stokes segmental radius to the bond length, that is,

$$\kappa = \frac{c_b b}{l} \tag{11}$$

Using $c_b = \langle c_b \rangle$ and the limiting value of b for $x = \infty$ (namely, b_∞), one obtains the values of κ_∞ shown in Table 2. They define the limiting segmental friction coefficients for the infinite chain. Alternatively, corresponding limiting values of the Stokes radius ($= \langle c_b \rangle b_\infty$) may be evaluated. Such values have been tabulated for PM and linear and cyclic poly(dimethyl siloxane) in various solvents [6].

As discussed, the increase in segmental friction with chain length produces the differences between the experimental curve for PM in benzene and the theoretical curves in Fig. 1. Similar but distinct differences would occur

TABLE 2 Values of c_b Required to Give Agreement with Experimental Values of D for PM in Various Solvents According to K-R and Oseen Hydrodynamics[a]

		c_b					
x	b (nm)	CCl₄/ 25°C/K-R	Benzene/ 25°C/K-R	Benzene/ 25°C/Oseen	Tetralin/ 22.2°C/K-R	Quinoline/ 22.2°C/K-R	Decalin/ 22.2°C/K-R
5	0.228	*0.314*	*0.267*	*0.225*	*0.160*	*0.148*	*0.135*
6	0.247	0.252	0.229	0.215	0.139	0.121	0.117
7	0.264	0.251	0.221	0.208	0.126	0.115	0.104
8	0.278	0.225	0.217	0.202	—	0.116	0.098
9	0.291	—	0.216	0.198	0.127	0.116	0.099
10	0.302	0.214	0.214	0.194	—	0.118	0.092
12	0.320	0.205	0.202	0.187	0.105	0.120	0.092
14	0.334	—	—	—	0.113	—	0.090
16	0.346	0.194	0.201	0.179	—	0.124	—
18	0.355	0.253	—	—	0.108	—	0.094
19	0.360	—	0.199	0.179	—	0.134	—
20	0.364	0.215	—	—	—	—	—
24	0.376	—	0.214	0.185	—	—	—
28	0.385	0.215	0.219	0.197	0.128	—	0.107
∞	0.438	—	—	—	—	—	—
$\langle c_b \rangle$		0.225	0.213	0.194	0.121	0.120	0.099
κ_∞		0.644	0.610	0.555	0.346	0.344	0.283
η_0(cP)		0.900	0.608	0.608	2.1	3.44	2.64
V_0/cm³/mol		96.5	88.9	88.9	136	118	159

[a] b, effective bond lengths of chains at 25°C on the basis of RIS statistics [3] ($b = (\langle r_0^2 \rangle/n)^{1/2}$); $\langle c_b \rangle$, average values of c_b over chain lengths studied excluding the values at $x = 5$; κ_∞, predicted limiting vaues of κ at infinite chain length; η_0, solvent viscosities; V_0, solvent molar volumes.

with PM in the other solvents. The conclusion to be drawn is that it is not possible to predict values of D for a given chain structure in a universal manner using rigid-body hydrodynamics in a continuum. The conclusion also extends to other polymer–solvent systems [6,8]. Further, the small differences found between the Oseen and K-R calculated values of D for PM in benzene mean that the hydrodynamic interactions are weak and the use of interaction tensors of higher order than the Oseen is not meaningful. In addition, they would still give a dependence on solvent.

The reductions in actual values of D below those expected on the basis of a constant value of κ or ζ in a given solvent must be due to differences between the average segmental and center-of-mass velocities, causing additional frictional forces. The rigid-body assumption puts these differences at zero. The deduced increase at small x in ζ or κ in proportion to b indicates an increasing difference between the two velocities up to a limiting value (characterized by κ_∞) as the center-of-mass velocity tends to zero. Such limiting behavior for local, segmental modes of motion relative to center-of-mass motion is expected from general considerations. The detailed interpretation of the phenomenon must lie with molecular dynamics computations, which, hopefully, would reflect the increased frictional losses. However, the computations must also reflect the structure and behavior of the solvent. A solvent continuum is not sufficient, as indicated by the variation of $\langle c_b \rangle$ or κ_∞ with solvent. The variation means that the segments are aware of solvent molecules, as may be expected from the similar sizes of the two entities.

Although the experimental results and calculations discussed refer to short chains, the changes in segmental friction with chain length have important consequences for infinite chains. This may be seen as follows [9]. Let D_x^0 be the diffusion coefficient for a chain of x segments calculated using the value of ζ $(=\zeta^0)$ at a short chain length (here, $x = 6$), which gave agreement with experiment at that chain length. Then the difference between the calculated and experimental curves in Fig. 1, for example, may be transformed to the fractional deviation

$$\Delta_x = \frac{D_x^0 - D_x}{D_x} \tag{12}$$

where D_x is the experimental diffusion coefficient. In the K-R approximation

$$\Delta_x = \frac{(1/x)(1/\zeta^0 - 1/\zeta)}{1/x\zeta + (1/6\pi\eta_0)(|R^{-1}|/x^2)} \tag{13}$$

where ζ is the value of the segmental friction coefficient required to give agreement with experiment.

Δ_x may also be written as

$$\Delta_x = \frac{1/r_{D^0 x} - 1/r_{Dx}}{1/r_{Dx}} = \frac{\psi_x^0 - \psi_x}{\psi_x} \tag{14}$$

where

$$\psi_x^0 = \frac{\langle s_{0,x}^2 \rangle^{1/2}}{r_{D,x}^0} \quad \text{and} \quad \psi_x = \frac{\langle s_{0,x}^2 \rangle^{1/2}}{r_{D,x}} \tag{15}$$

the ratios of the equilibrium radius ($\langle s_{0,x}^2 \rangle^{1/2}$) and the diffusion radii $r_{D,x}^0$ and $r_{D,x}$ based on ζ^0 and ζ. Equation (14) can be rearranged to give

$$\psi_x = \frac{\psi_\infty^0}{1 + \Delta_x} \tag{16}$$

and in the limit of infinite chain length,

$$\psi_\infty = \frac{\psi_\infty^0}{1 + \Delta_\infty} \tag{17}$$

where ψ_∞^0 is the value given by K-R theory for a constant value of ζ, namely, $8/3\pi^{1/2}$ $(= 1.505)$.

Figure 2 shows Δ_x versus x^{-1} for the solvents benzene and quinoline [9]. It can be seen that approximately linear behavior with distinct intercepts (Δ_∞)

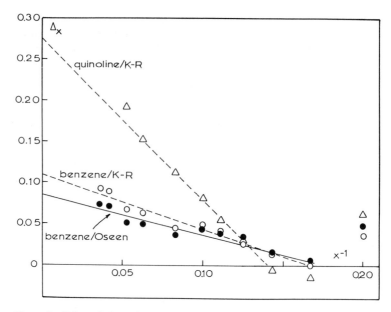

Figure 2   Polymethylene chains in benzene and quinolene at 25°C. Fractional decrease in diffusion coefficient (Δ_x) as a function of reciprocal chain length (x^{-1}) based on fitted values of κ at $x = 6$ (benzene/K-R: $\kappa_6 = 0.37$; benzene/Oseen: $\kappa_6 = 0.35$; quinoline/K-R: $\kappa_6 = 0.20$).

for the two solvents is defined. Unfortunately, the data for the other solvents in Tables 1 and 2 are not accurate enough for the present analysis. In this respect it should be remembered that the small differences $D_x^0 - D_x$ will emphasize experimental scatter. The data for benzene [11] and quinoline [7] are more recent than those for the other solvents [12,13].

The intercepts from Fig. 2 converted to values of ψ_∞ are given in Table 3, together with ψ_∞^0 and values derived independently [14,15] for high-

TABLE 3   Limiting Values of the Ratio of Equilibrium Radius to
Stokes–Einstein Diffusion Radius at Infinite Chain Length (ψ_∞)
for Various Polymer–Solvent Systems Together with
Solvent Molar Volume (V_0) and Viscosity (η_0)

System	ψ_∞	V_0 (cm³/mol)	η_0 (cP)
Rigid-body, unperturbed freely jointed chain in a continuum (ψ_∞^0)	1.505	0	—
PM/benzene/25°C/Oseen	1.38	—	—
PM/benzene/25°C/K-R	1.34	88.9	0.608
PS/θ-solvents	1.27	—	—
PMMA/θ-solvents	1.16	—	—
PM/quinoline/25°C/K-R	1.08	118	3.44

[a] PM, polymethylene; PS, polystyrene; PMMA, poly(methyl methacrylate).

molar-mass polystyrene and poly(methyl methacrylate) in θ-solvents, by comparing directly measured values of $\langle s_0^2 \rangle^{1/2}$ and r_D. In the analyses of data on poly(methyl methacrylate) and polystyrene, no distinction between solvents was found because molar-mass dependences starting from short chains were not investigated. The present, detailed evaluation of Δ_x on the basis of PM shows that such a distinction should be present.

Table 3 shows that ψ_∞ cannot be considered as a universal quantity. The effects of solvent, polymer structure, and chain length on ψ_x and ψ_∞ can be summarized with reference to Fig. 2 and Table 3. Figure 2 shows that the decrease in ψ_x with chain length is solvent dependent, being larger for a smaller value of κ, that is, for more freely moving segments. Values of ψ_∞ refer to equivalent impermeable behavior at infinite chain length. ψ_∞^0 refers to freely jointed chains with a Boltzmann distribution of rigid chain conformations (i.e., ζ constant) in a continuum. A decrease in ψ_∞ away from ψ_∞^0 is effectively an increase in $r_{D,\infty}$ relative to $\langle s_{0,\infty}^2 \rangle$ and implies less penetration of the molecular domain, giving a larger equivalent impermeable radius. The analysis presented here proposes the origin of this phenomenon as the increase in segment friction with chain length, the increase arising from the dynamic flexibility of the molecular conformations on the time scale of center-of-mass movements. The value of ψ_∞ (or, in general, ψ_x) will in general be dependent on polymer, solvent, and temperature, and theories and computations accounting for solvent structure are required for its prediction.

DIRECT MONTE CARLO SIMULATION OF CENTER-OF-MASS DIFFUSION

Introduction

MC simulations using Metropolis sampling use changes in the bond conformations of randomly chosen sections of polymer chains to produce new chain conformations. Such changes automatically produce center-of-mass movements in external space which can be recorded. There is no implication that the bond rotations chosen are dynamically correct, nor is there any real time scale automatically associated with the center-of-mass movements so generated, especially as the bond rotations are carried out sequentially, irrespective of chain length. However, provided that experimentally measured values of D exist, an effective time may be defined for a computer move through the Einstein relationship

$$D = \frac{\langle \Delta c^2 \rangle}{2\tau} \tag{18}$$

where $\langle \Delta c^2 \rangle$ is the directly computed mean-square displacement of the center-of-mass in one dimension per computer move. Using Eq. (18), center-of-mass diffusion may be modeled using relatively simple MC calculations rather than more complex molecular dynamics calculations. Such modeling can have applications where the gross movements of molecules need be compared— say, in porous media or during interactions with surfaces.

In the present section we consider preliminary results exploring the variation of τ with chain length for PM chains and the limitations of the proposed method. The values of D used are those that were analyzed in detail in the first part of the paper.

Computations and Experimental Data

Figure 3 shows the experimental values of D to be considered [7]. The linear portions of the curves have slopes of about -0.9. The computational results to be used refer to tetrahedral lattice chains with —CH$_2$— groups at lattice sites, a trans-gauche energy differences of 0 and $1kT$, and $g_\pm g_\mp$ pairs of bond conformations excluded. Poor to good solvent conditions were studied by changing the nearest-neighbor segment–segment energy from $-0.5kT$ to $+0.5kT$. Chains with numbers of segments (x) from 11 to 101 were used, and the elementary moves employed effectively changed the conformations of six consecutive bonds. $\langle \Delta c^2 \rangle$ was calculated over samples of 10^2 to 10^5 chain conformations by finding the mean-square displacement per computer move in each dimension and averaging the results. Thus

$$\langle \Delta c^2 \rangle = \frac{\langle \Delta x^2 \rangle + \langle \Delta y^2 \rangle + \langle \Delta z^2 \rangle}{3} \tag{19}$$

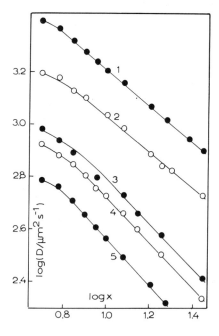

Figure 3 $\log D$ versus $\log x$ for poly-methylene chains in various solvents [7,11–13]: 1, benzene/25°C; 2, carbon tetrachloride/25°C; 3, tetralin/22.2°C; 4, decalin/ 22.2°C; 5, quinoline/25°C

Within the accuracy of the computations, no significant distinction could be drawn between the values of $\langle \Delta c^2 \rangle$ for different chain flexibilities and solvent conditions.

RESULTS AND DISCUSSION

Considering first the shortest chain length ($x = 11$), Table 4 shows the calculated values of $\langle \Delta c^2 \rangle$ and the derived values of ζ for the solvents of lowest and highest viscosity, benzene and quinoline. To better than an order of magnitude, $\tau \simeq 10^{-13}$ s.

The variation of $\langle \Delta c^2 \rangle$ with x is shown as the solid line in Fig. 4. The error bars on the computed points reflect the secondary effects of the different chain-flexibility and excluded-volume parameters used in the computations. The upper, dashed line is that predicted from Eq. (18) assuming that τ remains, for all chain lengths, at the value found for $x = 11$. That is, agreement with the experimental values of D is ensured at $x = 11$. The *ordinates* of the two lines are defined by the equations

$$\text{(dashed)} \quad \log\langle \Delta c_x^2 \rangle = \log D_x + \log 2\tau_{11} \tag{20}$$

and

$$\text{(solid)} \quad \log\langle \Delta c_x^2 \rangle = \log D_x + \log 2\tau_x \tag{21}$$

where x denotes number of segments. Further, the dashed line gives the same rate of increase of $\log\langle \Delta c_x^2 \rangle$ or $\log D_x$ with $\log x$ as that observed experimentally (Fig. 3). The values of $\langle \Delta c_x^2 \rangle$ computed directly (solid line) decrease more rapidly with x because in the computation each computer move occurs sequentially, whereas in reality, sections of chains move simultaneously, and the number of simultaneous movements increase with x.

The two lines in Fig. 4 may be analyzed as follows. From Eqs. (20) and (21), the difference in ordinate at a given value of x is

$$\log\frac{\tau_{11}}{\tau_x} = \log v_x \tag{22}$$

TABLE 4   Effective Times per Elementary Computer Move (ζ) for Undecane Obtained from Correlations, According to Eq. (18), Between Values of $\langle \Delta c^2 \rangle$ Computed Directly from MC Simulations and Experimental Values of D [a]

		$\tau(s)$	
$\epsilon g/kT$	$\langle \Delta c^2 \rangle$ (m^2)	Benzene	Quinoline
0	0.042×10^{-20}	1.3×10^{-13}	5.9×10^{-13}
1	0.045×10^{-20}	1.4×10^{-13}	6.3×10^{-13}

[a] $\epsilon g/kT$, values of trans-gauche energy difference used in the MC computations; values of $\langle \Delta c^2 \rangle$ are averages over the excluded-volume parameters used; D_{expt}/benzene/25°C = 1.59×10^{-10} m^2/s; D_{expt}/quinoline/25°C = 3.55×10^{-10} m^2/s.

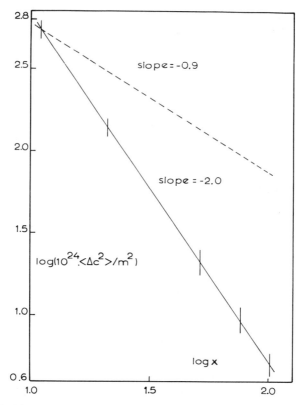

Figure 4   Mean-square center-of-mass displacement in one dimension per elementary computer move ($\langle\Delta c^2\rangle$) versus chain length (x): solid line, as computed directly, with moves occurring sequentially; dashed line, derived from Eq. (18) assuming that elementary computer moves occur simultaneously in the same time as for $x = 11$.

where v_x is the number of computer moves occurring in a chain of x segments in the real time for one computer move in a chain of 11 segments. The equations of the two lines are

$$\text{(dashed)}\quad \log\langle\Delta c_x^2\rangle = \log\langle c_{11}^2\rangle - 0.9(\log x - \log 11) \qquad (23)$$

$$\text{(solid)}\quad\;\; \log\langle\Delta c_x^2\rangle = \log\langle c_{11}^2\rangle - 2.0(\log x - \log 11) \qquad (24)$$

Thus the difference between them gives

$$\log v_x = 1.1 \log\frac{x}{11} \qquad (25)$$

Within the limits of accuracy of the computations, this result shows that to model the chain length dependence of D, the number of computer (6-bond) moves occurring simultaneously is approximately proportional to chain length (i.e., $v_x \propto x$).

To summarize, the present analysis shows that center-of-mass movements produced in MC calculations can be scaled to real time. The method appears reliable only in terms of chain-length dependence. From the computations, the smaller variations that may be expected to occur with chain flexibility and solvent quality are apparently not accurately reproduced. The study of such variations probably requires larger samples than those used in this preliminary work.

In general, one would expect v_x to be proportional to a positive power of x. The approximate proportionality to x found relates directly to the difference in the slopes of the lines in Fig. 4. For other systems, the experimental slope can be different. For example, a slope of -0.5 is found experimentally at short chain lengths for systems that behave impermeably due to stronger polymer–solvent interactions [6,7]. In addition, the slope for the directly computed results will depend on the chain model and the elementary movement used in the sampling. In this respect the tetrahedral lattice model is not a particularly accurate one for short PM chains in solvents. However, provided that both computational and experimental results exist for the chain lengths of interest, the law for scaling computer moves to real time can be defined.

An initial application of the approach presented here has been the use of the relationship between computed values of $\langle \Delta c^2 \rangle$ and D for PM chains in computer display [16] to depict visually the center-of-mass diffusion of chains of different lengths on correct relative time scales.

REFERENCES

1. M. Lal and R. F. T. Stepto, *J. Polym. Sci., Polym. Symp.* **61,** 401 (1977).
2. D. Rigby and R. F. T. Stepto, *Polymer* **28,** 423 (1987).
3. P. J. Flory, *Statistical Mechanics of Chain Molecules,* Wiley-Interscience, New York, 1969.
4. C. J. C. Edwards, A. Kaye, and R. F. T. Stepto, *Macromolecules* **17,** 773 (1984).
5. A. T. Clark and M. Lal, *Br. Polym. J.* **9,** 92 (1977).
6. C. J. C. Edwards and R. F. T. Stepto, in *Physical Optics of Dynamic-Phenomena and Processes in Macromolecular Systems,* ed. B. Sedlacek, Walter de Gruyter, West Berlin, 1985, p. 1.
7. I. J. Mokrys, D. Rigby, and R. F. T. Stepto, *Ber. Bunsenges. Phys. Chem.* **83,** 446 (1979).
8. C. J. C. Edwards, D. Rigby, and R. F. T. Stepto, *Macromolecules* **14,** 1808 (1981).
9. R. F. T. Stepto, *Makromol. Chem.* **190,** 549 (1989).
10. K. Dodgson, C. J. C. Edwards, and R. F. T. Stepto, *Br. Polym. J.* **17,** 14 (1985).
11. J. L. Hill and R. F. T. Stepto, *Preprints IUPAC International Symposium on Macromolecules,* Helsinki, 1972, Vol. 3, p. 325.
12. R. K. Dewan and K. E. van Holde, *J. Chem. Phys.* **39,** 1820 (1963).
13. R. Varoqui, M. Daune, and L. Freund, *J. Chim. Phys.* **58,** 394 (1961).
14. H. U. ter Meer, W. Burchard, and W. Wunderlich, *Colloid Polym. Sci.* **258,** 675 (1980).
15. W. Burchard and M. Schmidt, *Macromolecules* **14,** 210 (1981).
16. R. F. T. Stepto, in *Integration of Fundamental Polymer Science and Technology,* ed. L. A. Kleintjens and P. J. Lemstra, Elsevier, Amsterdam, 1986, p. 300.

Estimation of Diffusion Coefficients for Small Molecular Penetrants in Amorphous Polyethylene

S. TROHALAKI, A. KLOCZKOWSKI, and J. E. MARK
Department of Chemistry and the Polymer Research Center
University of Cincinnati
Cincinnati, Ohio 45221

D. RIGBY and R. J. ROE
Department of Materials Science and Engineering
and the Polymer Research Center
University of Cincinnati
Cincinnati, Ohio 45221

ABSTRACT

Molecular dynamics simulation has been performed with an amorphous assembly of n-alkane-like chains together with a low concentration of spherical Lennard-Jones molecules, in order to model the diffusion of small molecular penetrants in polyethylene. The chain molecules, previously shown to exhibit a number of realistic properties, were subject to potentials restricting bond lengths, bond angles, and dihedral angles. All polymer segments and penetrant particles also interacted according to truncated Lennard-Jones potentials. Lennard-Jones parameters appropriate for CO_2 were initially chosen for the free particles, and to investigate the effect of penetrant size, values of r^* appropriate to He and Ar were employed, holding all other parameters constant. Self-diffusion coefficients were evaluated for all three penetrant sizes at 240 K, and additionally, temperature was varied between 240 and 421 K for the largest and smallest penetrants. Simple Arrhenius behavior was found, with an activation energy independent of penetrant size, in agreement with recent experimental findings. Possible mechanisms for penetrant diffusion are discussed.

INTRODUCTION

Diffusion of gases in polymers is an important, and in many cases, controlling factor in a variety of applications, such as protective coatings, membrane separation processes, food packaging, and biomedical devices. The theoretical understanding of diffusion of gases in polymers is, however, very limited.

The diffusion process is formally described from a macroscopic viewpoint by Fick's two laws [1], which apply to homogeneous and isotropic polymers. Integration of the first law for the desired geometry and boundary conditions yields the total rate of diffusion of a penetrant gas through a polymer. Gas permeation is commonly described in terms of a "solution-diffusion" mechanism, and the diffusion of the penetrant in the polymer is the rate-determining step in the permeation process. When the solubility of the penetrant in the polymer is low, the diffusion coefficient may become independent of concentration.

During the last few decades, considerable effort has been made to develop a microscopic description of gas diffusion in polymers more detailed than Fick's continuum viewpoint. The various models developed to describe the diffusion of small molecules in polymers can be classified as either molecular models or free-volume models. Whereas free-volume models attempt to elucidate the relationship between the diffusion coefficient and the free volume of the system without consideration of a microscopic description, molecular models analyze specific penetrant and chain motions together with the pertinent intermolecular forces.

Molecular models commonly assume that fluctuating microcavities or "holes" exist in the polymer matrix. At equilibrium, a definite size distribution of microcavities is established on a time-average basis. A hole of sufficient size may contain a dissolved penetrant molecule, which can "jump" into a neighboring hole once it acquires sufficient energy. Diffusive motion results only when holes that have become vacant in this manner are occupied by other penetrant molecules.

Molecular models [2–7] include these characteristics largely to describe the Arrhenius behavior of diffusion coefficients observed experimentally at a temperature well above the glass transition:

$$D = D_0 \exp\left(-\frac{E_D}{RT}\right) \qquad (1)$$

where E_D is the apparent energy of activation for diffusion, D_0 is a constant, R is the universal gas constant, and T is the absolute temperature. The most advanced molecular theory of diffusion by Pace and Datyner [7] incorporates the main features of earlier models of Brandt [4] and Di Benedetto [5,6]. This theory, recently criticized by us [8], assumes that the polymer is composed of bundles of chains with four adjacent parallel chains forming tubes, and the activation energy of diffusion equals the minimum energy for symmetric separation of two chains to permit free passage of a penetrant molecule.

Free-volume models [9–13] are based on Cohen and Turnbull's theory [14] for diffusion in hard-sphere liquids in which the total free volume is assumed to have two contributions. One arises from molecular vibrations and cannot be redistributed without a large energy change, and the second is in the form of discontinuous voids. Diffusion is assumed to result from a redistribution of voids of free volume caused by random fluctuations in local density and is not due to a thermal activation process. An estimation of free volume for polymers based on rotational isomeric state model calculations has been proposed [15]. All the molecular and free-volume theories have been reviewed by Frisch and Stern [16].

The existing theories have not yet been powerful enough to predict diffusion coefficients, permeability, and selectivity for specific polymer-penetrant systems given only their chemical structures. There is a growing demand for better theoretical understanding of the diffusion of gaseous penetrants in polymers to achieve improvements in designing new, better membranes. Ultimately, a tailorlike preparation of an optimal membrane for a specific separation process would be possible. A better insight into this phenomenon might be gained by computer simulations. With the increasing power of available computer systems it has become possible to simulate polymer structure using Monte Carlo and molecular dynamics methods [17,18].

Because diffusion is a dynamic problem, the obvious approach in calculation of the diffusion coefficients is through molecular dynamics. The alternative stochastic Brownian dynamics method [19,20] requires the diffusion coefficients as a known quantity to perform simulations. Diffusion can be studied by either equilibrium or nonequilibrium molecular dynamics. The latter can be used to study steady-state diffusion as well as nonsteady-state problems [21]. Equilibrium molecular dynamics simulations of diffusion in simple binary mixtures have been performed by several authors [22–26].

Molecular dynamics of polymers is a difficult problem in itself and has become feasible only in the last few years. The first attempt was made by Ryckaert and Bellemans [27], who simulated liquid alkanes. All other molecular dynamics simulations of polymers [28–33] were based on alkanes, the most recent one by Rigby and Roe [28] utilizing molecules containing up to 50 methylene units. Longer chains (up to 200 monomers) were studied by extended Monte Carlo simulations of self-avoiding walks on a diamond lattice [34]. By cooling the system in a stepwise fashion, both the glassy and rubbery states can be studied with molecular dynamics [28]. This enables us to study diffusion coefficients as a function of temperature. The diffusion coefficient changes rapidly as the glass transition temperature is approached. We have performed an equilibrium molecular dynamics simulation of an asembly of n-alkane-like chains together with a low concentration of free one-center Lennard-Jones particles with parameters corresponding to CO_2 molecules. Also, to investigate the effect of the size of the free particles, simulations were performed employing values of the Lennard-Jones radii for the free particles appropriate for He and Ar, keeping all other parameters constant. In the present work simulations were performed well above the glass transition temperature.

The self-diffusion coefficients were determined for the CO_2 molecules, and for the hypothetical penetrant particles of the same mass and Lennard-Jones ϵ^* parameter as CO_2 but with varying r^*, at 240 K. Also, the self-diffusion coefficients were obtained for the largest and smallest free particles at temperatures ranging from 240 to 420 K, enabling calculation of activation energies. Mutual diffusion is a collective process and therefore is much more difficult to study than self-diffusion. The mutual diffusion coefficient is, however, approximately equal to the self-diffusion coefficient of the dilute component when the latter concentration is low, as is the case in this study, and the two become identical in the limit of vanishing concentration.

MODEL FOR SIMULATIONS

The simulation model, developed originally by Rigby and Roe [28], models individual chains as a sequence of 20 spherical CH_2 segments connected by springlike valence bonds subject to the potential-energy function

$$E_b = \tfrac{1}{2}k_b(l - l_0)^2 \tag{2}$$

where l_0 is the equilibrium bond length. The deformation of the bond angle θ between successive pairs of bonds from its tetrahedral value θ_0 is governed by a potential quadratic in $\cos \theta$,

$$E_\theta = \tfrac{1}{2}k_\theta(\cos \theta - \cos \theta_0)^2 \tag{3}$$

The dihedral angle ϕ, defined by three successive bonds, is constrained to lie mainly in the trans and gauche states by a threefold torsional potential

$$E_\phi = k_\phi \sum_{n=0}^{5} a_n \cos^n \phi \tag{4}$$

Nonbonded interactions between segments in different chains, between segments separated by more than three bonds along the chain backbone, between segments and free penetrant particles, and between penetrant particles are evaluated according to a truncated Lennard-Jones potential

$$E_{nb} = \begin{cases} 4\epsilon^* \left[\left(\dfrac{r^*}{r_{ij}} \right)^{12} - \left(\dfrac{r^*}{r_{ij}} \right)^{6} \right] + C & \text{for } r_{ij} \le 1.5r^* \\ 0 & \text{for } r_{ij} > 1.5r^* \end{cases} \tag{5}$$

The simulation consists of integration of the equations of motion of the individual segments and free particles

$$m_i \frac{d^2 \mathbf{r}_i}{dt^2} = \mathbf{F}_i = -\nabla_i E$$

where $\mathbf{r}_i$ is the position vector of the center of segment/particle i of mass m_i, $\mathbf{F}_i$ denotes the force on segment/particle i due to interactions with other segments/particles, and E is the total potential energy of the collection of segments comprising the contributions expressed by Eqs. (2) to (5). The method originally proposed by Verlet [35] was used for the numerical integration of the equations of motions.

$$\mathbf{r}_i(t + \Delta t) = -\mathbf{r}_i(t - \Delta t) + 2\mathbf{r}_i(t) + \frac{\mathbf{F}_i(t)(\Delta t)^2}{m_i} \tag{6}$$

and

$$\mathbf{v}_i(t) = \frac{\mathbf{r}_i(t + \Delta t) - \mathbf{r}_i(t - \Delta t)}{2\Delta t} \tag{7}$$

where $\mathbf{v}_i$ is the velocity of segment/particle i and Δt denotes the basic time step of the integration. From Eq. (6) it is evident that the derivation of the configuration of the system at time $t + \Delta t$ requires knowledge of its configuration at time $t - \Delta t$ as well as at time t.

As done in previous work [28], the parameters k_b, l_0, k_θ, and so on, defining the interaction potentials in Eqs. (2)–(5) were chosen such that the chain model will mimic a polyethylene molecule, and the segment effectively represents a CH_2 group. Following Helfand et al. [3], the spring constant k_b for bond-length stretching in Eq. (2) was weakened by a factor of about 7 from the realistic value. This was necessary because the use of a realistic value of k_b would have required too small a time step in the integration of equations of motion.

The simulation system consisted of a cubic box containing 500 CH_2 segments (25 chains) and four free penetrant particles. Surface effects were eliminated in the customary manner by employing periodic boundary conditions. The simulations were performed at constants N, V, and E with Δt set equal to 1.005×10^{-14} s. Each simulation was run for 100,000 time steps for a duration of 1005 ps. The Fortran programs for the simulation were run on the Cray Y-MP computer at the Ohio Supercomputer Center.

RESULTS AND DISCUSSION

The self-diffusion coefficient of a component in a mixture, D_α ($\alpha = 1, 2$), can be calculated from the Einstein relation

$$D_\alpha = \frac{1}{6N_\alpha} \lim_{t \to \infty} \frac{d}{dt} \sum_{i=1}^{N_\alpha} \langle [\mathbf{r}_i(0) - \mathbf{r}_i(t)]^2 \rangle \qquad (8)$$

where $\mathbf{r}_i$ is the position vector of particle, N_α is the number of particles of type α, and the angular brackets denote averaging over all choices of time origin. Equation (8) is equivalent to the formulation of D_α based on the velocity autocorrelation function [37,38].

Under each set of conditions, defined by the temperature, density, and value of r^*, we have performed between three and five consecutive simulations, each of duration 1005 ps, with coordinates saved every 25 ps. These data were used to compute mean-square displacements (MSDs) as a function of time, with averaging over all particles and time origins. Figures 1 and 2 show the resulting behavior for the polymer segments at temperatures of 240, 300, 361 and 421 K for systems in which the penetrant size, r^*, is fixed at 0.418 nm and 0.258 nm, respectively. The drawn curves represent mean values computed from the consecutive trajectories at a given ρ, T, r^*, while the error bars shown at intervals illustrate the standard deviations of the same data. The increase in relative error with time results almost entirely from the linear decrease in the number of time origins used in computing the MSD as time increases. In all cases, following an initial period of 30 to 50 ps during which chain segments move an average of 1 nm, essentially linear behavior is observed at long times. The curves for the two sizes of penetrant molecule are almost identical at the extreme temperatures, 240 and 421 K, while the agreement is less satisfactory at the intermediate temperatures. The corresponding MSDs of the penetrant particles are shown in Figs. 3 and 4, which again exhibit linear behavior, deteriorating only slightly at the longest times. In the

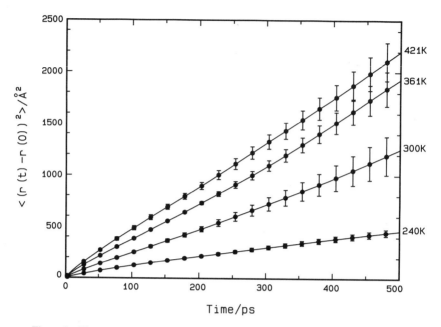

Figure 1   Mean-square displacements of polymer segments as a function of time in system containing 25 eicosane molecules and four Lennard-Jones penetrant particles. From top to bottom, curves were obtained at 421, 361, 300, and 240 K. Error bars depict standard deviations of data from three to five consecutive runs at each temperature. Penetrant Lennard-Jones parameters are $r^* = 4.18$ Å, $\epsilon^* = 1650$ J/mol.

case of penetrant, the initial curvature in the MSDs noted in Figs. 1 and 2 is no longer visible in Figs. 3 and 4.

Self-diffusion coefficients obtained from the data in Figs. 1 to 4, calculated according to Eq. (8), are listed in Table 1. These data are also plotted in the form of Arrhenius plots in Fig. 5. Activation energies E_D calculated from Fig. 5 lie in the range 6 to 7.5 kJ/mol for the penetrant and 6.8 to 7.5 kJ/mol for the polymer. Further, while the magnitude of D decreases with increasing size of the penetrant as expected, the activation energies for the diffusion of different-sized penetrants are comparable, the small difference observed being barely outside the range of uncertainty in the data (about 0.5 kJ/mol).

The present simulation results display a number of interesting features with regard to the activation energies for diffusion of penetrant and polymer. First, the temperature dependence of the diffusion coefficients is well represented by Arrhenius plots, indicating that the activation energies are fairly independent of temperature. Second, there is no obvious dependence of penetrant E_D on the size of the gas molecule when this is varied from 0.258 nm to 0.418 nm. Third, the activation energy for self-diffusion of polymer exceeds that of the penetrant only marginally, by some 500 J/mol. Finally, the absolute

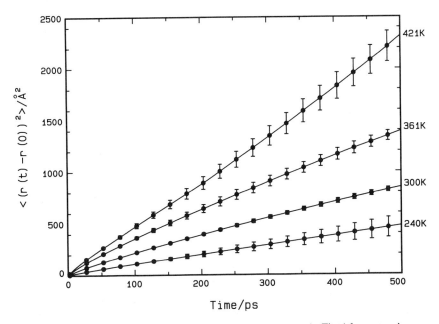

Figure 2 Mean-square displacements of polymer segments. As Fig. 1 for system in which the Lennard-Jones parameters of the penetrant are $r^* = 2.58$ Å, $\epsilon^* = 1650$ J/mol.

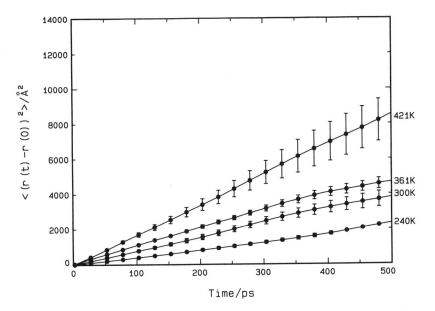

Figure 3 Mean-square displacements of penetrant particles for the system with $r^* = 4.18$ Å, $\epsilon^* = 1650$ J/mol. Temperature as in Fig. 1.

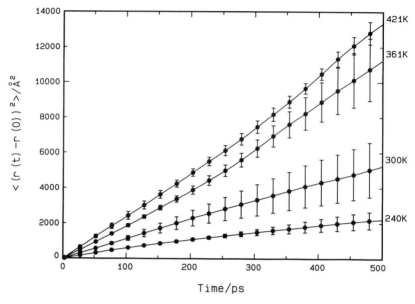

Figure 4 Mean-square displacements of penetrant particles for the system with $r^* = 2.58$ Å, $\epsilon^* = 1650$ J/mol. Temperatures are as in Fig. 1.

value of all activation energies is of the same order of magnitude as that associated with bond conformational transitions, which is estimated to be about 10.6 kJ/mol in the present systems. [39].

    Experimental determination of mutual diffusion coefficients of gases H_2, CO, and CO_2 in the alkanes C_{16}, C_{20}, and C_{28} at pressures of about 1.4 and 3.4

TABLE 1   Self-Diffusion Coefficients for Polymer Segment and Various Penetrants[a]

T (K)	$D_{polymer}$	$D_{penetrant}$
	$r^* = 0.418$ nm	
421	$0.7083 \pm .0362$	2.823 ± 0.257
361	$0.6013 \pm .0208$	1.781 ± 0.054
300	$0.3848 \pm .0431$	1.338 ± 0.093
240	$0.1556 \pm .0112$	0.716 ± 0.031
	$r^* = 0.258$ nm	
421	$0.7417 \pm .0416$	$4.387 \pm .125$
361	$0.4708 \pm .0125$	$3.399 \pm .179$
300	$0.2736 \pm .0097$	$1.867 \pm .405$
240	$0.1556 \pm .0611$	$0.8244 \pm .078$
	$r^* = 0.342$ nm	
240	$0.1280 \pm .0222$	$0.753 \pm .042$

[a] Units are Å^2/ps.

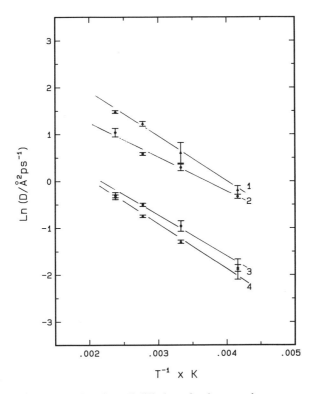

Figure 5   Arrhenius plots for self-diffusion of polymer and penetrant: curve 1, penetrant with $r^* = 2.58$ Å; curve 2, penetrant with $r^* = 2.58$ Å; curve 3, polymer in the same system as curve 2; curve 4, polymer in the same system as curve 1.

mPa and temperatures between 323 and 564 K was reported by Akgerman et al. [40–42]. Analysis of their data shows Arrhenius behavior in all systems, with essentially the same E_D value for the three penetrants. For example, E_D values in the C_{20} system are 13.5, 13.9, and 13.5 kJ/mol for H_2, CO, and CO_2, respectively. The effect of pressure is very small, and the difference in the diffusion coefficient between the two pressures is in most cases barely beyond the experimental error. These authors also studied diffusion of the alkanes C_8, C_{10}, C_{12}, and C_{14} in C_{16} [43]. The data again conform to simple Arrhenius behavior, with activation energies scattered in the range 14.7 to 15.3 kJ/mol (but with no clear systematic trends visible). This suggests that the activation energy for self-diffusion in pure C_{20} liquid might also be on the order of 15 kJ/mol. The results of experimental study of diffusion of gas molecules in polyethylene present a picture that is somewhat different from the diffusion in alkane systems. The activation energy of diffusion of CO_2 in the amorphous regions of both low-density and high-density polyethylenes (determined at temperatures 5 to 60°C) [44,45] lie in the range 35 to 38 kJ/mol, which is larger than the value in alkanes by more than a factor of 2. Moreover, with polyethylene, the E_D value depends on penetrant size, the dependence being described as proportional either to the diameter [44] or to the molecular volume [45], whereas with alkanes E_D is independent of penetrant size.

The activation energies for penetrant diffusion, 6 to 7.5 kJ/mol, found in this work by simulation are smaller by almost a factor of 2 than those, 13.5 to 13.9 kJ/mol, obtained experimentally with alkanes. The absolute values of the diffusion coefficients found by simulation are also somewhat higher; for example, for CO_2 at 361 K the simulation gives 17.8×10^{-9} m²/s, while the experimental value (extrapolated from higher-temperature data) [41] is 4.44×10^{-9} m²/s. The major factor most likely to have given rise to these discrepancies is the difference in liquid density. The simulated alkane system, originally studied by Rigby and Roe [28] and used directly in this work, exhibits a density lower than that observed experimentally (due to the truncation of the non-bonded Lennard-Jones interaction potential at a fairly short distance). Thus at 361 K the simulated system has a density of 0.576 g/cm³, whereas the experimental C_{20} system has a density of about 0.741 g/cm³ at 1.38 MPa pressure. The much larger free volume available in the simulated system is probably responsible for the reduced activation energy and increased diffusion coefficient. The much larger activation energy of gas diffusion in polyethylene compared to that in alkanes may probably be ascribed similarly to the higher density and smaller free volume in polyethylene.

Another subtle factor that might have contributed to the lower E_D value in the simulated system is conceivable. In our alkane models the bond stretching potential was weakened by a factor of 7 to reduce the bond stretching vibrational frequency and thereby to allow larger MD integration steps. Such a more flexible bond could have resulted [46] in a slightly enhanced conformational transformation and local segmental motions, which in turn would have allowed easier transport of penetrant gas molecules. Such a consideration is, however, merely speculative at this time and quantitative data to substantiate it are not available.

The difference in experimentally observed behaviors in gas diffusion between polyethylene and alkanes (i.e., the larger activation energy and the dependency on the penetrant size in polyethylene) is puzzling. One possible factor that gives rise to this difference is the presence of crystalline fractions in polyethylene, which restrict the movement of amorphous chains tied to them. The only other obvious difference is the much higher density of polyethylene.

More work is evidently needed to understand fully the mechanism of gas diffusion in amorphous polymers. The most important question raised in this work is on the apparently overriding effect of the density (hence the free volume) in determining the activation energy and penetrant size dependency. Molecular dynamics simulation offers a convenient tool to study such an effect, since in simulation, unlike in experimentation, the density of the system can easily be altered without altering other parameters at the same time. Similarly, it should be worthwhile to study the effect of penetrant size in a more dense system to see whether the penetrant size dependency arises when the "average free volume size" is comparable to or smaller than the penetrant gas molecules. The effect of the bond stretching potential on the activation energy of gas diffusion also has to be investigated. Some of these questions will be taken up in the continuation of the present study.

ACKNOWLEDGMENTS

This work was supported in part by NSF Grant DMR-8520921 awarded to RJR and Gas Research Institute Grant 5082-260-0666 awarded to JEM. We acknowledge computer time at the Ohio Supercomputer Center. Some early results were also obtained at the Pittsburgh Supercomputer Center.

REFERENCES

1. J. Crank. *The Mathematics of Diffusion,* 2nd ed., Clarendon Press, Oxford, 1975.
2. P. Meares, *J. Am. Chem. Soc.* **76,** 3415 (1954).
3. R. M. Barrer, *J. Phys. Chem.* **61,** 178 (1957).
4. W. W. Brandt, *J. Phys. Chem.* **63,** 1080 (1959).
5. A. T. DiBenedetto, *J. Polym. Sci.* **A1,** 3459, 3477 (1963).
6. A. T. DiBenedetto and D. R. Paul, *J. Polym. Sci.* **A2,** 1001 (1964).
7. R. J. Pace and J. Datyner, *J. Polym. Sci., Polym. Phys. Ed.* **17,** 437, 453, 465 (1979).
8. A. Kloczkowski and J. E. Mark, *J. Polym. Sci., Polym. Phys. Ed.* **27,** 1663 (1989).
9. H. Fujita, *Fortschr. Hochpolym.-Forsch.* **3,** 1 (1961).
10. H. L. Frisch, D. Klempner, and T. K. Kwei, *Macromolecules* **4,** 237 (1971).
11. S. A. Stern, S. S. Kulkarni, and H. L. Frisch, *J. Polym. Sci., Polym. Phys. Ed.* **21,** 467 (1983).
12. S. A. Stern, S.-M. Fang, and H. L. Frisch, *J. Polym. Sci., Part A-2* **21,** 201 (1972).
13. J. S. Vrentas and J. L. Duda, *J. Polym. Sci., Polym. Phys. Ed.* **15,** 403 (1977).
14. M. H. Cohen and D. Turnbull, *J. Chem. Phys.* **31,** 1164 (1959).
15. S. Trohalaki, L. C. DeBolt, J. E. Mark, and H. L. Frisch, *Macromolecules,* **23,** 813 (1990).
16. H. L. Frisch and S. A. Stern, *Crit. Rev. Solid State Mat. Sci.* **11,** 123 (1983).
17. M. P. Allen and D. J. Tildesley, *Computer Simulation of Liquids,* Clarendon Press, Oxford, 1987.
18. A. Baumgartner, *Annu. Rev. Phys. Chem.* **35,** 419 (1984).
19. M. Fixman, *J. Chem. Phys.* **69,** 1527 (1978).
20. E. Helfand, Z. R. Wasserman, and T. A. Weber, *J. Chem. Phys.* **70,** 2016 (1979).
21. W. G. Hoover and W. T. Ashurst in *Theoretical Chemistry,* Vol. 1, ed. H. Eyring and D. Henderson, Academic Press, New York, 1975.
22. G. Jacucci and I. R. McDonald, *Physica* **A80,** 607 (1975).
23. D. L. Jolly and R. J. Bearman, *Mol. Phys.* **41,** 137 (1980).
24. K. T. Toukubo and K. Nakanishi, *J. Chem. Phys.* **65,** 1937 (1976).
25. M. Schoen and C. Hoheisel, *Mol. Phys.* **52,** 33 (1984).

26. C. Hoheisel and R. Vogelsang, *Comput. Phys. Rep.* **8,** 1 (1988).

27. J. P. Ryckaert and A. Bellemans, *Chem. Phys. Lett.* **30,** 123 (1975); *Faraday Discuss. Chem. Soc.* **66,** 95 (1978).

28. D. Rigby and R. J. Roe, *J. Chem. Phys.* **87,** 7285 (1987); **89,** 5280 (1988).

29. J. M. R. Clarke and D. Brown, *Mol. Phys.* **58,** 815 (1986).

30. T. A. Weber, *J. Chem. Phys.* **69,** 2347 (1978); **70,** 4277 (1979).

31. T. A. Weber and E. Helfand, *J. Chem. Phys.* **71,** 4760 (1979); *J. Phys. Chem.* **87,** 2881 (1983).

32. W. L. Jorgensen, *J. Phys. Chem.* **87,** 5304 (1983).

33. D. Chandler, S. L. Hollengren, and Y. A. Montgomery, *J. Chem. Phys.* **73,** 3688 (1980).

34. K. Kremer, *Macromolecules* **16,** 1632 (1983).

35. L. Verlet, *Phys. Rev.* **159,** 98 (1967).

36. E. Helfand, Z. R. Wasserman, and T. Weber, *Macromolecules* **13,** 526 (1980).

37. J. O. Hirschfelder, C. F. Curtiss, and R. B. Bird, *Molecular Theory of Gases and Liquids,* Wiley, New York, 1954.

38. P. J. Hansen and I. R. McDonald, *Theory of Simple Liquids,* Academic Press, New York, 1976, p. 239.

39. R. J. Roe and D. Rigby, unpublished work.

40. M. A. Mathews, J. B. Rodden, and A. Akgerman, *J. Chem. Eng. Data* **32,** 319 (1987).

41. J. B. Rodden, C. Erkey, and A. Akgerman, *J. Chem. Eng. Data* **33,** 344 (1988).

42. J. B. Rodden, C. Erkey, and A. Akgerman, *J. Chem. Eng. Data* **33,** 450 (1988).

43. M. A. Matthews, J. B. Rodden, and A. Akgerman, *J. Chem. Eng. Data* **32,** 317 (1987).

44. A. S. Michaels and H. J. Bixler, *J. Polym. Sci.* **50,** 413 (1961).

45. R. Ash, R. M. Barrer, and D. G. Palmer, *Polymer* **11,** 421 (1970).

46. E. Helfand, Z. R. Wasserman, T. A. Weber, J. Skolnick, and J. H. Runnels, *J. Chem. Phys.* **75,** 4441 (1981).

18

Applications of a Pseudokinetic Algorithm for Polymer in the Continuum

CHRISTIAN M. LASTOSKIE and WILLIAM G. MADDEN
Central Research Department and Polymer Products Department
Experimental Station
E.I Du Pont de Nemours and Company
Wilmington, Delaware 19880

ABSTRACT

A pseudokinetic simulation technique, previously employed for polymers on the lattice, is extended to a limited class of polymer models in the continuum. The method invokes a rapid exchange of mass among the chains on the time scale of the motion of individual beads and consequently explores the conformations of individual chains much more rapidly than in conventional simulations. The continuum models are designed so that the specified kinetic processes can proceed athermally. The distribution of molecular weights obtained is independent of thermodynamic state and identical to the results obtained using the corresponding lattice algorithm. The intra- and intermolecular radial distribution functions calculated using the pseudokinetic algorithm are found to be in quantitative agreement with those obtained from a conventional simulation. The chain centers-of-mass radial distribution function is obtained for the first time in a polymer melt. The statistical precision of this quantity is enhanced considerably by the pseudokinetic elements of the algorithm. The properties of melt polymer confined between hard parallel plates are also examined. The structural detail on the length scale of individual beads is much richer than on the lattice, but global properties of the chains are found to be very similar to earlier lattice results. The properties of a melt polymer slab in the continuum are presented for the first time. No new short-range structure is observed, and the global reorganization of the chains near the free melt interface is again in close agreement with lattice results. The equilibrium density in the interior of the melt slab is unexpectedly low. This may be a result of the truncation of the interbead potential.

INTRODUCTION

The computer simulation of polymers is plagued by a broad range of relaxation times, ranging from subpicosecond intramolecular vibrations to glassy dynamics which are painfully slow on the time scale of human activity. In any dynamical computer simulation, one must therefore choose a model that samples configuration space efficiently on the time scale of interest, whether the method chosen involves the rigorous motion of the molecular dynamics method or the coarse-grained motions of suitable Monte Carlo algorithms. This often means that the model chosen implicitly preaverages motion on short time scales and that one simply declines to extend the simulation run to still longer times. However, even purely dynamical analyses presume that an average is taken over a large number of independent equilibrium starting configurations. In dense polymeric systems, such averages are achieved either by examining very large samples, containing many spatially uncorrelated microdomains, or by running very long simulations, in which configurations widely separated in time may be regarded as temporally uncorrelated and statistically independent.

When the dynamics of a model are not themselves at issue, it is wholly unnecessary to sample configurations on any time scale relevant to real polymers. The pioneering single-chain simulations of Wall and co-workers [1] and of Rosenbluth and Rosenbluth [2] produced new chain configurations via self-avoiding random walks, and the successive configurations thus achieved were statistically uncorrelated. The computer effort required to produce such configurations is thus unrelated to any real polymeric motion. However, the well-known attrition problem limits these methods to very short chains or to dilute systems. For long chains or high density, polymer Monte Carlo methods are usually based on perturbed configuration methods similar in spirit to those used for small molecule liquids. The earliest use of such methods was by Verdier and Stockmayer [3] who applied it to single-chain dynamics on the lattice. In the ensuing years, a dizzying array of on- and off-lattice variations have been employed, nearly all of which share the feature that successive acceptable configurations differ from one another only locally, and many successful Monte Carlo transitions are therefore required to rearrange the overall configuration of even a single chain. At high density, a significant fraction of all successful moves are partial reversals of earlier moves, and the net rate of configurational readjustment is further diminished. The local nature of movement in these traditional polymer Monte Carlo simulations deliberately echos the expected motion in real polymers and thus provides for a coarse-grained dynamical interpretation of the Monte Carlo sequence. If such an interpretation is at all valid, the effective rate at which such simulations sample the configuration space of a collection of densely packed chains is ultimately determined by the perversely slow dynamics of real polymeric media.

In 1978, Olaj et al. [4] introduced a lattice simulation method designed to sample polymer configurations on a completely occupied lattice, where no mass movement is possible. Mansfield [5] later independently constructed a similar simulation technique. The essential idea is that when reactions are possible, chemical equilibrium is part and parcel of an overall thermodynamic equilibrium. If the energy of the transition state (col) is much higher than that of typical equilibrium configurations, only the relative magnitudes of the appropriate rate constants contribute to the final equilibrium properties. For polymer simulations, the reactions of interest are those in which chains abstract mass from one another. Some of the commonly postulated processes are similar to those known to occur in polymer melts; others are quite arbitrary. On the fully occupied lattice, the rate constants for mass interchange define the only meaningful time scale in the simulation. In practice, the detailed kinetics are regarded as irrelevant, and only static equilibrium averages are taken. For this reason, Madden [6] dubbed the reactive elements of the method "pseudokinetic." The technique has been used to confirm the accuracy of the Flory theorem for melt polymer [4,5] and to explore the conformations of chains at the interface between amorphous and crystalline domains [5].

Because the pseudokinetic events can produce frequent and dramatic changes in the global properties of individual chains, Madden [6,7] coupled pseudokinetics to ordinary mass motion for application to inhomogeneous polymer melts on a lattice. A detailed analysis of the configurational properties of such systems requires a spatial partitioning of the simulation sample into small slabs parallel to the interface of interest. The rapid reorganization of the chains via the pseudokinetic mechanisms leads to vastly improved statistical precision within these small subsamples. The method was applied to a polymer film adsorbed on an unstructured planar surface, to polymer confined between flat plates, and to bulk polymer [6,7]. Tests against simulations with the pseudokinetic elements suppressed revealed that properties which depend on the global configurations of chains (e.g., the radius of gyration) are sampled two orders of magnitude more efficiently at a volume fraction of 0.75 when the pseudokinetic events are included [7]. This sampling advantage must necessarily increase at higher volume fractions because all lattice methods that rely solely on the motion of individual beads become completely ineffectual as the volume fraction approaches unity. Using this technique, it has proven possible to execute reliable simulations of dense lattice polymer on a microcomputer of modest capacity (i.e., a MicroVAX II). For many purposes, useful results can be produced on a daily basis.

Traditionally, lattice models are applied to polymers because the discrete nature of the space and the corresponding quantization of the volume fluctuations allow for a more rapid sampling of polymer configurations than would be possible at high density in the continuum. The difficulties associated with the simulation of dense polymer melts in the continuum are well documented. Such simulations usually require the use of either a supercomputer or a nearly dedicated mainframe. Clearly, the incorporation of pseudokinetics into a continuum simulation is potentially even more advantageous than on a lattice. We develop here a pseudokinetic algorithm suitable for application to a limited, but important, class of polymer models in the continuum. The introduction of these techniques makes possible the simulation of polymer in the continuum on the current generation of workstations (e.g., a DECstation 3100). A preliminary description of this algorithm has appeared elsewhere [8].

MODEL SYSTEMS

Pseudokinetic mechanisms are easily accommodated on a lattice with nearest-neighbor interactions only because each kinetic event leads to no net change in the numbers of either the covalent or cohesive contacts among the beads and is thus strictly athermal. For an arbitrary model polymer in the continuum, it is not possible to specify useful athermal kinetic mechanisms because the energetics of covalent bonds are generally very different from that of the cohesive interactions. However, if the model is constructed so as to ensure that there is a significant spatial domain over which the two kinds of interactions are identical, it becomes possible to exchange mass among the chains athermally. Several suitable hard-core model systems are shown in Fig. 1a and b. These are variations of a tethered chain, with or without cohesive forces. Each includes a covalent potential $v(r)$ acting between bonded neighbors, and a cohesive potential $u(r)$ acting between all other pairs of beads. For convenience, the minimum of $v(r)$ is made identical to that of $u(r)$. Of course, molecular stability always arises from the fact that bonding potentials are much more negative than normal intermolecular interactions. In this work the total number of covalent bonds is preserved throughout the simulation, and the value given to the minimum in $v(r)$ represents a thermodynamically irrelevant shift in the zero of energy. When the cohesive forces are either hard sphere or square well, athermal reactions may occur over the full breadth of the corresponding covalent potential. From a practical point of view, this has the advantage that the covalent potential may be made wide enough to ensure that the pseudokinetics proceeds efficaciously.

It is also possible to devise suitable models with fully continuous potentials. Two such models, based on a Lennard-Jones fluid, are shown in Fig. 1c

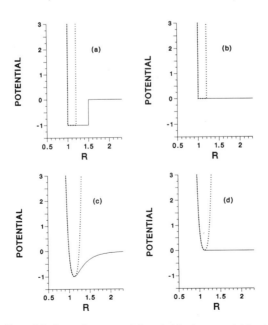

Figure 1 Potentials for polymer models suitable for pseudokinetic simulation.
———: $\mathbf{u}(r)$; $\cdots\cdots$: $\mathbf{v}(r)$.

and d. In Fig. 1c, the cohesive interactions are given by a Lennard-Jones potential,

$$u(r) = 4\epsilon\left[\left(\frac{\sigma}{r}\right)^{12} - \left(\frac{\sigma}{r}\right)^{6}\right]$$ (1)

and the covalent potential is obtained by reflecting the cohesive potential about its minimum at $r = r_m$,

$$v(r) = \begin{cases} u(r) & r < r_m \\ u(2r_m - r) & r_m < r < 2r_m \\ \infty & r > 2r_m \end{cases}$$ (2)

In Fig. 1d, a corresponding construction is made from a purely repulsive interaction known in the liquids literature as the WCA reference potential [9]:

$$u(r) = \begin{cases} 4\epsilon\left[\left(\frac{\sigma}{r}\right)^{12} - \left(\frac{\sigma}{r}\right)^{6}\right] + \epsilon & r < r_m \\ 0 & r > r_m \end{cases}$$ (3)

The covalent potential is again constructed according to Eq. (2). At either high density or high temperature, the underlying structure of the WCA reference fluid is nearly identical to that of the parent Lennard-Jones fluid. For either model, athermal kinetics will be possible only when both the bond to be formed and the bond to be broken are less than r_m (here, $2^{1/6}\sigma$). Since the repulsive core prevents close approach of two bead centers, the effective range over which kinetics can occur is roughly 0.9σ to 1.12σ. At melt densities, it transpires that this range is large enough to permit the kinetics to proceed at a reasonable rate. All results reported here are for either the Lennard-Jones model or for the WCA analog. When the Lennard-Jones model is used, the potential is cut off (without shift) at 2.5σ.

SIMULATION ALGORITHM

In the lattice simulations of Madden [6,7], mass movement took place via the reptation-like step proposed by Wall and Mandel [10]. Thus it was convenient to include pseudokinetic steps which were also end initiated. An end bead was selected, as was a neighboring site. If the neighboring site was empty, a reptation was attempted. Otherwise, one of four kinetic processes was attempted. As a result, the movement of beads on the lattice was intimately intermingled with the kinetic processes. At low-volume fractions, bead motion was more frequent, but at high-volume fractions, the kinetic events were dominant. Despite this, the resulting distribution of molecular weights proved completely independent of temperature, volume fraction, or sample confinement. This is a strong indication that the details of the molecular configurations do not affect the pseudokinetics and makes plausible the contention that pseudo-kinetics do not themselves bias the sampling of molecular configurations. As noted above, it was possible, on the lattice, to confirm the latter assertion via exhaustive tests [7]. Because of the difficulties associated with the conventional simulation of melt polymer, direct tests of the pseudokinetic methods for continuum models will be more limited in scope. Thus indirect assessments, such as the insensitivity of the distribution of molecular weights to thermodynamic parameters and geometrical constraints, are especially important.

In the continuum version of the pseudokinetic algorithm, the movement of mass and the pseudokinetics are compartmentalized. The algorithm proceeds via an alternating sequence of movement and pseudokinetic cycles:

Movement Cycle

Beads are selected randomly and subjected to random Cartesian displacements as in conventional Monte Carlo. Moves are accepted according to the usual canonical criterion. Because of the large size of the system, a cell method is used for keeping track of possible interacting neighbors. The edge length of a bookkeeping cell is normally set in the neighborhood of one bead diameter. This process is continued for L_m beads selections followed by entry into the kinetic cycle.

Kinetic Cycle

The pseudokinetic processes employed in this work are identical to those used in the lattice simulations [6,7]: (1) an end bead attacks an interior bead of a different chain, abstracting that portion of the other chain on a randomly chosen side of the victim bead; (2) an end bead attacks an interior bead on the same chain, forming a new covalent bond, and severs whichever covalent bond of the victim bead prevents ring formation; (3) an end bead attacks an end bead of a different chain, forming a transient superchain which is then severed at a randomly chosen bond; and (4) an end bead attacks the other end of the same chain forming a transient ring which is then opened at a randomly chosen bond.

Ends are randomly chosen as attacking beads, and potential victims are randomly selected from nearby cells, which include all beads within some *stalking radius* of the attacking end. The stalking radius is set slightly in excess of the athermal attack range for the model. If a suitable victim is found within the attack range, the appropriate kinetic event is attempted and a new attacking end is selected. For each attacking end, a maximum of 50 potential victims are stalked before the search is abandoned. Typically, there are only about 25 beads in the cells examined, so there is a fair probability that any suitable victims present will be discovered. At moderate to high densities, a suitable victim bead is ultimately found for 80 to 95% of all attacking ends, and between 12 and 20 potential victims are examined on average before one is found within the attack radius. An attack is allowed to proceed to conclusion only if the covalent bond to be broken is also shorter than the athermal attack distance. For the hard-core models of Fig. 1, all covalent bonds satisfy this condition, but for the continuous models, only about half the covalent bonds are suitably compressed. A new attacking end is selected after each attempted reaction, regardless of the outcome. The pseudokinetic sequence proceeds for L_k end bead selections, followed by a return to the movement cycle.

In most runs, the kinetic cycle was entered four times in every pass (with a pass defined as one attempted mass movement per bead), and in each kinetic cycle, L_k was between 10 and 30 times the number of end beads. These choices may not be optimal for the simulation conditions presented here. The duration of the movement sequence must be long enough to create new attacker–victim pairs. These will arise less frequently at low densities. Long kinetic sequences make sense only if the new ends just formed have suitable victims other than those to which they had previously been covalently bonded. Otherwise, a large number of kinetic reversals will occur, and little statistical advantage will be gained for all the additional effort. For the continuous models of Fig. 1c and d at melt densities, two to three suitable victims surround a typical end bead. For the hard-core models of Fig. 1a and b, the covalent potential may be adjusted to include additional victims. This condition is density dependent, and there may be a percolation transition associated with the ability of the kinetic processes alone to transport ends throughout the sample. Fortunately, the execution of a pseudokinetic event is nearly as rapid in the continuum as on the lattice, so one can afford to enter the kinetic cycle frequently and attempt a large number of attacks with little impact on the overall execution time of the simulation. Even with the low value of L_m and high values of L_k employed here, the pseudokinetics consume only about 10 to 15% of the computer time in the simulation.

The pseudokinetics necessarily results in an ensemble of microstates with a distribution of polymer molecular weights. For comparison with theories for monodisperse samples, this proves to be a disadvantage. Synthetic polymer melts—even samples subjected to repeated fractionation—do possess a distribution in molecular weights. Thus the pseudokinetic simulation methods provide a means for exploring the effects of polydispersity on the measured properties. The number-averaged molecular weight, M_N, is invariant from microstate to microstate because the number of chains is preserved in the kinetic processes described above. If each interior bead is equally likely to be attacked by an end, the expected distribution of molecular weights is the "most probable distribution" [11], for which $M_w/M_N = 2.0$ and $M_z/M_w = 1.5$. This is, in fact, observed for unrestricted pseudokinetics on the lattice. If kinetic events are allowed to succeed only when the product chains have molecular weights lying in the range $M_L = M_N - \Delta$ and $M_H = M_N + \Delta$, the lattice simulations produced a square mound distribution (to within 1 part in 1000 for M_w/M_N). Figure 2 shows that the off-lattice algorithm behaves in

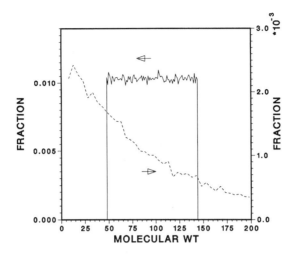

Figure 2 Fraction of chains as function of chain length (molecular weight). - - - -: unrestricted kinetics ($M_w/M_N = 1.93$); ———: kinetics restricted to give products in the range $48 < M < 144$ ($M_w/M_N = 1.085$).

precisely the same way. As in the lattice studies, the distributions do not depend on thermodynamic parameters or on the spatial boundary conditions placed on the sample. Other methods for biasing the distribution have been suggested [4,5] and a variation of the semigrand methods of Kofke and Glandt [12] should also be applicable. With some trial-and-error adjustment of parameters, virtually any distribution of molecular weights should be accessible via a pseudokinetic algorithm.

Ordinarily, one would simulate a polydisperse polymer sample by creating a large number of finite realizations with chain lengths selected by sampling some specified a priori distribution of molecular weights. A conventional simulation would then be performed on each realization, with an equilibration sequence followed by a production run. The pseudokinetics may be regarded as providing an athermal mechanism for moving from one realization to another "on the fly." Because the realization achieved at the end of the kinetic sequence has an energy identical to that of the realization just before entering the sequence, the two states are equally probable and are correctly weighted with respect to one another. The crucial assumption in the pseudokinetic procedure is that the mass distribution in the microstate thus produced is typical of an equilibrium configuration for the new molecular weight distribution and that no relaxation is required after an attack. It is readily established on the lattice that the transition probabilities for a forward and reverse reaction are identical in a completely athermal simulation and are independent of the details of the bead configurations. Thus microscopic reversibility is enforced and the essential issue is whether the process is ergodic. As long as the portion of the algorithm that moves mass is itself ergodic, the overall algorithm ought to be as well. However, this does not *guarantee* that a faithful sampling of chain conformations can be achieved on time scales shorter than that of melt diffusion. The chain conformations accessible to the pseudo-kinetics might be statistically skewed by the local distribution of attackers and victims. This must happen at very low density, where the pseudokinetics eventually becomes "diffusion limited."

Although we employ a Monte Carlo procedure to move mass within the system, the model we have chosen is suitable for molecular dynamics simulation as well. When two beads have penetrated within an attack radius, it is impossible for an outsider examining their local motions to determine whether those beads are interacting via the intermolecular or the intramolecular forces. As a consequence, the phase-space trajectories for realizations with different molecular weight distributions will be rigorously coincident for times comparable to the vibrational period of a covalent bond. At melt densities, an enormous number of realizations will be simultaneously coincident. Thus one may halt a molecular dynamics run, enter a pseudokinetic sequence, and return to the molecular dynamics motion without any readjustment of velocities and without any kinks or cusps in the overall trajectory. Neither energy nor momentum conservation will be in any way affected by the crossover to a new realization. Of course, one may not sensibly attempt dynamical averages which span any pseudokinetic events. However, the postkinetic and prekinetic samples are very different in their connectedness, and if a new time origin is established after each kinetic sequence, one may achieve a faster exploration of the effect of initial starting configurations on dynamical averages. This will prove useful only if the time scale of the dynamical process of interest is much shorter than that of the conformational relaxation of the chains by normal mechanisms.

RESULTS

Figure 3 shows the bead–bead radial distribution function (RDF) obtained for the Lennard-Jones polymer model at a density $\rho^* = N_{bead} \sigma^3/V = 0.75$ and a temperature $T^* = k_B T/\epsilon = 2.50$ and compares it with that of a collection of Lennard-Jones monomers at the same thermodynamic state. The polymer sample contained 72 polymer chains confined to a box with periodic boundary conditions in all directions with $M_N = 96$ and with pseudokinetic constraints such that $48 < M < 144$. The constraints on the molecular weights lead to an ensemble with $M_w/M_N = 1.085$. The polymer simulation included 800 passes and about 33,000 successful attacks per chain end. Despite the "loose" nature of the covalent potential, the polymer sample exhibits a sharper first peak in the RDF and an enhanced secondary structure when compared against the monomeric fluid. This is as expected for such a simple, polymer model, with its high degree of radial symmetry. The persistent secondary structure may not

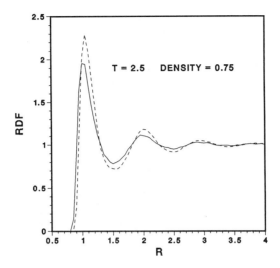

Figure 3 Radial distribution functions. ————: monomeric Lennard-Jones fluid;
– – – – –: Lennard-Jones polymer model (cf. Fig. 1c).

be present when the length of a covalent bond differs substantially from σ or
when other structure breaks the coordination symmetry. In Fig. 4, the polymer
bead–bead radial distribution function is decomposed into intermolecular,
intramolecular, covalent, and noncovalent contributions. All RDFs were obtained by binning bead separations in the usual fashion and analyzing the
results relative to the expected probability density for pairs of ideal gas par-

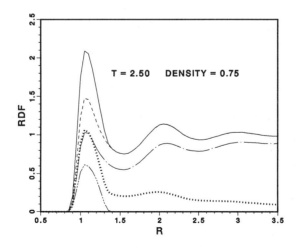

Figure 4 Decomposition of the radial distribution function for Lennard-Jones
polymer. ————: full RDF; – – – – –: noncovalent RDF; —·—·—: intermolecular
RDF; ······: intramolecular RDF; —··—··: covalent RDF.

ticles. The distribution functions so calculated are therefore *mean* RDFs because no distinctions are made on the basis of chain length or position of a bead on the chain.

Of particular interest are the intramolecular bead–bead RDF, which includes all correlations among beads on the same chain, and the intermolecular bead–bead RDF, which counts correlations only among beads on different chains. A recent theory by Schweitzer and Curro [13], based on the RISM integral equation of Chandler and Andersen [14], assumes a particular form for an intramolecular function, $\omega(r) = \rho g^{[\text{INTRA}]}(r)$. Their equation then predicts $g^{[\text{INTER}]}(r)$ using a suitable closure borrowed from the literature of small-molecule liquids. Schweitzer and Curro assume that, in the melt, the function $\omega(r)$ does not depend on density or temperature and can be approximated by some simple, relatively unstructured form. These authors have considered several possible approximations for $\omega(r)$, including a sum of Gaussians suggested by the Flory ideality theorem for bulk melts [11]. None of those approximations for $\omega(r)$ include any detailed short-range structure (apart from a possible absolute exclusion of beads from the core region). Figure 4 shows that there is indeed short-range structure in $g^{[\text{INTRA}]}$ (and hence in ω), but that structure does not persist beyond about two bead diameters. It is likely that the longer-range behavior of $\omega(r)$ is correctly represented by Schweitzer and Curro's Flory-like approximation, but the implications of the short-range structure are less obvious. For some properties, such as the thermodynamics of isochoric mixing, the short-range errors may cancel, as they seem to do in crude lattice and cell theories. For an accurate prediction of the equation of state, however, the short-range details may prove more important.

Because $g^{[\text{INTRA}]}(r)$ and $g^{[\text{INTER}]}(r)$ clearly exhibit both short- and long-range structure and because they are relatively easy to calculate, they may be used to assess directly the effects of the pseudokinetics on the sampling. Figure 5 compares $g^{[\text{INTRA}]}(r)$ and $g^{[\text{INTER}]}(r)$ for the WCA polymer of Fig. 1d obtained from two separate simulations, with and without pseudokinetics. The sample in each simulation was made up of 96 chains with $M_N = 96$, $M_w/M_N = 1.08$, $T^* = 3.00$, and $\rho^* = 0.75$. The calculations with the pseudokinetics suppressed are shown for a single realization of the molecular weight distribution. Repetitions on different realizations give results in quantitative agreement with those shown in the figure. The superb agreement between the two sets of distribution functions supports the assertion that the pseudokinetics do not perturb the structure on the length scales of either individual beads or chains. [The root-mean-square (rms) radius of gyration for these samples is about 5.3.]

In the pseudokinetic algorithm, the centers of mass of the chains move dramatically as mass is exchanged among them. As a result, one can obtain good statistics on the center-of-mass radial distribution function. This quantity is shown for a Lennard-Jones polymer at $T^* = 2.50$ and $\rho^* = 0.75$ in Fig. 6, apparently for the first time here. Although it is fundamentally impossible to

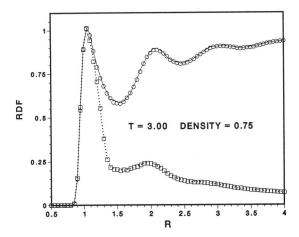

Figure 5 Intra- and intermolecular RDFs for WCA polymer (cf. Fig. 1d). Connected curves give the results obtained using the pseudokinetic algorithm. Symbols give the results obtained from a conventional Monte Carlo simulation.

measure this quantity (because the center of mass of the chain cannot be tagged), the center-of-mass RDF provides additional insight into the nature of the correlation hole in the melt. Very few chains have nearly coincident centers of mass, so statistics are poor below about 1σ. The center-of-mass RDF is not influenced by local detail, and the continuum result is therefore nearly identical to that observed in the lattice studies [15]. The correlation hole should deepen as the density is reduced. This expectation has already been confirmed in lattice studies [15].

Figures 7 to 10 report analyses for a simulation of a Lennard-Jones polymer melt confined between two hard walls perpendicular to the z coordinate and separated by 30σ. The sample consisted of 96 chains with $M_N = 96$,

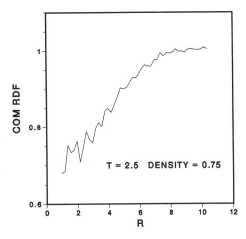

Figure 6 Center-of-mass RDF for Lennard-Jones polymer.

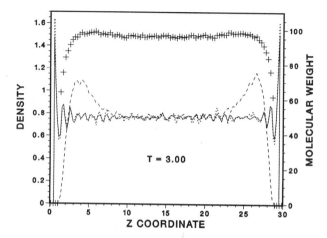

Figure 7   Distribution of Lennard-Jones polymer confined between parallel plates.
————: bead density; · · · · ·: renormalized end density; — — — —: renormalized centers-of-mass density; + + + +: molecular weight.

$M_N/M_w = 1.085$, $T^* = 3.00$, and $\rho^* = 0.75$. Equilibrium averages were taken for 2000 mass moves per particle. Each bead was assumed to have an additive underlying hard-sphere diameter of 1.0σ and thus could approach the walls no closer than 0.5σ. Figure 7 compares the density of beads, the density of chain ends, and the density of chain centers of mass across the sample. To facilitate shape comparisons among these quantities, the absolute end densities are multiplied by $M_N/2$ and the absolute center-of-mass densities are multiplied by M_N. In a bulk system, the renormalized densities would thus be identical to the mean bead density. In an inhomogeneous system, qualitative differences

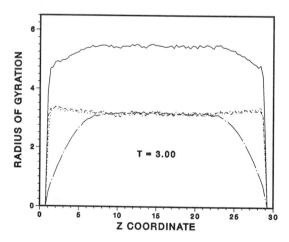

Figure 8   Contributions to R_g as a function of position for Lennard-Jones polymer confined between parallel plates. ————: R_g; — — — —: X_g; · · · · · ·: Y_g; — · — · —: Z_g.

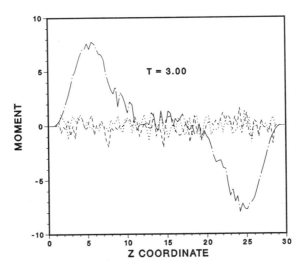

Figure 9 Third moments of the chains as a function of position for Lennard-Jones polymer confined between parallel plates. $----$: $\langle \delta x^3 \rangle$; $\cdots\cdots$: $\langle \delta y^3 \rangle$; $-\cdot-\cdot-$: $\langle \delta z^3 \rangle$.

among these quantities in the neighborhood of an interface may be more readily discerned. The essential differences between the results of a lattice simulation and those of a continuum simulation are apparent in Fig. 7. Both the bead density and the end density show the pronounced layering, well known for small molecule liquids and observed previously for simple, purely

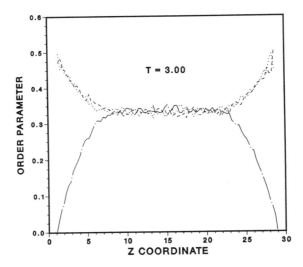

Figure 10 Mean-square direction cosines of the end-to-end vector for Lennard-Jones polymer confined between parallel plates. $----$: S_x; $\cdots\cdots$: S_y; $-\cdot-\cdot-$: S_z.

repulsive polymer models in the continuum [16–20]. No lattice model can adequately describe such behavior. Note that, at contact, the (renormalized) end density in the continuum is systematically higher than the bead density [16]. In the lattice studies, the reverse was found [6,7].

The center-of-mass density shown in Fig. 7 is also strongly affected by the presence of the walls. However, it is very similar to that observed in the lattice work. Chains in the neighborhood of the interface are asymmetrically collapsed, producing an enhanced center-of-mass density within one radius of gyration of the wall. Tests on the lattice showed that this structure scaled precisely with the bulk radius of gyration (and hence with $M_N^{1/2}$). A similar scaling has been observed in the continuum [21]. Note that the centers of mass in Fig. 7 are accumulated in bins of 0.4σ. The only previous simulation for chains of this length, a bravura calculation by Kumar et al. [16], employed standard simulations methods, a very short-ranged potential and required a bin size of 2σ to achieve adequate statistics. (Because conventional simulations require such long runs to achieve good statistics for global properties of chains, the *bead* density is often obtained with far better precision than that shown here [16].) Also shown in Fig. 7 is the variation of molecular weight across the interface. The segregation of short chains in the neighborhood of the interface is well understood in terms of an avoided entropy reduction by longer chains when collapsed against the wall. Although this molecular weight segregation can affect the center-of-mass density and other global chain properties in the neighborhood of the interface, comparison with the results obtained for monodisperse samples shows that the segregation effects are usually inconsequential for samples with this narrow a distribution of molecular weight.

The rms radius of gyration is defined by

$$R_g = (X_g^2 + Y_g^2 + Z_g^2)^{1/2} \tag{4}$$

where

$$X_g = \langle \delta x^2 \rangle^{1/2} = \left\langle \frac{1}{M} \sum_i (x_i - x_{cm})^2 \right\rangle^{1/2} \tag{5}$$

In Eq. (5) the sum is over all beads within a chain and the average is over all chains and over many microstates. Analogous equations define the other Cartesian contributions Y_g and Z_g. In an inhomogeneous system, these quantities may be partitioned according to the position of the center of mass of the chain. Figure 8 shows how these four quantities vary across the confined polymer sample. Once again the results are reminiscent of those found in the lattice studies. X_g and Y_g are perturbed only very slightly in the neighborhood of the interface, while Z_g is dramatically reduced. Those chains whose centers of mass are in the immediate neighborhood of the interface are increasingly flattened and approach two-dimensional behavior. Although X_g and Y_g seem to deviate slightly from ideal melt behavior as one approaches the interface, their reduction due to the segregation effect acts in opposition to the splaying one would expect [7,20] in a strictly two-dimensional system (where the Flory ideality theorem does not hold). The contribution of chain fluctuations perpendicular to the interface, Z_g, is a dramatic function of position and dominates the positional dependence of R_g. This behavior is overwhelmingly a consequence of chain collapse and is again in close agreement with previous results for lattice polymers.

The third moments of the mass distribution in the chains are given by

$$\langle \delta q^3 \rangle = \left\langle \frac{1}{M} \sum_i (q_i - q_{cm})^3 \right\rangle \tag{6}$$

where ($q = x, y, z$). Like all odd moments, these quantities must vanish in an isotropic environment. However, the contributions of individual chains fluctuate wildly in sign and magnitude, and the precision of the third moments provides for a very sensitive test of the algorithm.[†] In an inhomogeneous system, the $\langle \delta q^3 \rangle$ may be accumulated according to the position of the chain centers of mass, and their characters will help to establish whether a bulk domain is present. For a sample confined between flat plates, the third moments must be antisymmetric with respect to the midpoint of the sample. Figure 9 displays the variation of the $\langle \delta q^3 \rangle$ across the interface. The x and y moments fluctuate about zero in a manner similar to that found in bulk samples. The z moment has a positive peak near $z = 0$ and a corresponding negative peak near $z = 30$. In each case this illustrates the asymmetric collapse of surface chains toward the center of the sample. Note that as in the lattice studies, the bulklike domain (where all three moments fluctuate about zero) is somewhat more narrow than the bulk region observed in Figure 8. This was also observed in the lattice studies. In general, one expects higher moments to be more sensitive to the presence of a wall, and the spatial domain over which such moments are affected will be correspondingly larger. By these increasingly stringent measures, a rigorously bulklike region will develop only when the interfaces are separated by distances much larger than the length of an extended chain. However, the domain established by the uniformity of the second and third moments is sufficiently bulklike for most purposes.

Figure 10 shows the variation of the mean-square direction cosines of the end-to-end vector, $S_q = \langle \cos^2 \phi_q \rangle$ ($q = x, y, z$). We do not construct a conventional dimensionally dependent order parameter because we find it more appealing to observe the crossover between the three-dimensional bulk value of $\frac{1}{3}$ in the center of the sample and the two-dimensional bulk value of $\frac{1}{2}$ for S_x and S_y as the chains become flattened against the walls. For those chains with centers of mass very near the interface, S_z becomes increasingly small. For all

---

† At the extreme edge of an interface, data accumulated on a per chain basis can become unreliable because of low chain counts. These data are occasionally omitted from the figures when they are so widely scattered that their inclusion would be confusing or misleading.

the morphological quantities described here, it is important to keep in mind that very few chains actually lie flat against the interface and that most of the mass in close proximity to the interface is associated with chains whose center of mass is more distant and whose properties are less dramatically perturbed by it. The mean-square direction cosines define a bulk region in the center of the sample of about the same extent as that observed in Fig. 8.

We have also performed what we believe to be the first simulation of a free melt slab in the continuum. Figures 11 to 14 present analyses similar to those undertaken for polymer confined between hard walls. The results shown are for a Lennard-Jones polymer made up of 96 chains, with $M_N = 96$, $M_w/M_N = 1.085$ at $T^* = 2.25$. In a preliminary report [8], we presented the results for a sample adsorbed on a flat adhesive surface. The temperature of that simulation was 1.50, and we noted that the very slow relaxation suggested glassy behavior. Subsequent efforts seem to confirm this. The present result was obtained by heating the earlier sample to a very high temperature ($T^* = 10$) until surface melting was observed. The sample was then displaced from the wall and relaxed at $T^* = 2.25$. Thus as equilibrium was approached, the outer surface sharpened, while the surface previously in contact with the wall became more diffuse. Equilibrium averages were then taken for 4800 passes. The equilibrium density ($\rho\sigma^3$) in the center of the slab is 0.63. This is a bit low by the standards of small-molecule liquids near their triple points. It is well known [22], however, that the use of truncated potentials leads to misleadingly low predictions for the bulk density in inhomogeneous simulations of this sort, and this probably holds true for polymers as well. It may also be that typical melt temperatures for this model lie closer to 1.5 than to 2.25. Further studies are under way to explore these possibilities.

For the free surface of the melt, neither the overall bead density nor the

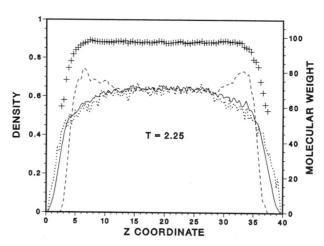

Figure 11   Distribution of a Lennard-Jones polymer in a free melt slab. Key as in Fig. 7.

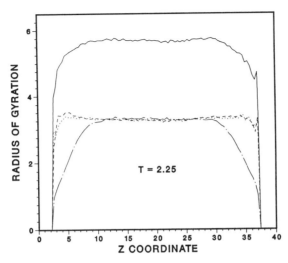

Figure 12   Contributions to R_g as a function of position for Lennard-Jones polymer in a free melt slab. Key as in Fig. 8.

density of end beads shows any oscillatory structure. There is, however, an excess of ends on the dilute side of the interface and a corresponding depletion of ends on the more concentrated side. As in the simulation between hard walls, this is accompanied by a segregation of low-molecular-weight chains into the interface. While segregation of short chains in the neighborhood of the interface might be expected to contribute to the surface excess of ends, the effect would be an enhancement of the end density throughout the interface.

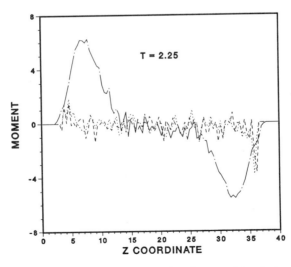

Figure 13   Third moments of the chains as a function of position for Lennard-Jones polymer in a free melt slab. Key as in Fig. 9.

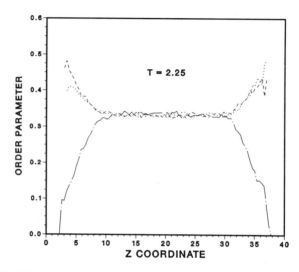

Figure 14   Mean-square direction cosines of the end-to-end vector for Lennard-Jones polymer in a free melt slab. Key as in Fig. 10.

The depletion of ends on the high-density side of the interface confirms that the origin of this phenomenon lies elsewhere. This behavior is quite similar to the interfacial structure observed in the lattice studies of a free melt surface. There the peculiar character of the end density was found to be preserved as the distribution of molecular weights was made more narrow.

The center-of-mass density peaks at about one radius of gyration away from the surface and is very similar to the corresponding result in Fig. 7. Similar behavior is observed in Figs. 12 to 14 for the radii of gyration, the third moments, and the mean-square direction cosines. All the figures show a bulklike region in the center of the slab with appropriate structuring at the interfaces. The amplitude of the structure is always somewhat less pronounced than that of polymer between hard walls. Although the chains at the extremes are flattened, they are never strictly two dimensional. In fact, a simulation of infinite duration would produce a minute concentration of polymer gas, which would itself have three-dimensional bulklike character. Thus the direction cosine would smoothly return to a mean value of $\frac{1}{3}$ and the third moment would return to zero in the polymer gas. The radius of gyration would rise to a new constant value, different from that in the melt. No such evaporation is to be expected in this finite run, but some results for chain configurations at the extremes of the interfaces may presage this behavior. Unfortunately, statistics are too poor there (even with the assist from the pseudokinetics) to make any definitive characterization.

CONCLUSIONS

The pseudokinetic techniques of Olaj et al. [4] and of Mansfield [5] can be applied to a limited but important class of models in the continuum. The distribution of molecular weights obtained in the simulations is found to approach the most probable distribution for condensation polymer when no restrictions are placed on the molecular weights and a square mound distribution when symmetric constraints are enforced. As in earlier lattice studies, the distribution of molecular weights is found to be independent of the thermodynamic state of the system. Tests against simulations without the pseudokinetics in operation show that neither the intramolecular nor the intermolecular distribution function is affected by the introduction of the kinetic events. The pseudokinetic elements do not appear to enhance the rate of mass relaxation, but do provide for a marked improvement in the statistical precision of global properties of individual chains. This is illustrated in the bulk by calculating the center-of-mass radial distribution function. Applications to polymer melt confined between parallel plates and to a free-melt polymer slab confirm that the expected enhancement in statistical precision is indeed achieved.

While new detail on the length scale of individual beads is present in the continuum model, all global properties of chains are very similar to those observed previously in lattice simulations. No new features appear in the continuum. This confirms the longstanding presumption that when one is interested only in polymer properties characterized by the radius of gyration, it is entirely unnecessary to work with a continuum model. For some applications, simulations of continuum models for polymer melts require an excessive amount of computer time to obtain the properties of interest, and lattice simulations provide a more intelligent and effective utilization of available resources. However, the quantitative prediction of many properties of real polymers does require more realistic models. It is unlikely that pseudokinetic techniques will be applicable to any atomically detailed model for polymers with complex local structure. But the models under consideration here can provide useful information that bridges the gap between the length scales for which the lattice models are adequate and those which require a detailed representation of the individual monomer units.

REFERENCES

1. F. T. Wall, L. A. Hiller, and J. Wheeler, *J. Chem. Phys.* **22,** 1036 (1954).
2. M. N. Rosenbluth and A. W. Rosenbluth, *J. Chem. Phys.* **23,** 356 (1955).
3. P. H. Verdier and W. H. Stockmayer, *J. Chem. Phys.* **36,** 227 (1962).
4. O. F. Olaj, W. Lantschbauer and K. H. Pelinka, *Chem. Kunstst. Aktue* **32,** 199 (1978); O. F. Olaj and W. Lantschbauer, *Makromol. Chem. Rapid Commun.* **3,** 847 (1982).

5. M. L. Mansfield, *J. Chem. Phys.* **77,** 1554 (1982); *Macromolecules* **16,** 914 (1983).

6. W. G. Madden, *J. Chem. Phys.* **87,** 1405 (1987).

7. W. G. Madden, *J. Chem. Phys.* **88,** 3934 (1988).

8. C. Lastoskie and W. G. Madden, *Polym. Prepr. Am. Chem. Soc., Div. Polym. Chem.* **30,** 39 (1989).

9. J. D. Weeks, D. Chandler, and H. C. Andersen, *J. Chem. Phys.* **54,** 5237 (1971).

10. F. T. Wall and F. Mandel, *J. Chem. Phys.* **63,** 4592 (1975).

11. P. J. Flory, *Principles of Polymer Chemistry,* Cornell University Press, Ithaca, N.Y., 1953.

12. D. A. Kofke and E. D. Glandt, *Mol. Phys.* **64,** 1105 (1988).

13. K. A. Schweitzer and J. Curro, *Macromolecules* **20,** 1928 (1987).

14. D. Chandler and H. C. Andersen, *J. Chem. Phys.* **57,** 1930 (1972).

15. W. G. Madden, unpublished.

16. S. K. Kumar, M. Vacatello, and D. Y. Yoon, *J. Chem. Phys.* **89,** 5206 (1988).

17. R. Dickman and C. K. Hall, *J. Chem. Phys.* **89,** 5206 (1988).

18. A. Yethirag and C. K. Hall, preprint.

19. K. F. Mansfield and D. Theodorou, *Polym. Prepr., Am. Chem. Soc., Div. Polym. Chem.* **30,** 76 (1989); *Macromolecules* (in press).

20. I. Bitsanis and G. Hadziioannou, *Polym. Prepr., Am. Chem. Soc., Div. Polym. Chem.* **30,** 78 (1989).

21. S. K. Kumar, M. Vacatello, and D. Y. Yoon, *Macromolecules* (in press).

22. J. S. Rowlinson and B. Widom, *The Molecular Theory of Capillarity,* Clarendon Press, Oxford, 1982.

19

Computer Simulation
of Models
for Rubber Elasticity

J. H. WEINER and J. GAO
Division of Engineering and Department of Physics
Brown University
Providence, Rhode Island 02912

ABSTRACT

Recent computer simulation studies of dense polymer systems have revealed strong coupling on the atomic level between the covalent interactions responsible for the chain bonds and the noncovalent (e.g., excluded-volume) interactions. To describe these two types of interactions on an equal footing, we have introduced a new concept called the intrinsic atomic level stress. We have performed the computer simulation by the method of molecular dynamics of a model network of tetrafunctional freely rotating chains with nodes at the sites of a diamond lattice. The simulation shows that the intrinsic atomic stresses in the network are independent of extension and almost the same as those in the corresponding melt. This leads to a new physical picture of the stress in such systems; that is, the macroscopic stress in a deformed network may be regarded as arising from the orientation by the deformation of the intrinsic atomic stress systems.

INTRODUCTION

If a rubberlike solid is regarded as a collection of atoms, two principal types of interactions must be noted. First, of course, are the covalent interactions responsible for the topological structure of the long-chain molecules. In addition, there are important noncovalent interactions, including those representing excluded volume effects, which prevent the collapse of the network.

An assumption that has frequently been made in theories of rubber elasticity, particularly in earlier treatments of the subject [1], is that the covalent and noncovalent interactions may be regarded as two decoupled systems. The spherically symmetric noncovalent interactions are then assumed to be capable of contributing only a hydrostatic pressure to the stress in the system, while the nonhydrostatic stress in a deformed rubberlike solid is taken as due solely to the covalently bonded chains, treated as entropic springs.

Some more recent theories of rubber elasticity have treated aspects of the noncovalent interactions, for example, through their effects on junction fluctuation [2], entanglements [3], or chain confinement [4]. However, in all of these approaches, the entropic spring concept retains a central role.

By means of the molecular dynamics simulation of idealized atomistic models of polymer systems, we have been reexamining the assumption of decoupled covalent and noncovalent interactions [5–9]. A key tool in this work is the virial stress theorem [5], which treats these two types of interactions on an equal footing. We find that on the atomic level, these two types of interactions are far from decoupled. Rather, there are important mutual effects of one type of interaction on the other, and these increase with system density. As a consequence, at the liquid-like packing fractions typical of real systems, we find that the nonhydrostatic contribution to the stress made by the non-covalent interactions is at least comparable in magnitude to the covalent contribution.

In recent work [9], for example, we have performed the molecular dynamics simulation of a tetrafunctional network, based on a diamond lattice, utilizing a freely jointed chain model with the noncovalent interactions represented by the purely repulsive part of a Lennard-Jones potential. There we found that at high packing fractions the nonhydrostatic contribution to the stress made by the noncovalent interactions is almost four times the covalent contribution.

Here we describe further molecular dynamics simulations, continuing to use the tetrafunctional network based on a diamond lattice, but with freely rotating chain models (i.e., with controlled bond angles) and with the addition of an attractive tail to the noncovalent potential.

MODEL

The chain model we employ is a freely rotating chain with N bonds and $N + 1$ atoms with positions denoted by $x_i, i = 0, 1, \ldots, N$. The covalent potential $V_c(x_0, x_1, \ldots, x_N)$ serves to keep bond lengths close to b and bond angles close to θ_0 by the use of stiff springs as follows:

$$V_c(x_0, x_1, \ldots, x_N) = \sum_{\beta=1}^{N} u_{cb}(r^\beta) + \sum_{\beta=1}^{N-1} u_{ca}(\cos \theta_\beta) \tag{1}$$

where

$$u_{cb}(r) = \tfrac{1}{2}\kappa_b(r - b)^2 \tag{2}$$

is the covalent bond potential and

$$u_{ca}(\cos \theta) = \tfrac{1}{2}\kappa_\theta(\cos \theta - \cos \theta_0)^2 \tag{3}$$

is the covalent angle potential; $r^\beta = x_\beta - x_{\beta-1}$, $r^\beta = |r^\beta|$ and $\cos \theta_\beta = -r^\beta \cdot r^{\beta+1}/r^\beta r^{\beta+1}$.

The noncovalent interaction is through a truncated Lennard-Jones potential,

$$u_{nc}(r) = \begin{cases} 4\epsilon\left[\left(\dfrac{\sigma}{r}\right)^{12} - \left(\dfrac{\sigma}{r}\right)^6\right] + \epsilon & \text{for } r \leq r_0 \\ 0 & \text{for } r > r_0 \end{cases} \tag{4}$$

where r is the distance between any pair of atoms.

To form a network, as in our previous work [9], tetrafunctional junctions are taken initially as fixed at the sites of a diamond lattice. Chains with $N = 50$ bonds connect these junctions, with 16 chains in the basic cell, which has edges oriented in the x_1, x_2, and x_3 directions; periodic boundary conditions in these directions are employed as is customary in molecular dynamics. Typically, the end-to-end distance of chains in the undeformed reference configuration has a value of r/Nb between 0.1 and 0.15. Care is taken in the preparation of the initial state of the network to ensure that no permanent chain entanglements are introduced, since our purpose is to study the nature of covalent-noncovalent interplay in the simplest setting. For the values of the parameters used in the noncovalent potential, chains cannot pass through each other and the initial network topology is conserved in subsequent thermal motion or deformation of the system.

The model network is subjected to a uniaxial constant volume deformation in the x_1 direction by subjecting all of the network junctions fixed in the cell faces to that deformation. Network junctions in the interior of the cell are permitted to undergo thermal motion.

INTRINSIC ATOMIC STRESSES

We have, in a recent publication [10], introduced the concept of intrinsic atomic level stresses, and we utilize them in the interpretation of the results of the present simulation. While we refer the reader to Ref. 10 for complete details, the basic idea may be described briefly as follows: Let $t_{ij}, i, j = 1, 2, 3$, be the macroscopic stress tensor in the system, referred to the fixed reference or laboratory frame x_1, x_2, x_3. Then it is a consequence of the virial stress theorem that t_{ij} may be written as

$$\frac{v t_{ij}}{kT} = \sum_{\beta} \langle \sigma_{ij}(\beta) \rangle \tag{5}$$

where v is the volume of the basic cell, k is Boltzmann's constant, T is temperature, the summation is over all atoms β in the basic cell, carats denote time averages, and the definition of $\sigma_{ij}(\beta)$ in terms of covalent and noncovalent interactions is given in Ref. 10. In view of Eq. (5), we may refer to $\langle \sigma_{ij}(\beta) \rangle$ as the atomic stress at atom β or, equivalently, as the contribution which atom β makes to the macroscopic stress t_{ij}. The atomic stress $\langle \sigma_{ij}(\beta) \rangle$ is referred, as is t_{ij}, to the fixed laboratory frame x_1, x_2, x_3. Therefore, much of the physical significance of $\langle \sigma_{ij}(\beta) \rangle$ is lost because of the large-amplitude thermal motion which occurs in polymer systems. For example, in a polymer melt, where there are no long-time hindrances to large rotations, $\langle \sigma_{ij}(\beta) \rangle$ must be hydrostatic, that is, of the form $-\pi(\beta)\delta_{ij}$, where δ_{ij} is the Kronecker delta ($\delta_{ij} = 1$ for $i = j$ and $\delta_{ij} = 0$ for $i \neq j$) and $\pi(\beta)$ is the contribution to the pressure made by atom β.

For this reason we have found it useful at each time step to refer $\sigma_{ij}(\beta)$ to a local coordinate system $\bar{x}_r(\beta), r = 1, 2, 3$, which moves so that it always bears a fixed relation to the covalent bonds attached to atom β. The components of the atomic stress referred to $\bar{x}_r(\beta)$ are denoted by $\bar{\sigma}_{rs}(\beta)$ and $^*\sigma_{rs}$ denotes the time average of $\bar{\sigma}_{rs}(\beta)$ averaged over all of the atoms of the system. We refer to $^*\bar{\sigma}_{rs}$ as intrinsic atomic stresses. It is convenient to choose the coordinates of $\bar{x}_1(\beta)$ and $\bar{x}_2(\beta)$ in the plane of the covalent bonds attached to atom β, with $\bar{x}_1(\beta)$ bisecting the angle between them. Under these conditions it follows from symmetry considerations that the tensor $^*\bar{\sigma}_{rs}$ will have only diagonal nonzero components.

RESULTS

In addition to the parameter values already given, the molecular dynamics simulations were performed with $\sigma = 1.5b$, $\kappa_b b^2/kT = 200$, $\kappa_\theta/kT = 600$, $\epsilon/kT = 0.66$, $r_0 = 1.5\sigma$, and with volume of the basic cell $(18.6b)^3$. Other conditions of the calculations and the procedure followed were the same as in Ref. 10.

Results for the macroscopic stress difference, $t_{11} - t_{22} = t_{11} - \frac{1}{2}(t_{22} + t_{33})$ as determined from the simulation of the model in uniaxial, constant-volume, deformation with extension λ are shown in Fig. 1. The large error bars are a reflection of the sluggishness of the network model due to bond angle restrictions and to the junctions. Nevertheless, the simulations show the large variation in the stress to be expected in a network with relatively short chains.

The intrinsic atomic stresses, $^*\sigma_{11}$, $^*\sigma_{22}$, and $^*\sigma_{33}$, for the same series of computations on the deformed network show strikingly different behavior; their values, shown in Fig. 2, are almost completely independent of the extension λ in the range studied. Furthermore, simulations were performed of the

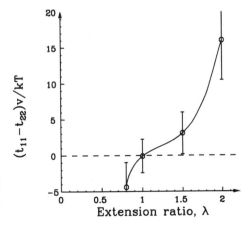

Figure 1 Macroscopic stress difference, $t_{11} - t_{22} = t_{11} - \frac{1}{2}(t_{22} + t_{33})$ as determined from molecular dynamics simulation of tetrafunctional network model in uniaxial constant-volume deformation.

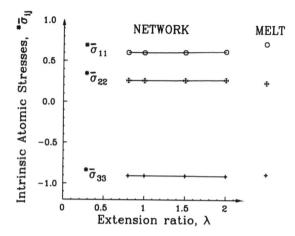

Figure 2 Intrinsic atomic stresses $^*\sigma_{11}$, $^*\sigma_{22}$, and $^*\sigma_{33}$ as determined from molecular dynamics simulation of tetrafunctional network model in uniaxial constant-volume deformation and for corresponding melt. Estimated error in calculated values of $^*\sigma_{ij}$ is less than 1%.

polymer melt corresponding to the network, one formed from the network by releasing the chains from all junctions while retaining the pressure constant. The intrinsic atomic stresses were computed for this melt and are also shown in Fig. 2; they are seen to have almost the same values as in the network.

CONCLUSIONS

The physical picture of the nature of stress in polymer networks and melts that emerges from the molecular dynamics simulations reported here and from our earlier work is very different from the usual one. We find that each atom of the system is the bearer of an intrinsic stress system whose principal values are relatively independent of whether the given system is a melt or a network, and in the latter case, they are also independent of the degree of extension. The principal directions of the intrinsic atomic stress system, however, bear a fixed relation to the covalent bonds attached to the atom in question. Therefore, in the melt where segment orientation is unrestricted, the time average of the intrinsic atomic stresses lead to a purely hydrostatic state of the macroscopic stress. In the network, however, deformation results in segment orientation and the resulting biased orientation of the intrinsic atomic stress systems results in a nonhydrostatic macroscopic state of stress.

ACKNOWLEDGMENTS

This work has been supported by the Gas Research Institute (Contract 5085-260-1152). The computations were performed on the Cyber 205 and on the ETA 10, both at the John von Neumann Center, Princeton, with time provided by the JVNC National Allocation Committee.

REFERENCES

1. See, for example, H. M. James and E. Guth, *J. Chem. Phys.* **11,** 455 (1947).
2. P. J. Flory and B. Erman, *Macromolecules* **15,** 800 (1982).
3. M. Doi and S. F. Edwards, *J. Chem. Soc., Faraday Trans. 2* **74,** 1802 (1978).
4. R. J. Gaylord and J. F. Douglas, *Polym. Bull.* **18,** 347 (1987).
5. J. Gao and J. H. Weiner, *Macromolecules* **20,** 2520 (1987).
6. J. Gao and J. H. Weiner, *Macromolecules* **20,** 2525 (1987).
7. J. Gao and J. H. Weiner, *Macromolecules* **21,** 773 (1987).
8. J. Gao and J. H. Weiner, *Macromolecules* **22,** 979 (1989).
9. J. H. Weiner and J. Gao, *Proceedings of 5th IFF-ILL Workshop, Molecular Basis of Polymer Networks,* Jülich, West Germany, 1988 (in press).
10. J. Gao and J. H. Weiner, *J. Chem. Phys.* **90,** 6749 (1989).

20

Computer Simulation of Hairpin Dynamics In Wormlike Main-Chain Nematic Polymer Liquid Crystals

D. R. M. WILLIAMS and M. WARNER
Cavendish Laboratory
Madingley Road
Cambridge CB3 0HE
United Kingdom

ABSTRACT

In a previous paper [1] we introduced a simple model for the dynamics of a single hairpin in nematic wormlike main-chain polymer liquid crystals. In this paper the model is generalized to the multihairpin case and the results of a computer simulation based on the Langevin equation are presented. It is shown that the decay of n hairpins may be described by a simple system of master equations.

INTRODUCTION

In this chapter we address the dynamics of a particular type of "defect" that can occur in polymer liquid crystals (PLCs) that have a flexible nematic backbone. The dynamics of PLCs in general is an important problem because it should eventually lead to an understanding of their complicated rheological behavior. The complexity and subtlety of flexible main-chain nematic PLCs come from the interplay between nematic alignment and the polymeric tendency to maximize entropy. A first-order phase transition separates the dominance of the former at low temperatures and the latter at high temperatures. The conformations of chains at lower temperatures are stretched out along the ordering direction [2] and are predicted to become rodlike eventually, albeit with thermally generated transverse wiggles [3]. Their wormlike (rather than the usual Gaussian) character then becomes of the essence to consider. The polymers considered here consist of stiff nematic monomeric units joined together by spacers to form a reasonably stiff chain overall. The chain stiffness implies that the polymer pays an energy penalty for bending, and the nematic elements mean that the polymer pays an energy penalty for swimming across the nematic direction. The question arises as to how such a wormlike chain moves in the nematic field presented by its neighbors. We shall answer this question for relatively well-ordered low-temperature nematic melts.

We represent the polymer of length L in two dimensions by the trajectory of the angle it makes with the x-axis, $\theta(s)$, where s is the arc length measured from one end of the polymer. The nematic director is assumed to point in the $\pm x$ direction. The energy may be written as a functional of $\theta(s)$:

$$U[\theta(s)] = \frac{1}{2} \int_0^L ds \left[\epsilon \left(\frac{d\theta(s)}{ds} \right)^2 + 3aS \sin^2 \theta(s) \right] \tag{1}$$

Here S is the order parameter $\langle P_2[\cos \theta(s)] \rangle$ (where the average is taken over all worms for all values of s) and ϵ and a are the elastic bend and nematic mean field coupling constants respectively. Equation (1) is a good representation for three-dimensional nematic worms at temperatures where the meandering away of the chain from the nematic director is expensive (compared with $k_B T$) and hence the variation of the ϕ coordinate when θ varies from 0 to $\pi/2$ is small. The obvious configuration for a polymer molecule to have to minimize its energy is for it to lie entirely along the nematic direction. However, at finite temperatures entropic considerations will be important. There will be many small excursions $\theta(s) \neq 0$ away from $\hat{x}$, but more important, if the worm is flexible enough, it will undergo rapid reversals in direction to form hairpins, as suggested by de Gennes [4]. The energetic cost of a hairpin is counterbalanced by the entropy of where it can be placed along the chain's length. Such defects drastically affect chain dimensions (the first one halves the x-dimension) and are predicted to have a significant influence on the dielectric response [5]. It is also likely that these hairpins play an important part in the dynamics of chains and hence in the rheology of main chain (nematic backbone) PLCs (MCPLCs). They may also characterize the dynamics of how dielectric and nonlinear optical response builds up.

We have chosen here to work in two dimensions because hairpins are essentially two-dimensional objects. To study the dynamics of hairpin creation and destruction, we first need to find the static equilibrium trajectories for the hairpins. This is done for the single- and multihairpin cases in the following section. Next, the dynamics of a single hairpin are discussed and it is shown that in the low-temperature limit the destruction of a single hairpin is brought about by the shuffling of length between the two arms of the hairpin until one arm is sufficiently short that the bend can unwind. Then this "diffusing pulley" model is generalized to the case of many hairpins on a single chain and the friction matrix and mobility matrix for a multihairpin system are calculated. This enables a set of n Langevin equations to be written down for the motion of n hairpins. These equations then form the basis of a computer simulation. In the final section it is shown that a set of simple master equations for the probability of finding n hairpins in a chain at time t can be written down which mimic the more complicated Langevin approach very well.

STATICS FOR THE SINGLE- AND MULTIHAIRPIN CASES

In this section we review what is known about the statics of hairpins. To calculate the trajectories for main-chain nematic polymers in mechanical equilibrium, one needs to apply Euler–Lagrange minimization to the energy in Eq. (1). This was first done by de Gennes [4] for the case of a single hairpin in an infinitely long chain using the coordinate system shown in Fig. 1. The equation that needs to be solved is

$$\epsilon \frac{d^2 \theta(s)}{ds^2} = 3aS \sin \theta(s) \cos \theta(s) \tag{2}$$

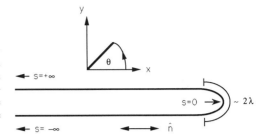

Figure 1 Coordinate system used for calculating the trajectory of a single hairpin in an infinitely long chain. The chain bends over a distance of a few λ, the "hairpin length," and for the remainder of the trajectory is aligned very closely with the nematic director $\hat{n}$.

with free end boundary conditions $d\theta(s)/ds = 0$. This is a pendulum equation and the solutions will thus in general be elliptic functions. However, for the special case of an infinitely long chain the solution is particularly simple and has the form

$$\tan\frac{\theta(s)}{2} = \exp\left(\frac{s}{\lambda}\right) \tag{3}$$

In this equation we have defined a "hairpin length" of

$$\lambda = \sqrt{\frac{\epsilon}{3aS}} \tag{4}$$

This is roughly the length over which the hairpin bends. One can also calculate the energy of an infinite hairpin, which is

$$U_h = 2\sqrt{3aS}\,\epsilon \tag{5}$$

Both the hairpin energy and the hairpin length represent the geometric means of the nematic and bending influences. Statistical mechanics based on simply inserting Eq. (5) into a Boltzmann distribution appears to agree well with the results of a more extensive calculation [3]. From Fig. 1 it is clear why solution (3) was called a hairpin. Along both arms of the chain the worm is aligned closely with the nematic director and along a short length of a few λ's undergoes a well-defined bend. The shape of this bend represents a balance between the nematic term in the energy equation [Eq. (1)], which forces the bend to be as sharp as possible so that the chain points against the director for only a short length, and the elastic term, which forces the bend to be as slow as possible.

The case of a hairpin or hairpins in a chain of finite length is discussed elsewhere [1]. There it is shown that in a chain of length L there can exist only a finite number of hairpin solutions. Thus there is the single hairpin solution, a two-hairpin solution, and so on, up to the n_{max} hairpin solution with

$$n_{max} = \left[\frac{L}{\pi\lambda}\right] \tag{6}$$

where $[x] =$ "the largest integer $\leq x$." As expected, these solutions have the hairpins equally spaced along the chain. However, these solutions are the ones that satisfy the conditions of mechanical equilibrium exactly. In practice there are many more solutions which are close to mechanical equilibrium (close in the sense that they take a long time to move). These solutions have any number of hairpin bends placed almost anywhere along the chain. An example is shown in Fig. 5. The only two conditions that these solutions must satisfy is that two hairpin bends are not closer than a few λ's and that similarly, no hairpin bend is less than a few λ's from the chain ends. If either of these conditions is violated, potential effects drive the system rapidly away from its original configuration. In the first case the two close hairpin bends will annihilate one another, and in the second case the hairpin bend will drop off the end of the chain. These processes are discussed further later.

Thus the hairpin length λ not only represents the length scale on which the chain bends to form a hairpin but also gives a measure of how much arc length must separate two hairpins for them to be effectively noninteracting. From Eq. (3) it is clear that the effect of the hairpin bend decays away with distance exponentially from the bend, and indeed it has been argued [1] that the nearest-neighbor interactions between hairpin bends separated by an arc length r is given roughly by

$$U_{nn} \simeq -4U_h \, \exp\!\left(-\frac{r}{\lambda}\right) \tag{7}$$

It is this exponential decay of the hairpin–hairpin interaction that allows us to ignore potential effects in the dynamics of multihairpin systems until the hairpins approach each other (or the chain ends) closely.

DYNAMICS OF A SINGLE HAIRPIN

In this section we discuss a model for the dynamics of the destruction of a single hairpin in a chain of finite length [1]. Simple energy arguments and more complicated arguments based on the eigenmodes of oscillation of a hairpin in a wormlike chain are used to justify the model. To discuss the motion of a single hairpin in a chain, we first model the chain of length L as a series of N stiff discrete links of fixed length $b = L/N$ linking $N + 1$ beads as shown in Fig. 2. The coordinates we use to describe the chain are its center of mass $(\bar{x}, \bar{y})$ and the angles θ_i made by each monomer with the x-axis. The nematic director is chosen to lie along the x-axis. A discrete version of the energy equation [Eq. (1)] is used:

$$U(\boldsymbol{\theta}) = \frac{\epsilon}{2b} \sum_{i=1}^{N-1} (\theta_{i+1} - \theta_i)^2 + 3aSb \sum_{i=1}^{N} \sin^2(\theta_i) \tag{8}$$

We entirely neglect the effect of entanglements in order to simplify the problem to that of a single chain moving in the mean field of the surrounding chains. By doing this we can ignore the motion of the center of mass, whose trajectory at finite temperature will be merely a random walk. We thus restrict

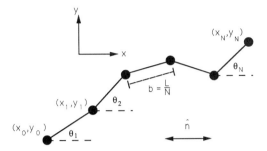

Figure 2 Coordinate system used for the N-link chain consisting of $N + 1$ beads. The bond length b is the same for each link. The coordinates (x_i, y_i) are used only in calculating the Rayleighian and are replaced by the θ_i's and the center of mass in the equations of motion.

our attention to the θ_i's. Naturally, the effects of inertia are neglected because these are important only for polymers on very short time scales.

Full Equation of Motion

The approach used here is first to write down the mobility matrix $\mathcal{M}_{ij}(\boldsymbol{\theta})$ for an N-monomer chain defined by

$$\dot{\theta}_i = -\sum_{j=1}^{N} \mathcal{M}_{ij}(\boldsymbol{\theta}) \frac{\partial U}{\partial \theta_j} \tag{9}$$

In practice it is not possible to write down the mobility matrix explicitly. However, one can calculate its inverse [the friction matrix $\mathcal{F}_{ij}(\boldsymbol{\theta})$] using Rayleigh's dissipation function

$$\mathcal{R} = \frac{1}{2} \frac{\mu L}{N+1} \sum_{i=1}^{N+1} \dot{x}_i^2 + \dot{y}_i^2 \tag{10}$$

and the equation of motion for a system dominated by friction,

$$\frac{\partial \mathcal{R}}{\partial \dot{\theta}_i} = -\frac{\partial U}{\partial \theta_i} \tag{11}$$

Here μ is a friction constant per unit length. Once one has the friction matrix it is a simple matter, by adding random (Brownian) forces, to write down a Langevin or Smoluchowski equation [6] to describe the dynamics of a hairpin. However, at low enough temperatures for the hairpins to be considered as identifiable entities, the time scale for hairpin destruction can be calculated analytically from a simple model (see below) and is very long—basically scaling as L^3. This implies that a direct simulation of the motion of an N-monomer hairpin is not very practical. One can improve matters somewhat by using the x_i's and y_i's directly as coordinates and introducing forces to keep the bond lengths constant, but this still does not allow the simulation of the destruction of a very long hairpin.

Eigenmode Analysis

Because the time scales for the full problem are so long, we will be look-ing at the much simpler and more numerically tractable problem of the eigen-modes of a single hairpin in a wormlike chain. It has been shown [1] that all hairpin solutions for a finite continuous chain are unstable, and provided that the hairpin length remains larger than a few monomer sizes, this is also true for an N-link chain. However, the period associated with the unstable mode of motion grows at least exponentially with the length of the chain, and hence chains that are more than a few hairpin lengths and which are in a single hairpin configuration may be considered to be quasi-stable. This observation leads us to consider the eigenmodes of the linearized equation of motion. We take a chain in a hairpin configuration θ_0 and perturb it slightly so that $\theta = \theta_0 + \alpha$. The perturbation α is then expanded in eigenmodes ψ^σ

$$\alpha(t) = \sum_{\sigma=1}^{N} a^\sigma(t)\psi^\sigma \tag{12}$$

Here the ψ^σ are defined to be the simultaneous eigenvectors of the friction matrix and the matrix formed from the second derivatives of the energy equation (8) evaluated at the hairpin trajectory θ_0, so that

$$\omega^\sigma \sum_{j=1}^{N} \mathcal{F}_{ij}(\theta_0)\psi_j^\sigma = \sum_{j=1}^{N} \frac{\partial^2 U}{\partial\theta_i \partial\theta_j}\bigg|_{\theta=\theta_0} \psi_j^\sigma \tag{13}$$

The eigenvectors are chosen to have normalization

$$\sum_{i=1}^{N} \sum_{j=1}^{N} \psi_i^\mu \mathcal{F}_{ij}(\theta_0)\psi_j^\nu = \delta^{\mu\nu} \tag{14}$$

The details of how the friction matrix $\mathcal{F}_{ij}$ is calculated can be found in our previous paper [1]. We now use the amplitudes of the eigenmodes, the a^σ's, as our coordinates. The energy expressed in terms of these coordinates is

$$U(\mathbf{a}) = \frac{1}{2} \sum_{\sigma=1}^{N} \omega^\sigma a^{\sigma 2} \tag{15}$$

and the mobility matrix for these coordinates is simply the identity matrix. When there are no random forces acting on the chain, the amplitudes thus obey the simple time evolution $a^\sigma = a_0^\sigma \exp(-\omega^\sigma t)$. In the presence of random forces the a^σ's independently undergo random walks in parabolic potentials.

Numerical calculation shows that for a chain with L equal to or greater than a few λ's, all the eigenfrequencies are large and positive except for the first. This is negative (except when $b \simeq \lambda$), but it becomes very small as L/λ is increased. Typical plots of the eigenmodes superimposed on the hairpin state are shown in Fig. 3. From these it is clear that the first mode involves the hairpin moving along its own length and undergoing virtually no change in energy. Hence it has an eigenfrequency close to zero. However, the next two

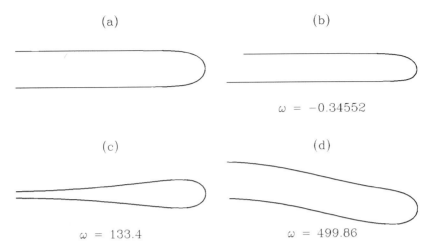

Figure 3 "Ground state" hairpin solution (a) and first three eigenmodes (b) to (d) (superimposed on the ground state) for a 100-link chain. The eigenfrequencies listed [corresponding to those in Eq. (13)] are expressed in terms of the dimensionless variable $\epsilon/\mu L^4$. The parameters chosen here are $L = 1$, $\epsilon = 1$, and $L/\lambda = 35$. Note that the first excited state clearly represents a hairpin shifting length from one arm to the other and the eigenvalue is negative and small, indicating that the eigenmode is barely unstable. The scales on the x and y axes have been made unequal for purposes of clarity.

modes (and all higher modes) involve both a large nematic and a large bending energy penalty and hence decay rapidly. Thus at long times the high-energy modes become unimportant and the hairpin diffuses along its own length. This is the method by which hairpins will be destroyed at low temperatures. Of course, once the hairpin has moved a significant distance along its own length, the chain will no longer be in a state that is a solution of the equations of mechanical equilibrium, and the linearized approximation will break down. However, it is clear that even when this occurs, the hairpin will still move by diffusing in this manner because any other motion involves a large energy penalty.

The importance of the eigenexpansion carried out above is that it includes the effects of the wormlike character of the chain. If these were ignored, energy considerations would lead one to guess that a single hairpin would be destroyed by shuffling length between its arms (i.e., moving along its own length). This is because [1] the energy change in moving length from one arm to the other is exponentially small in L/λ. It is clear that this is still the case even when the full correlation effects of each segment of the chain being linked are included.

We thus argue that at low temperatures the destruction of a single hairpin will proceed by the diffusion of length between the arms of the hairpin. This will proceed until one of the arms is only a few λ's long, when potential

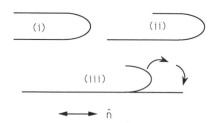

Figure 4 Two stages involved in the destruction of a single hairpin with large L/λ. In (i) we see a symmetric hairpin. In (ii) random forces have shuffled the length between the arms. In (iii) the shortest arm has become less than a few hairpin lengths long and the hairpin bend falls off the chain end because of potential effects.

effects will take over and the hairpin will rapidly unwind, as shown in Fig. 4. The temperature must be low enough so that the higher-order eigenmodes or "wiggles" are not prevalent in the motion of the hairpin. The conditions under which this is the case do not appear to be very stringent and are discussed elsewhere [1].

Simplified Model for Hairpin Motion

The notion (established by the eigenmode analysis above) of the shuffling of length between the arms of a single hairpin allows us to produce a simple mathematical model for the destruction of such a hairpin. One assumes that the chain consists of two perfectly straight arms of lengths s_1 and $L - s_1$. The coordinates used in this model are s_1 and the center of mass $\bar{x}$. The hairpin bend is assumed to have zero length. One can calculate the Rayleighian in these coordinates

$$\mathscr{R} = \frac{\mu}{2} \left\{ L\dot{\bar{x}}^2 + 4\dot{s}_1^2 \left[L - s_1 - \frac{1}{L}(L - s_1)^2 \right] \right\} \tag{16}$$

which gives a mobility in s_1 space of

$$\mathscr{L} = \frac{1}{4\mu} \left(\frac{1}{s_1} + \frac{1}{L - s_1} \right) \tag{17}$$

This mobility has a very simple form that can be interpreted as being formed from contributions algebraically added from each of the arms. This mobility can form the basis for the study of the destruction of a single hairpin either using the Smoluchowski equation or a computer simulation based on the Langevin equation. We do not do either here, preferring to concentrate on the more general problem of the motion of n hairpins. However, we note that this model predicts that the time for destruction of a single hairpin will scale like

$$\tau_d \propto \frac{\mu L^3}{k_B T} \tag{18}$$

Here the factor of $\mu L / k_B T$ comes from the inverse of the diffusion constant, and there is another factor of L^2 because the chain must diffuse over a length proportional to L.

DIFFUSING PULLEY MODEL FOR MULTIHAIRPIN DYNAMICS

In this section we extend the simple model of single-hairpin motion discussed at the end of the preceding section to the case of many hairpins on the same chain. This involves using the Rayleighian approach to calculate the friction matrix for the n-hairpin case. Although in theory a knowledge of the friction matrix is sufficient for the purposes of computer simulation by the Langevin approach, it would be more elegant and faster to have its inverse (the mobility matrix). It will be shown that the n-hairpin diffusing pulley model is open to an alternative physical interpretation which enables us to calculate the mobility matrix directly and hence provide a reasonably simple and explicit form for the Langevin equations.

The obvious generalization of the foregoing model to the multihairpin case is to consider a chain with n hairpins and $n + 1$ straight segments aligned along the nematic director as shown in Fig. 5. The curved pieces of chain (which are, in effect, the pulleys) are assumed to have negligible length. The coordinates used are the center of mass of the chain ($\bar{x}, \bar{y}$) and the lengths of the segments s_i with $i = 1, \ldots, n$. The length of the final segment s_{n+1} is not independent of all the others because the length of the chain must be conserved, so that

$$s_{n+1} = L - \sum_{i=1}^{n} s_i \qquad (19)$$

When the only external field acting on the chain is a nematic field, these coordinates suffice to describe the simple model. However, when a field that has a polarity (say, an electric field) is applied, one needs to introduce another coordinate ξ, which takes a value $+1$ when the first segment is pointing "up" and -1 when it is pointing "down." The hairpin bends, which are treated as points, are assumed to diffuse freely along the chain but with a mobility matrix that depends on the lengths s_1 to s_n. In this model the center of mass is assumed to diffuse freely and will be ignored in most of the work discussed hereafter.

Figure 5 Coordinate system used in the diffusing pulley model for the n-hairpin case. The lengths between the hairpin bends are assumed to be perfectly straight and aligned along the nematic director. The bends themselves are assumed to be of negligible length but for clarity are shown here as being finite.

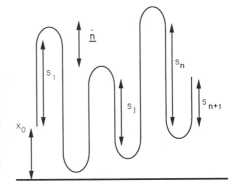

To calculate the friction matrix for the diffusing pulley system, we use the Rayleighian method. We first need to calculate the Rayleighian from

$$\mathcal{R} = \frac{\mu}{2} \int_0^L ds [\dot{x}^2(s) + \dot{y}^2(s) + \dot{z}^2(s)] \tag{20}$$

where $(x(s), y(s), z(s))$ is the position of a segment of the chain at arc length position s. In the model adopted here the chain is assumed to have no extension in the y or z directions, so these variables disappear from Eq. (20). If x_0 is the x position of one end of the chain, we define $\beta(s)$ via

$$x(s) = x_0 + \beta(s) \tag{21}$$

From Fig. 5 it is easy to show that $\beta(s)$ is

$$\beta(s) = (-1)^{p+1}s + \sum_{i=1}^{p-1} [(-1)^{i+1} + (-1)^p]s_i \qquad \text{for } s \in \left(\sum_{i=1}^{p-1} s_i, \sum_{i=1}^{p} s_i \right) \tag{22}$$

The dissipation function or Rayleighian is [from Eq. (20)]

$$\mathcal{R} = \frac{1}{2}\mu \left(L\dot{x}^2 + \int_0^L ds\,\dot{\beta}^2(s) - \frac{1}{L}\left[\int_0^L ds\,\dot{\beta}(s) \right]^2 \right) \tag{23}$$

To calculate the friction matrix $\mathcal{A}_{ij}$ for the s variables, one uses

$$\mathcal{A}_{ij} \dot{s}_j = \frac{\partial \mathcal{R}}{\partial \dot{s}_i} = -\frac{\partial U}{\partial s_i} \tag{24}$$

This yields a matrix of the form

$$\mathcal{A}_{kj} = \mu\{\mathcal{B}_{kj} + \mathcal{C}_{kj} + \mathcal{D}_{kj}\} \tag{25}$$

where

$$\mathcal{B}_{kj} = [1 + (-1)^{k+j}]\Theta(\max(k+1, j+1)) \tag{26}$$

$$\mathcal{C}_{kj} = [(-1)^{k+1} + (-1)^{j+1}]\Lambda[\max(k+1, j+1)] \tag{27}$$

and

$$\mathcal{D}_{kj} = -\frac{1}{L}[(-1)^{j+1}\Theta(j+1) + \Lambda(j+1)][(-1)^{k+1}\Theta(k+1) + \Lambda(k+1)] \tag{28}$$

The two functions $\Theta(k)$ and $\Lambda(k)$ are simple combinations of the lengths and are defined by

$$\Theta(k) = \sum_{i=k}^{N+1} s_i \qquad \text{and} \qquad \Lambda(k) = \sum_{i=k}^{N+1} (-1)^i s_i \tag{29}$$

The complicated nature of the mobility matrix arises from the fact that the motion of any given hairpin is linked to that of all the other hairpins.

The friction constant for center-of-mass variable $\bar{x}$ from Eq. (23) is simply μL, which is a constant independent of the internal chain coordinates. In the presence of random forces but in the absence of entanglements, the center of mass will therefore follow a random walk, provided that the potential is independent of the center of mass, as the nematic and bending potentials are.

It would be useful to have an analytic form for the inverse of $\mathcal{A}_{ij}$, the mobility matrix $\mathcal{L}_{ij}$. Finding this directly from Eq. (25) looks very difficult. However, by interpreting the diffusing pulley model as a purely one-dimensional system, an explicit expression for $\mathcal{L}_{ij}$ can be found. To see this we first write down the Langevin equations for the s variables. In a finite time interval Δt the change in the variable s_i is

$$\Delta s_i = \sum_{j=1}^{n} \mathcal{L}_{ij} f_j(t) \sqrt{\Delta t} + k_B T \sum_{j=1}^{n} \frac{\partial \mathcal{L}_{ij}}{\partial s_j} \Delta t \qquad (30)$$

where the $f_i(t)$ are Gaussian random variables satisfying

$$\langle f_i(t) \rangle = 0 \qquad \text{and} \qquad \langle f_i(t) f_j(t') \rangle = 2\mathcal{A}_{ij} k_B T \delta_{tt'} \qquad (31)$$

One can show from Eqs. (30) and (31) that if Δt is small,

$$\langle \Delta s_i \Delta s_j \rangle = 2k_B T \Delta t \, \mathcal{L}_{ij} \qquad (32)$$

We now take the model shown in Fig. 5 and stretch it out along a line as shown in Fig. 6. Now the hairpin bends in Fig. 5 are assumed to have no length, so the model in Fig. 6 may be interpreted as that of a series of sticks $1, 2, \ldots, n+1$ of lengths $s_1, s_2, \ldots, s_{n+1}$ which are continually colliding with each other and transferring length to and from neighboring sticks. In a small time step Δt, each stick undergoes a random walk in x space with a diffusion constant inversely proportional to its length. After each time step the sticks are readjusted so that any overlap is divided equally between them as shown in Fig. 7. This allows one to write down equations for the change in length of each stick after a small time Δt. These are

$$2\Delta s_1 = -\sqrt{\frac{\gamma}{s_1}} f_1 + \sqrt{\frac{\gamma}{s_2}} f_2 \qquad (33)$$

and

$$2\Delta s_i = \sqrt{\frac{\gamma}{s_{i-1}}} f_{i-1} - \sqrt{\frac{\gamma}{s_{i+1}}} f_{i+1} \qquad i = 2, \ldots, n \qquad (34)$$

where $\gamma = 2k_B T \Delta t / \mu$ and the f_i's are Gaussian random variables satisfying

$$\langle f_i \rangle = 0 \qquad \text{and} \qquad \langle f_i f_j \rangle = \delta_{ij} \qquad (35)$$

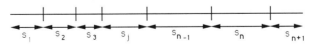

Figure 6 Here the diffusing pulley model shown in Fig. 5 has been stretched out into one dimension to form the "stick model." Each interbend length s_j may be thought of as a stick that is continually colliding and exchanging length with the surrounding sticks.

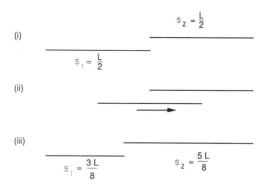

Figure 7 Stick model for a chain with a single hairpin. In (i) the two arms or sticks are of equal length. In (ii) the left stick jumps to the right by $L/4$ under the influence of thermal forces. In (iii) the jump is complete and the overlap length between the sticks is redistributed equally between them. The sticks are shown as being displaced vertically for clarity.

Using Eqs. (34) and (35), one can obtain expressions for the correlation functions $\langle \Delta s_i \, \Delta s_j \rangle$ which when they are compared to Eq. (32) give an explicit form for the mobility matrix:

$$\mathcal{L}_{11} = \frac{1}{4\mu}\left(\frac{1}{s_1} + \frac{1}{s_2}\right) \tag{36}$$

$$\mathcal{L}_{1j} = \mathcal{L}_{j1} = \frac{1}{4\mu}\left[\frac{1}{s_2}(\delta_{j,3} - \delta_{j,1}) - \frac{1}{s_1}\delta_{j,2}\right] \qquad j = 2,\dots,n \tag{37}$$

and

$$\mathcal{L}_{ij} = \frac{1}{4\mu}\left[\delta_{i,j}\left(\frac{1}{s_{i-1}} + \frac{1}{s_{i+1}}\right) - \frac{1}{s_{i-1}}\delta_{i-1,j+1} - \frac{1}{s_{i+1}}\delta_{i+1,j-1}\right] \qquad i,j = 2,\dots,n \tag{38}$$

It can be verified that Eq.(38) is indeed the inverse of the friction matrix (25). These equations allow us to form the derivatives of the mobility matrix needed in Eq. (30). Along with a redefinition of the random forces, this produces the simplified set of Langevin equations for multihairpin dynamics

$$\Delta s_i = g_i(t) + \frac{k_B T}{4\mu}\left[(\delta_{i,2} - \delta_{i,1})\frac{1}{s_i^2} + \delta_{i,n}\frac{1}{s_{n+1}^2}\right] \tag{34}$$

Here the $g_i(t)$ are Gaussian random variables satisfying

$$\langle g_i(t) \rangle = 0 \qquad \text{and} \qquad \langle g_i(t)g_j(t') \rangle = 2\mathcal{L}_{ij} k_B T \delta_{tt'} \tag{39}$$

To simulate the motion of an n-hairpin system it is merely now a matter of starting with some initial configuration of lengths between the hairpins and use Eq. (34) to iteratively calculate the lengths after each time step. This is done until two hairpins approach close to each other so that they annihilate

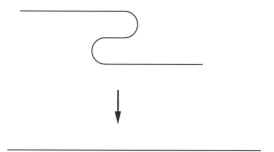

Figure 8 Process of internal annihilation of two hairpins. The two bends approach each other to within a few hairpin lengths and then annihilate because of potential effects.

(Fig. 8) or one falls off the end and is destroyed (Fig. 4). Both these events are driven by potential effects and are assumed to occur on time scales that are instantaneous compared to the diffusive time scales in the rest of the problem. The first type of event reduces the number of hairpins by two and the second type by one. For simulation purposes it is necessary to choose a small length d_{cul} as the annihilation length for the hairpins. This length is chosen to be much less than L/n. Once the number of hairpins has been reduced, a new mobility matrix needs to be calculated which has a different number of elements to the previous matrix. It is this complication of a varying number of variables in the problem that makes computer simulation all the more necessary. We have started the system in a given configuration with $\mathcal{N}$ hairpins with the lengths s_1 to $s_{\mathcal{N}+1}$ chosen from a uniform distribution. The system is then allowed to evolve and the number of hairpins decreases, at first rapidly and then more slowly. After a given number of time steps the initial number of hairpins is restored and the system is again allowed to evolve. This is done many times and the results averaged to reduce the noise in the data. We stress here that there is no hairpin creation allowed in this simulation; the hairpin bends merely move until they either fall off the end of the chain or are annihilated internally. This simulation provides a great deal of information about the distribution of the lengths $s_1, s_2, \ldots$. Here we prefer to concentrate on a simpler quantity, namely $P(n, t)$, the probability of finding n hairpins at time t.

In Figs. 9 and 10 can be seen the results of the computer simulation. The system was started with $\mathcal{N} = 30$ hairpins and the parameters chosen were $\mu = 0.5, L = 1.0$, and $k_B T = 1.7$. One unit of time on the graphs is 4.436×10^{-4} in the same units. Each "run" of the program was for 100 time units and data were averaged over 1000 runs. In Fig. 9 can be seen the average number of hairpins $\langle n(t) \rangle = \sum_{j=0}^{\mathcal{N}} j P(j, t)$ versus time. Naturally this decays from 30 toward zero because there is no hairpin creation, only annihilation. In Fig. 10 the full distribution function $P(n, t)$ versus n is plotted for various times. The distribution decays rapidly from its initial sharp peak at $n = 30$ to a "Gaus-

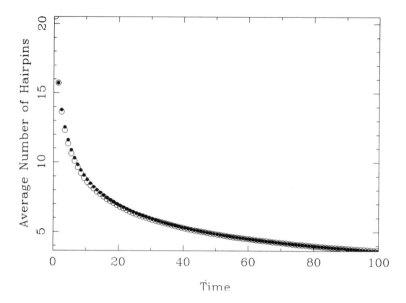

Figure 9   Average number of hairpins $\langle n(t) \rangle$ versus time for a chain originally with 30 hairpins. The open circles represent the data from the Langevin simulation described earlier. The closed circles are from results from the master equation approach. The uncertainties in the Langevin data are not visible on this scale.

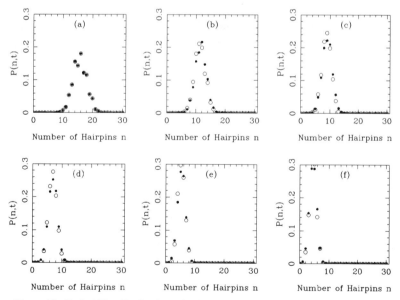

Figure 10   Probability distribution $P(n, t)$ for the number of hairpins versus n for a number of times. Again the open circles are from the Langevin computer simulation and the closed circles are the master equation solution. The times are (a) $t = 2$, (b) $t = 5$, (c) $t = 10$, (d) $t = 20$, (e) $t = 40$, and (f) $t = 60$.

sian" type of bump at $t = 2$. This bump then sharpens and moves toward $n = 0$. Of course, as t goes to infinity the bump becomes a sharp peak at $n = 0$.

MASTER EQUATION MODEL FOR MULTIHAIRPIN DYNAMICS

In the preceding section we showed how the probability of finding n hairpins in a given chain evolved with time based on a Langevin equation approach to the problem. It is of interest to enquire if this information could be gained from a simpler set of equations which involve only $P(n, t)$ and which ignore the length distributions. To write down an equation for $\partial P(n, t)/\partial t$ we adopt a pre-averaging approach. We would like to know how long it takes for a given hairpin bend in an n-hairpin chain to be destroyed. On average such a hairpin bend has to diffuse an arc distance of roughly $\langle d \rangle = L/(n + 1)$ before it meets another hairpin or falls off the chain end. We hypothesize here an average diffusion constant for this process which depends only on the average length of an arm, that is, $L/(n + 1)$. This diffusion constant must be proportional to $k_B T/\mu$ and is thus chosen to be $\langle D \rangle = k_B T (n + 1)/\mu L$. The time for a given hairpin in a given chain to be destroyed is then proportional to

$$\tau = \frac{\langle d \rangle^2}{\langle D \rangle} = \frac{\mu L^3}{k_B T (n + 1)^3} \tag{40}$$

When only hairpin destruction is allowed there are in general four processes which contribute to the time rate of change of $P(n, t)$. These are:

(a) $n + 2$ hairpins $\rightarrow$ n hairpins

(b) $n + 1$ hairpins $\rightarrow$ n hairpins

(c) n hairpins $\rightarrow$ n $-$ 2 hairpins

and

(d) n hairpins $\rightarrow$ n $-$ 1 hairpins

Processes (a) and (c) are caused by two hairpins colliding internally, whereas processes (b) and (d) are caused by a single hairpin bend falling off the end of a chain. Processes (a) and (b) cause $P(n, t)$ to increase, whereas (c) and (d) cause it to decrease. In a given chain with n hairpins there are only two "external" segments where a bend can fall off the end, whereas there are $n - 1$ "internal" segments that can be annihilated as a result of hairpin collisions. Thus the contribution to $\partial P(n, t)/\partial t$ from processes of type (c) will be roughly $(n - 1)/2$ times as large as that from type (d). If we examine only the contribution from (c), we can say that

$$\frac{\partial P(n, t)}{\partial t} = -\Omega \alpha (n + 1)^3 (n - 1) P(n, t) + \text{other terms} \tag{41}$$

Here $\Omega = k_B T / \mu L^3$, and α is a constant, positive numerical factor. The term $P(n, t)$ comes from the fact that the rate of decay must be proportional to the number of chains with n hairpins. If we include the contributions from all four processes, the equation becomes

$$\frac{1}{\Omega} \frac{\partial P(n, t)}{\partial t} = -\alpha(n + 1)^3(n - 1)(1 - \delta_{0,n})P(n, t) - \beta(n + 1)^3(1 - \delta_{0,n})P(n, t)$$
$$+ \alpha(n + 3)^3(n + 1)(1 - \delta_{\mathcal{N},n})(1 - \delta_{\mathcal{N}-1,n})P(n + 2, t) \qquad (42)$$
$$+ \beta(n + 2)^3(1 - \delta_{\mathcal{N},n})P(n + 1, t)$$

Here β is another positive numerical constant. The δ's occurring in this equation are there to ensure that the equation remains valid for the special cases where n takes extreme values. Thus, for instance, there is no process (c) or (d) when $n = 0$. The equation is valid for $n = 0$ to $n = \mathcal{N}$ and so really represents a string of coupled first-order linear differential equations with constant coefficients which could be written in matrix form $\dot{P} = AP$, where P is the vector $(P(0, t), \ldots, P(\mathcal{N}, t))$ and A is a constant matrix. Formally, this equation may be solved trivially to yield $P(t) = P_0(t) \exp(At)$, although in practice to evaluate $\exp(At)$ requires a knowledge of the eigenvalues of A. Here we solve the equations numerically. The equations contain two arbitrary constants α and β which are independent of any of the variables in the problem, such as temperature or $\mathcal{N}$.

We are now in a position to compare the results for $P(n, t)$ obtained from the set of master equations (42) with the results from the Langevin equation. The comparison may be seen in Figs. 9 and 10. The Langevin and master equation distribution functions $P_L(n, t)$ and $P_M(n, t)$ were made to agree at a particular time t_s (in this case $t_s = 2$) by setting $P_M(n, t_s) = P_L(n, t_s)$. The parameters α and β were chosen to give a good agreement between the distributions at later times. Here we chose $\alpha = 1.0$ and $\beta = 2.0$. The results are not very sensitive to the choice of β because for n large there are many more internal hairpins then external ones, and hence internal annihilation dominates over bends falling off the end of the chain. It is clear from Fig. 9 that the master equations predict rather accurately the average number of hairpins, for the Langevin simulation and master equation solution lie virtually on top of one another. A more stringent test of Eq. (42) is made in Fig. 10 by comparing the actual distributions. In Fig. 10(i), $P_M(n, t)$ was made to agree with $P_L(n, t)(t_s = 2)$ and in Fig. 10(ii) to (vi), we can see that the master solution mimics the shape of $P_L(n, t)$ even if it does not always accurately reproduce $P_L(n, t)$ for specific values of n.

CONCLUSIONS

In this chapter we have produced a model for the dynamics of hairpin defects which are likely to be important in the rheology and dielectric response of nematic polymer liquid crystals which have a semiflexible backbone. This model is the "diffusive pulley," whereby straight lengths of chain aligned along the nematic director are shuffled between hairpin bends. It has been put on a reasonably firm basis by reference to energy arguments and more important, by using the eigenmodes for a wormlike chain in a nematic field. The friction and mobility matrices for the n-hairpin case have been calculated and these have been used to simulate the motion of the many hairpins in a single chain using a series of Langevin equations. These equations were used to produce graphs of the distribution function $P(n, t)$, the probability of finding n hairpins on a given chain at time t. It was then shown that this distribution could reasonably be modeled accurately by a much cruder and simpler master equation approach. The main conclusion from this approach is summarized in master equation (42). The work here is clearly incomplete for it does not include the possibility of hairpin creation. However, by detailed balance and using the equilibrium distribution for the number of hairpins calculated [5] one can easily include terms for hairpin creation in the master equation.

ACKNOWLEDGMENT

The authors wish to thank Keith Gelling for useful discussions.

REFERENCES

1. D. R. M. Williams and M. Warner, *J. Phys. France* **51,** 317 (1990).
2. J. F. D'Allest, P. Sixou, A. Blumstein, R. B. Blumstein, J. Teixeira, and L. Noirez, *Mol. Cryst. Liq. Cryst.* **155,** 581 (1988).
3. M. Warner, J. M. F. Gunn, and A. B. Baumgärtner, *J. Phys.* **A18,** 3007 (1985).
4. P.-G. de Gennes, in *Polymer Liquid Crystals,* ed. A. Ciferri, W. R. Krigbaum, and R. B. Meyer, Academic Press, New York, 1982, Chap. 5.
5. J. M. F. Gunn and M. Warner, *Phys. Rev. Letts.* **58,** 393 (1987).
6. M. Doi and S. F. Edwards, *The Theory of Polymer Dynamics,* Clarendon Press, Oxford, 1986.

21

Polynematic Phases and Conformations: Theory and Simulation

CLIVE A. CROXTON
Department of Mathematics
University of Newcastle
Newcastle, New South Wales 2308
Australia

ABSTRACT

The configurational and phase transitional properties of a polynematic molecule are modelled in terms of a semi-flexibility coupled linear sequence of rods. The properties are determined on the basis of the iterative convolution approximation, and the results compared with a parametrically identical Monte Carlo simulation. The agreement between the two analyses is generally good to excellent. Both approaches indicate that the nematic-isotropic transition migrates to lower temperatures with increasing flexibility and shows a *reversal* back to higher temperatures prior to attaining full flexibility. The rod-coil and order-disorder transitions appear largely independent, particularly for fully flexible sequences. In all cases the "rod-coil" transition amounts to no more than a discontinuous elongation of the molecule, and only for the stiffest systems can a rod state be said to develop. A rationalization of the phase and conformational behavior of these polynematic systems is presented in terms of the thermal activation of hairpin states.

INTRODUCTION

Considerable interest has developed in the extension of our understanding of conventional low-molecular-weight mesogenic systems to polymeric mesophases which offer the possibility of intriguing new phenomena associated with the additional internal degrees of freedom available within the molecule itself. Of the two principal classifications of liquid-crystal (LC) polymers, main chain and side chain, the latter are probably of more interest to the LC materials chemist as alternative substrates for electro-optical displays, while a major component of the former classification is biological in origin, consisting of alternating sequences of helical and flexible polypeptide: indeed, Jähnig [1] was one of the first to discuss such systems in the context of the nematic environment of membrane structure. A second form of backbone structure consists of a linear sequence of stiff and less stiff sections, or a uniformly stiff (but not rigid) sequence that is characterized by a persistence length l, and as such is termed a wormlike chain having its flexibility distributed homogeneously along its length. Flory [2] modeled such a system as a sequence of flexibly connected rigid rods each of length l, in which case chain flexibility is localized at the junction points; subsequent mesogenic behavior was then treated as for conventional low-weight rigid-rod systems whose LC characteristics are determined by their axial ratio.

Analyses of the wormlike chain in the presence of an external potential presented by Jähnig, and more recently by ten Bosch et al. [3], were essentially perturbative in their approach, and as such were incapable of describing important qualitative characteristics of the system, such as the singular approach to the rod limit at low temperatures, with the associated development of hairpin states (de Gennes [4]). A treatment that embodies both the weak and strong limits of nematic potential has been developed by Warner et al. [5], who find that the problem is equivalent to the diffusion of the head of a unit tangent vector to the chain over a unit sphere, and may be described by a spheroidal wave equation whose eigenvalues and eigenvectors are determined in the limits of weak and strong nematic potential. More recently, Vroege and Odijk [6] have investigated induced chain rigidity on the basis of a nonlinear integrodifferential equation first formulated by Khokhlov and Semenov [7], again for wormlike chains interacting via hard-core repulsions in the second virial approximation.

In a variety of earlier papers (Croxton [11,12]) the iterative convolution (IC) description of polymer configuration was presented. In those papers the configurational properties of fully self-interacting polymer sequences were determined for a wide variety of polymer geometries (linear, ring, star, ladder) under various degrees of confinement (surfaces, occlusions), both in the absence of solvent and as a function of solvent quality. In addition to the geometrical properties, a priori determinations of the polymer scattering function were also presented on the basis of the IC technique. In all cases the only input was the specification of the intersegmental pair potential $\Phi(i, j)$ between each pair of segments i, j within the system, which may itself be homo- or heterogeneous. The objective of the IC treatment was to provide a unified approach to polymer structure and scattering, and certainly, comparison with Monte Carlo estimates of the various quantities appears to support the IC technique as a successful theory of polymer configuration.

In this chapter we make a direct comparison of the iterative convolution (IC) [15] and Monte Carlo (MC) [21] analysis of such polynematic systems *on the same parametric basis,* providing an unequivocal appraisal of both the technique and the results of the two approaches. The IC development of what might be termed the "correlation function approach" to the description of the configurational properties of polymers derives essentially from its successful application in the statistical theory of liquids. We provide an opportunity to present a coherent description of its application to polynematic macromolecules both in theory and simulation.

THEORY

Here we extend the iterative convolution technique [11] to the description of linear semiflexible nematic chains in which the coupling of the internal degrees of freedom causes a change in chain statistics and the development of orientational order. As such, the present analysis represents the *orientational* counterpart to previous IC descriptions which were based essentially on spatially dependent interactions between the segments. We model the chain as a sequence of unit rods (Fig. 1a) flexibly connected such that the bend potential between adjacent rods is given by

$$\Phi(\theta_{i,i\pm1}) = \begin{cases} 0 & \theta_{i,i\pm1} \leq \theta_{max} \\ +\infty & \theta_{i,i\pm1} > \theta_{max} \end{cases} \tag{1}$$

Clearly, the specification of the range θ_{max} specifies the flexibility of the sequence, becoming a stiff rod when $\theta_{max} = 0$, and totally flexible when $\theta_{max} = \pi$. The unit rods provide a natural measure of the persistence length l, and to this extent our model coincides with Flory's description [2].

Since we are concerned essentially with the *orientational* features of the system, in particular the orientation of the coupled rods with respect to an external director field, it is convenient to represent the system as a bunch of unit vectors as shown in Fig. 1b. If we designate the director by the vector 0, then given that the system is cylindrically symmetric about 0 [$\Phi(\phi_{oi}) = $ const., where Φ_{0i} is the rotational displacement of the ith rod about the vector 0], we need only to determine the angular distribution functions $Z(\theta_{ij}|N)$. The distribution $Z(\theta_{0j}|N)$ describes the orientation of the jth rod with respect to the director 0, while $Z(\theta_{ij}|N)$ describes the relative orientation of rods i and j, given of course, that they are in a coupled sequence of N rods subject to an external director field. Coupling between rods to the director field is discussed below (Eq. (3)].

The *order parameter S* for the set of N rods in the sequence with respect to the "external" vector "0" is defined in terms of the orientational average $\langle \cdot \rangle$ along the chain

$$S = \frac{1}{N}\sum_{j=1}^{N} S_j = \frac{1}{N}\sum_{j=1}^{N} \left\langle \frac{3\cos^2\theta_{0j} - 1}{2} \right\rangle$$

$$= \frac{1}{N}\sum_{j=1}^{N} \int_{1}^{-1} \left(Z(\cos\theta_{0j}|N) \frac{3\cos^2\theta_{0j} - 1}{2} \right) (d\cos\theta_{0j}) \tag{2}$$

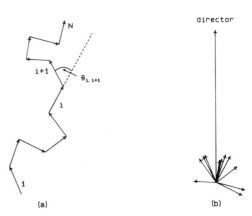

Figure 1 (a) Linear nematic polymer comprising N unit rods, each of which has its orientation with respect to an external director field 0 specified by (θ_i, ϕ_i). (b) Equivalent representation of the polynematic sequence shown in (a).

where S_j is the order parameter of the jth rod, and with a knowledge of the $Z(\cos\theta_{0j}|N)$ the order parameter may then be readily determined. This is a trivial extension of the conventional definition of the orientational order parameter for a single rod [10]. If we do not distinguish between parallel and antiparallel alignment with respect to the director, S varies between $+1$ and $-\frac{1}{2}$, corresponding to perfect alignment along and perpendicular to the director field, respectively. $S = 0$ corresponds to a random distribution of the rods with respect to the director: $Z(\cos\theta_{0j}) = $ const.† A clear distinction should be made between the "external" order parameter arising from the degree of rod alignment with respect to an external director and the "internal" order parameter relating to the alignment of the rods with respect to the long molecular axis of the sequence. (Note that by molecular axis we do *not* mean the chain backbone contour.) The distinction is perhaps clearest for a fairly stiff sequence in which there is an intrinsically high degree of internal alignment of the rods along the molecular axis and a correspondingly high value of the internal order parameter. The external order parameter with which we are concerned here, however, may nevertheless remain small or even zero for random orientations of the molecular axis with respect to the external field.

It may be argued that while the internal order parameter is sensitive to chain stiffness, the external function with which we are concerned here is only weakly dependent upon the stiffness of the chain. One can imagine progressively stiffening a linear sequence of rods and observing the order parameter. Ultimately, the chain reduces to a single rigid rod for which stiffness has no significance, and we must conclude in this case that stiffness and chain length can have no bearing on the external order parameter S. Similar reasoning may be applied to a perfectly flexible chain resulting in the same conclusion. We shall return to this point later. In the case of simple rigid linear nematics an association between the molecular axis, the director, and the order parameter is more easily made. Here, however, molecular flexibility introduces an additional complication giving rise to some dispute in the literature as to whether there is or is not close identification of the director with the molecular axis (see, e.g., de Gennes [4], Warner et al. [5], and Meyer [8]). Meyer [8] catalogs numerous qualitative differences between the two possibilities, the most notable being the projected divergence of the splay constant with molecular length in the case of close coincidence of the director and the molecular axis. We investigate this aspect of the system below in the course of our Monte Carlo simulations.

It should be said that Warner et al. specifically distinguish between the two possibilities, and emphasize that the molecular axis (chain backbone) in

---

† Note that for a uniform, spherically symmetric distribution of the end points of the vectors corresponding to a random distribution we require that $Z(\cos\theta_{0j}) = $ const. rather than $Z(\theta_{0j}) = $ const.

their wormlike model is *not* to be identified with the director. Nevertheless, the chain is visualized as meandering about the director, and it does appear that the mean molecular axis *does* remain in close registry with the director. This being the case, with no prospect of an orientational distribution of the mean molecular axes with respect to the director, Warner et al. are in fact determining the internal order parameter discussed above, and this will be sensitive to chain rigidity. To that extent these authors are calculating a different order parameter to that being considered here.

We work in the mean-field approximation of Maier and Saupe [9], who in their original presentation assumed that the stability of the nematic phase arises from the dipole–dipole part of the anisotropic dispersion forces, which in conjunction with an assumption of cylindrical symmetry about the preferred axis and in the absence of polarity and steric interactions yields

$$\Phi(\theta_{0i}) = \frac{-AS}{2V^2}(3\cos^2\theta_{0i} - 1) \tag{3}$$

for the orientation-dependent part of the potential energy of the ith rod with respect to the director. V is the molar volume, A is a constant taken to be independent of pressure, volume, and temperature, and S is as defined above. In fact, the change of volume across the nematic–isotropic transition is so small (Chandrasekhar [10]) that for the present purposes we may regard it as constant and absorb it into the parameter A; moreover, we prefer to work in terms of the reduced potential parameter $A^* = A/kT$ throughout. Other workers [16] have also presented justifications for the use of the original Maier–Saupe representation of the orientational potential adopted here, although we should point out that Vroege and Odijk [6] have recently criticized this approach.

It is not appropriate here to review the iterative convolution approach; for details of the technique we refer the reader to the literature (Croxton [11]). Basically, the technique permits the description, in terms of the spatial probability distributions $Z(ij|N)$, of the configurational properties of self-interacting polymer systems in which the only input specification is the set of pairwise interactions operating between each pair of monomers, i, j. Thus systems of nonlinear geometry, the incorporation of boundaries and other confinements, solvents, external fields, and so on, may be readily described, and as such the theory provides a unified statistical mechanical approach to the description of configurational and scattering properties of homo- and heteropolymeric systems. However, all previous applications of the technique have been based on central pairwise *radial* interactions, while this is its first application to orientationally dependent functions. Essentially, the IC integral equation relates the spatial probability distribution $Z(ij|N)$ between any two monomers i and j within the N-mer in terms of the propagation of correlation via all other monomers in the sequence. It may then be shown [11] that to a good approximation

$$Z(ij|N) = H(ij) \prod_{\substack{k=0 \\ \neq i,j}}^{N}{}' \int Z(ik|N)Z(kj|N)\, dk \tag{4}$$

$$H(ij) = \exp\left(-\frac{\Phi(ij)}{kT}\right) \tag{5}$$

where $\Phi(ij)$ is the interaction potential operating between segments i and j and Π' is the geometric mean of the $N-2$ convolution integrals arising from $1 \leq k \neq i,j \leq N$. A qualitative understanding of Eq. (4) may be gained by regarding $Z(ij|N)$ as comprising the "direct" Boltzmann distribution $H(ij)$, mediated in the field of the neighboring segments given by the geometric mean of the product of the normalized convolution integrals. For the present problem the spatial distributions take the form of angular functions and the IC equation takes the form†

$$Z(\cos\theta_{ij}|N) = H(\cos\theta_{ij}) \prod_{\substack{k=0 \\ \neq i,j}}^{N}{}' \int Z(\cos\theta_{ik})Z(\cos\theta_{kj})\, d\mathbf{k} \tag{6}$$

where $d\mathbf{k} = d\phi_{0k}\, d(\cos\theta_{0k})$ and the chain–director function

$$H(\cos\theta_{0j}) = \exp\left(A^*S\,\frac{3\cos^2\theta_{0j}-1}{2}\right) \tag{7a}$$

where the asterisk designates the reduced potential multiplier $A^* = A/kT$. A clear distinction should be made between chain–director functions $H(0j)$ and chain–chain functions, $H(ij)$. The chain–chain functions $H(ij)$ are rather more subtle in this orientational problem than in the purely spatial systems considered previously. In fact, the $H(ij)$ must first be determined as the appropriate *convolution of distributions* implied by the flexibility defined in Eq. (1) for the $|i-j|$ flexible joints between the two rods in question. Thus for two rods i and j their relative angular distributions are correctly given as the convolution of the intervening angular distributions:

$$H(\cos\theta_{ij}) = \int H(\cos\theta_{i,j-1})H(\cos\theta_{j-1,j})\, d(\mathbf{j}-\mathbf{1}) \tag{7b}$$

Although the deviation from a simple arithmetic addition and renormalization of the rectangular distributions is not great, Eq. (7b) is nevertheless technically correct, and since it is readily evaluated is used throughout these IC calculations (Fig. 2). The $H(ij)$ functions depend sensitively on both chain flexibility [the "bend" potential, Eq. (1)] and their relative separation, $|i-j|$. This is a fundamental difference for this system compared with homogeneous

---

†Note that for a uniform, spherically symmetric distribution of the end points of the vectors corresponding to a random distribution we require that $Z(\cos\theta_{0j}) = $ const. rather than $Z(\theta_{0j}) = $ const.

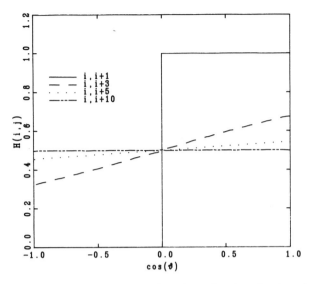

Figure 2 Direct interaction function $H(\cos \theta_{ij})$ as a function of various chain contour separations $|i - j|$ in a sequence having stiffness $\theta_{max} = \pi/2$ [Eq. (1)].

spatially dependent (as opposed to orientationally dependent) systems in which all $H(ij)$ are identical, regardless of i, j.

The vectors involved in the convolution integral are illustrated in Fig. 1c. The integral is evaluated numerically and iteratively for all orientations of the vectors i, j, k. It should be noted that the stiffness of the sequence implies maximum permissible angular ranges for the angles θ_{ik}, and so on, and these are determined as the cumulative sum of adjacent flexibilities specified in Eq. (1). Thus the maximum relative angular range between rods i and k is

$$\theta_{ik\,max} \leq |i - k|\theta_{max} \tag{8}$$

where θ_{max} is the maximum relative angular displacement of adjacent rods. Evaluation of the convolution integral [Eq. (6)] is practicable in real space because of the short range of the functions: the substantially longer-ranged integrals in previous applications of the IC technique have been handled by fast Fourier transform techniques.

A further complication in this IC analysis arises from the fact that one of the principal parameters to be determined in this investigation is the order parameter S, yet this function features explicitly in the specification of the orientational interaction, Eq. (3). In fact, S has to be determined self-consistently in the course of the iterative solution of the coupled integral equations, and we seek convergence of both S and the angular distribution functions, the former being related to the latter via Eq. (2).

The complete set of chain–director $[Z(\cos \theta_{0i}|N)]$ and chain–chain $[Z(\cos \theta_{ij}|N), i \neq j \neq 0]$ orientational distributions are determined via Eq. (6)

on the basis of which the order parameter [Eq. (2)] and the mean-square chain length $\langle R_N^2 \rangle$ may be determined. This latter quantity is defined as

$$\langle R_N^2 \rangle = N + 2 \sum_{i<j} \langle \cos \theta_{ij} \rangle \tag{9}$$

Similarly, the mean-square radius of gyration $\langle S_N^2 \rangle$ of the sequence may be determined:

$$\langle S_N^2 \rangle = \frac{1}{(N+1)^2} \sum_{i \le j} \langle R_{ij}^2 \rangle$$

$j = i$ is included because $\langle R_{ij}^2 \rangle$ is the mean-square length of the subchain consisting of links (not segments) i through j.

RESULTS AND DISCUSSION

In this section we present the results of the IC calculations [15], although frequent comparisons will, of course, be made with other analyses wherever possible. Unfortunately, however, there is some diversity in the literature regarding the basic parametric description of the problem, even including the specification of the mean-field interaction operating within the system, and accordingly, qualitative comparisons are only possible in the majority of cases. Clearly, direct comparison of the IC results with the MC data would appear to form the most appropriate basis for discussion, and this we shall do later in this chapter. We anticipate the comparison, however, with the observation that in general the qualitative agreement between the present calculations and the MC simulations is good to excellent, and the IC technique again appears to provide a sound semiquantitative description of this polymeric system.

In Fig. 3 we show $Z(\cos \theta_{0i}|N)$ as a function of reciprocal temperature A^* for $i = 5, N = 10$ for chains of stiffness $\theta_{max} = \pi/2$. This function corresponds to the orientational distribution of segment 5 within the 10-mer relative to the director field: distributions $Z(\cos \theta_{0i}|10)(i \ne 5)$ for other segments in the sequence show a marginal relaxation in their form as the ends of the chain are approached, but do not differ significantly from the distributions shown here. The most striking feature of these distributions is their very sharp transition from a uniform spherically symmetrical orientational distribution for $A^* \le 4.0$ to a strongly bimodal function at $A^* \ge 4.5$, suggesting a reduced reciprocal nematic–isotropic transition temperature $A_{NI}^* = (T_{NI}^*)^{-1}$ somewhere between these two values. Such distributions are fully consistent with the development of hairpins, although the latter cannot be explicitly resolved on the basis of the IC technique. Indeed, at large A^*, corresponding to temperatures well below the transition, the orientational distribution reflects the almost perfect alignment of the monomers along the director field for all stiffnesses, while at higher temperatures an almost uniform orientational distribution confirms that

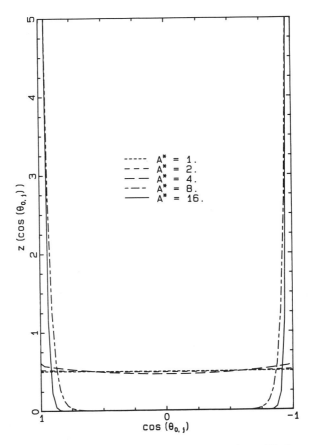

Figure 3 Normalized orientational distribution function $Z(\cos\theta_{0i}|N)$ determined on the basis of the IC approximation as a function of A^*, for $i = 5$, $N = 10$. Note the transition in the form of the distribution over the range $4.0 < A^* < 4.5$. Distributions are shown for a chain flexibility of $\theta_{max} = \pi/2$.

a thermotropic nematic–isotropic transition has taken place. These features are confirmed on the basis of the Monte Carlo determinations, although the latter estimates suggest a somewhat more relaxed orientational distribution for a given specification of N, A^*, and θ_{max}. This discrepancy is attributed to the approximate nature of the IC technique.

These orientational distributions may now be used to estimate the orientational averages $\langle\cos^2\theta_{0j}\rangle$, $\langle\cos\theta_{ij}\rangle$. On the basis of a knowledge of $\langle\cos^2\theta_{0j}\rangle$ the order parameter of the chain with respect to the external director may be determined, as defined in Eq. (2). In Fig. 4 we plot the family of order parameter curves as a function of chain stiffness and reciprocal temperature for $N = 10$. The curves exhibit a number of distinctive features, in particular a pronounced order–disorder transition in the orientational order parameter of

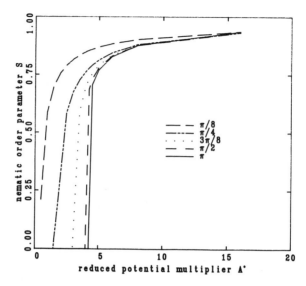

Figure 4 Order parameter curves S as a function of reciprocal temperature A^* and chain stiffness $0.125\pi \leq \theta_{max} \leq \pi$ for chains of length $N = 10$ in the IC approximation.

the system over a relatively narrow reciprocal temperature range. The transition appears to be first order and located at $A^* \sim 4.5$, which concurs very closely with the original Maier–Saupe estimate for nematic liquid crystalline systems for which $A_{NI}^* = 4.541$ [10].

As we have discussed already, a priori considerations suggest that the location of the transition should be independent of stiffness, and indeed, to anticipate the results of our Monte Carlo simulations, this does appear to be the case to within statistical error for the particular choice of flexibility given in Eq. (1). The somewhat more substantial spread in the location of the transitions on the basis of the IC technique must therefore be considered an artifact of the numerical analysis. Certainly, we could find no basis for its elimination. It should be said, however, that this conclusion is in marked contrast to the results of Warner et al. [5] and subsequently, Wang and Warner [14], who find that the location of the transition varies as the square root of the elasticity modulus ϵ or stiffness: $kT_{NI} \sim \sqrt{\epsilon}$. No comment is made by these authors regarding perfectly flexible ($\epsilon = 0$) or perfectly rigid ($\epsilon = \infty$) systems, particularly since these systems should reduce to the known result for a single rigid rod. Moreover, the result of Warner et al. appears at variance with the Monte Carlo results reported later in this chapter. However, the Warner result does appear more consistent with an estimate of the *internal* order parameter, that is, the degree of alignment of the chain along the mean molecular axis, although they frequently reassert that there is *not* a rigid identification of the molecular axis and the director in their treatment, and that their order parameter is determined with respect to the director. They also conclude that the transition is length dependent, again at variance with the IC and MC determinations.

The discontinuous behavior of the order parameter at the transition temperature strongly suggests that the phase transition is first order, even for the relatively short chains investigated here. This appears to concur with the conclusions of Warner et al. [5], but is in disagreement with ten Bosch et al. [3], who obtain a second-order transition. Again, these qualitative features are confirmed by the Monte Carlo determinations. Warner et al. discuss two limiting cases for the evaluation of the order parameter at the transition corresponding to a weak and a strong nematic limit. They find the discontinuity ΔS_{NI} to be 0.755 and $\frac{1}{3}$, respectively, while in both cases the transition temperature T_{NI} varies as the square root of the stiffness parameter. Unfortunately, both estimates are only to leading order and there is no guarantee that higher terms will not modify their conclusions. Indeed, further analysis by Wang and Warner [14] shows that higher terms may be significant. Certainly, both our IC and MC determinations are inconsistent with this conclusion.

The $\langle \cos \theta_{ij} \rangle$ yield the mean-square end-to-end length $\langle R_N^2 \rangle$ [Eq. (9)] and radius of gyration $\langle S_N^2 \rangle$, both of which are seen to depend sensitively on chain stiffness and reciprocal temperature A^* (Figs. 5 and 6). It is appropriate to observe from the outset that both these quantities depend fundamentally on the orientation, parallel or antiparallel, of the constituent rod segments. In this respect we anticipate some qualitative differences from the behavior of the order parameter reported above, and the dependence of these quantities on the development of hairpin states.

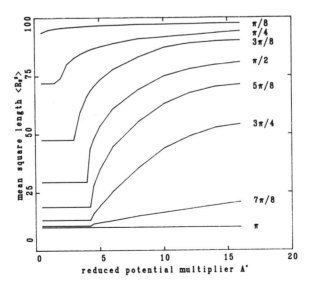

Figure 5 Mean-square end-to-end length $\langle R_N^2 \rangle$ as a function of A^* and chain stiffness $0.125\pi \leq \theta_{max} \leq \pi$ ($N = 10$) in the IC approximation.

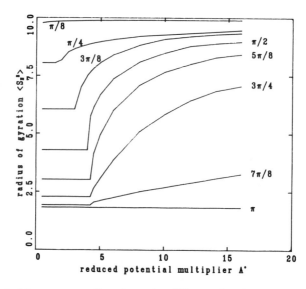

Figure 6   Mean-square radius of gyration $\langle S_N^2 \rangle$ as a function of A^* and chain stiffness $0.125\pi \leq \theta_{max} \leq \pi$ ($N = 10$) in the IC approximation.

At high temperatures, corresponding to small values of the reduced potential multiplier A^*, the mean-square length of the sequence corresponds to that of a constrained random walk. Thus for a chain of $N = 10$ segments $\langle R_N^2 \rangle = 100(10)$ for $\theta_{max} = 0$, (π) corresponding to a perfectly stiff and perfectly flexible chain, respectively. For these two limiting flexibilities it is straightforward to show that $\langle R_N^2 \rangle$ is independent of A^*, and this is confirmed by the IC calculations shown in Fig. 5.

For chains of intermediate flexibility the sequences show what is probably best described as an "elongation" with increasing A^*, coinciding with the order–disorder transition. Even for the highest values of A^* investigated, the system cannot be said to have achieved a rodlike state on the basis of estimates of $\langle R_N^2 \rangle$ and $\langle S_N^2 \rangle$ (Figs. 5 and 6). Certainly, the elongation is a sensitive function of chain stiffness: For the most flexible system the chain shows *no* elongation with increasing A^*, yet shows the most pronounced order–disorder transition. This behavior we are able to reconcile in terms of the development and entrapment of hairpins, and discuss their role on the basis of our Monte Carlo investigations later in this chapter.

At large flexibilities the role of hairpins becomes crucial. Their presence permits the development of a high degree of orientational order while substantially reducing the mean-square length of the system. With decreasing temperature these high-energy hairpins may be frozen out, the more readily the less flexible the system is. Hairpins are identified by $\cos\theta_{0j} \cos\theta_{0j+1} < 0$ at some point along the chain† for a given configuration, and unfortunately, we have not been able to explicitly investigate the development of hairpins on the basis of the IC technique: We can only surmise their existence on the basis of the observed behavior of S and $\langle R_N^2 \rangle$. Nevertheless, we discuss the contributions from hairpins below.

Results determined on the basis of the IC approximation are displayed in Fig. 5, from which we see that $\langle R_N^2 \rangle$ shows a discontinuous increase at the reduced isotropic–nematic transition temperature $T_{NI}^* = (A_{NI}^*)^{-1}$. Unlike the order parameter S, the mean-square length shows a sensitive dependence on flexibility: as mentioned above, this we attribute to the differing orientation dependence of S and $\langle R_N^2 \rangle$.

Of particular interest is the perfectly flexible chain $\theta_{max} = \pi$, which shows a strong first-order transition in the order parameter, while $\langle R_N^2 \rangle$ retains its random walk value regardless of the reduced temperature. It is a straightforward exercise to show that for perfectly flexible sequences of perfectly aligned rods ($S = 1$), the mean-square length is N, just as it is for the random walk result at $S = 0$. This initially surprising result is confirmed by the Monte Carlo calculations and illustrates the subtle role that stiffness plays in determining S and $\langle R_N^2 \rangle$. It is appropriate at this point to observe that such behavior is fully consistent with the development of hairpin states within the system, and a qualitative discussion from this point of view is given below. Although Khokhlov and Semenov [7] do not determine the order parameter for their semiflexible macromolecules, they do make passing reference to the fact that for perfectly flexible molecules the mean-square length varies insignificantly with external field, although they do report a discontinuous transition in the mean-square projection of the end-to-end distance vector on the orientation axis for such systems.

The mean-square radius of gyration $\langle S_N^2 \rangle$ shows a similar dependence to $\langle R_N^2 \rangle$ upon temperature and stiffness, as might be expected (Fig. 6). Although the scattering function for such chains cannot be calculated on the basis of the IC approximation, we nevertheless anticipate that the transition should be characterized by large anomalous scattering in the Guinier region with the discontinuous increase in $\langle S_N^2 \rangle$ at A_{NI}^*.

THE ROLE OF HAIRPINS

The order parameter and length transitions may be usefully discussed in terms of hairpins, or chain reversals; indeed, de Gennes [4] has already presented a qualitative discussion in these terms. Unfortunately, as we have already ex-

† The identification of $\cos \theta_{0i} \cos \theta_{0i+1} < 0$ as a hairpin does not necessarily require a sharp reversal in chain direction. Even for the stiffest chains capable of forming a hairpin (albeit in a large number of steps) there must be some point along the contour at which $\cos \theta_{0i} \cos \theta_{0i+1} < 0$.

plained, hairpin reversals cannot be identified on the basis of the IC analysis. Nevertheless, a qualitative discussion in terms of hairpins is possible, and their explicit contribution to the chain configurations will be presented below on the basis of a Monte Carlo analysis [21]. The system is regarded as a one-dimensional constrained random walk (constrained by the chain's intrinsic stiffness) linking a number n_H of hairpins. Clearly, the intrinsic stiffness of the chain implies a bending energy penalty for chain reversals executed in a small number of steps, while the external potential disfavors transverse excursions across the nematic field. Evidently, there is an energy E_H associated with the formation of a hairpin, the magnitude depending on the number of steps required to complete the reversal, the intrinsic chain flexibility, and the external potential multiplier A^*S. The probability of formation of a hairpin may be expressed in Boltzmann form and their development discussed in terms of thermal activation. Only at high temperatures (small A^*) can hairpin states be activated with significant probability, and only then in systems of sufficient flexibility. With decreasing temperature (increasing A^*), the hairpins are frozen out and the system makes its transition from coil to rod and the chain length increases accordingly. The rate at which the hairpins are eliminated from the system with decreasing temperature depends, of course, on E_H. For highly flexible chains the reversal can be effectively executed in two steps, incurring a very low energy penalty: Accordingly, such systems show a very gradual rod–coil transition (Fig. 5). Indeed, for perfectly flexible sequences, chain reversals along the director carry no energy penalty and cannot be thermally eliminated. Monte Carlo simulation appears to confirm this proposal and will be reported in some detail below.

MONTE CARLO SIMULATION

Given the somewhat idealised representation of the semiflexible macromolecule in the theoretical treatments, the most appropriate comparison with "experiment" is perhaps made in terms of a three-dimensional off-lattice Monte Carlo simulation where exactly the same parametric specification as for the theoretical calculations is used. We have previously reported in this chapter the results of an iterative convolution (IC) description of semiflexible chain molecules [15], and we now present the results of a Monte Carlo simulation of such a system providing a direct assessment of the IC analysis, and a basis for comparison with other theoretical treatments. We observe from the outset that the qualitative agreement between the IC description and the current Monte Carlo estimates is excellent, and quantitative agreement between the IC and MC estimates is generally good.

The various theoretical analyses of the rod–coil transition in semiflexible linear sequences are generally made as a function of reduced temperature or chain concentration, and although these treatments differ as to how they model the chain flexibility (i.e., whether the flexibility is localized at the rod junctions or distributed homogeneously along its length), all treatments appear to confirm the existence of a transition from the coil to an extended or "rod" state with decreasing temperature or increasing concentration. Whether the order–disorder and rod–coil transitions are distinct, and to what extent the stiffness and its distribution along the chain contour determine the nature of the transitions, represent some of the primary objectives of the Monte Carlo simulation apart from providing a direct basis for the assessment of the iterative convolution estimates previously reported. Associated with this phenomenon there will, of course, be consequences for the mean-square end-to-end length $\langle R_N^2 \rangle$ of the linear sequence of N rods and the mean-square radius of gyration $\langle S_N^2 \rangle$. Monte Carlo simulations of such systems have been reported by Bluestone and Vold [17], de Vos and Bellemans [18], Baumgartner and Yoon [19], and Khalatur et al. [20] invariably on the basis of chain concentration, which is only indirectly related to the present parametric description. Relatively little attention has been given to estimates $\langle R_N^2 \rangle$, $\langle S_N^2 \rangle$, the orientational distributions, the development of hairpins, and length and flexibility dependence, all of which are important aspects of the rod–coil transition and of considerable theoretical and experimental interest. De Gennes [4] has implicated the role of hairpins in the rod–coil transition, and accordingly, in this study we make a detailed assessment of their contribution to the geometrical properties of the system and their role in the reconciliation of a variety of transition phenomena which otherwise appear superficially inconsistent.

In the present three-dimensional continuum Monte Carlo analysis the sequence is modeled on the same parametric basis as for our earlier iterative convolution (IC) description [15]. That is, the chain is considered to consist of sequentially connected unit rods, each of which experiences an orientational potential with respect to an external director field, and a bend potential developed at the junction between sequentially adjacent rods. In this Monte Carlo study we consider two forms of bend potential:

1. A "rectangular" bend potential, in which adjacent rods are perfectly flexibly connected, but have a restricted relative angular range θ_{max}:

$$\Phi(\theta_{i, i \pm 1}) = \begin{cases} 0 & \theta_{i, i \pm 1} < \theta_{max} \\ +\infty & \theta_{i, i \pm 1} \geq \theta_{max} \quad 1 \leq i \leq N - 1 \end{cases} \tag{10a}$$

Thus, for a perfectly rigid rod, $\theta_{max} = 0$, while for a perfectly flexible chain, $\theta_{max} = \pi$ (Fig. 1). It is important to recognize that for this specification of chain flexibility the bend energy of the system is *zero* for all configurations satisfying the $\theta < \theta_{max}$ criterion. This choice of bend potential provides a useful means of assessment of the relative contributions of bend, flexibility, and orientational effects, particularly when taken in comparison with the harmonic bend potential, discussed below. The "rectangular" potential was used throughout the theoretical IC calculations.

2. A harmonic bend potential

$$\Phi(\theta_{i,i\pm1}) = k(1 - \cos\theta_{i,i\pm1})^2 \tag{10b}$$

between sequentially adjacent rods was also investigated, where k is a reduced stiffness constant or modulus of rigidity. In this model the bend energy is zero only for collinear configurations or when $k = 0$, representing a perfectly flexible system.

Excluded volume processes are neglected in this simulation; however, this is not expected to modify substantially the conclusions drawn regarding the orientational features of the system. An external director field, designated as "unit vector 0" is aligned along the z-axis, and the ith unit rod vector in the chain interacts with the director via a cylindrically symmetric mean field Maier–Saupe potential [9]:

$$\Phi(\theta_{0i}) = -A^*S \frac{3\cos^2\theta_{0i} - 1}{2} \qquad 1 \le i \le N \tag{11}$$

$A^* = A/k_BT$ is a reduced potential multiplier assumed independent of pressure and volume [10], A is the strength of the quadrupolar mean field and varies from substance to substance, θ_{0i} is the angle between the ith rod and the director, and S is the nematic order parameter determined with respect to the external director:

$$S = \frac{1}{N} \left\langle \frac{\displaystyle\sum_{i=1}^{N} \exp(-\Phi_i)S_i}{\displaystyle\sum_{i=1}^{N} \exp(-\Phi_i)} \right\rangle \tag{12}$$

where the order "instantaneous" parameter of the ith rod in a given chain configuration is

$$S_i = \frac{3\cos^2\theta_{0i} - 1}{2}$$

and where the angular brackets $\langle\cdot\rangle$ denote a configurational average over the chain. The reduced energy of the ith rod in a given chain configuration is

$$\Phi_i = \frac{1}{2}(\Phi_{i,i-1} + \Phi_{i,i+1}) + \Phi_{0i}$$

representing the sum of the bend [Eqs. (10a) and (10b)] and orientational [Eq. (11)] energy components, respectively.

It is particularly important to distinguish between the external order parameter considered here and the internal order parameter of the rods defined with respect to the principal molecular axis of the system. For a relatively stiff chain the latter order parameter may remain high even at the lowest values of A^* (high temperatures), while the external order parameter may assume much lower values, reflecting the orientational isotropy of the molecule. We determine both functions in the course of this investigation. Careful attention should be paid in comparing our results with other analyses in which the distinctions between director, molecular axis, and chain backbone are less precisely drawn, with some inevitable confusion in the resulting conclusions [4,5,14]. It follows that if we do not distinguish between parallel and anti-parallel alignment along the director, S varies between 1 and $-\frac{1}{2}$, corresponding to perfect alignment along and perpendicular to the director, respectively, while for a random distribution of the rods with respect to the director, $S = 0$.

Simulation of semiflexible chains of the kind described above presents an additional difficulty to those usually encountered in Monte Carlo simulation, and this arises from the fact that the chain–director interaction [Eq. (11)] is a function of the (unknown) order parameter S. Clearly, the order parameter must be determined self-consistently with the interaction function Eq. (2), and this is achieved as follows. An initial estimate for the value of the order parameter, say S_0, is chosen. (To expedite convergence, we generally take $S_0 = 0$ for $A^* < 4.5$ and $S_0 = 1$ for $A^* > 4.5$.) This particular choice is made on the basis of the known Maier–Saupe result for isolated single rods and the iterative convolution results reported above. However, in the vicinity of $A^* \sim 4.5$, *both* trial values ($S_0 = 0, 1$) are adopted for a given value of A^* when we find a pair of equilibrium values for the order parameter reflecting the instability of the system and a sensitivity as to whether the transition is approached from the high as the low side of A_{NI}^*. Then a linear, random, three-dimensional sequence is put down and the order parameter of that configuration determined on the basis of Eq. (3). Equation (11) with the estimated value S_0 is used to provide a weighting function $W(A^*, S_0) = \sum_{i=1}^{N} \exp(-\Phi_i)$ for that configuration,† and a mean first refinement of the order parameter, say S_1, over a batch of 500 configurations is calculated using weighted Boltzmann statistics. (Note that Φ_i contains both bend and orientational contributions to the total energy of the configuration.) In this technique, rather than generate configurations with a Boltzmann frequency by means of Metropolis sampling, we generate *random* configurations and weight them with the appropriate Boltzmann factor. The two techniques may be shown to be equivalent in the limit of a large number of generated configurations.

Batch means are treated as single measurements and accumulated until the confidence interval for S_1 is less than a predetermined limit; then the new value S_1 is compared with the original estimate, S_0. If the two values are consistent (our criterion is that they do not differ by more than 0.005), we begin to accumulate statistics for chains using the equilibrated S_1; if not, we substitute the new value, S_1, as our working estimate and repeat the process,

---

† The equivalence of random configurations, but weighted by the Boltzmann factor and the conventional Metropolis approach to Monte Carlo simulation, is exploited here. It enables a complete parametric set of the potential multipliers A^* to be investigated on the basis of one complete configurational run.

producing a further refinement, S_2, and so on. For the first estimate S_1 the limit on the confidence interval is fairly large (generally five times the final difference allowed between input and output S), to allow faster movement toward a more closely equilibrated value of S. As the value of S equilibrates, the limit on the confidence intervals is adjusted downward, ending at half the difference allowed between input and output S, or 0.0025.

We point out that we use *batch* means rather than individual configuration values for determining confidence intervals throughout the programs because the theoretical basis for the estimate of confidence intervals presumes that the data are measurements normally distributed about a true mean value [22]. The various quantities we measure may or may not satisfy this condition if individual configuration values are used. There is an optimum batch size, however: If the batch sizes are too small, they may not yield a normal distribution of means, while if they are too large (for a given investment of computer time), the statistical data base may become inadequate. We found that increasing batch size from 100 to 500 configurations yielded an increase in our estimate of mean-square length for $\theta_{max} = 3\pi/8, N = 100$ by almost 50%, while increasing batch size further from 500 to 1000 configurations per batch yielded only a further 1% increase in the estimated mean-square length. We therefore chose batch sizes of 500 configurations as representing the optimum size, given the constraints of time and adequacy of the statistics.

Once the equilibrated value of S has been determined it may be used in Eq. (11) and the statistics allowed to accrue in the course of estimation of the configurational properties of the system. Thus the mean-square separation between the centers of rods p and q is

$$\langle R_{pq}^2 \rangle = \left\langle \frac{\exp\left(-\sum_i \Phi_i\right) R_{pq}^2}{\exp\left(-\sum_i \Phi_i\right)} \right\rangle \tag{13}$$

and the mean-square radius of gyration

$$\langle S_N^2 \rangle = \frac{1}{(N+1)^2} \sum_{p \geq q} \langle R_{pq}^2 \rangle \tag{14}$$

are also determined on the basis of the batched statistics described above, the simulations continuing until the errors are <2% for the longest chains investigated ($N = 100$).

Hairpin reversals along the director are identified by the product $\cos \theta_{0i} \cos \theta_{0i+1} < 0$. Note that this criterion does *not* necessarily require a sharp reversal of the chain backbone: Even for stiff chains in which chain reversals along the director are necessarily spatially extended structures, there must arise at some point along the contour the condition $\cos \theta_{0i} \cos \theta_{0i+1} < 0$. By averaging over the batches of configurations we are able to determine the average number of hairpins $\langle n_H \rangle$ as a function of chain length, stiffness, and reciprocal temperature and to determine their role in the rod–coil and order–disorder transitions.

Finally, we identify an internal order parameter representing the degree of alignment of the rods along the principal molecular axis of the system. This axis is identified as that which minimizes $\sum_i R_i^2$, the sum of the perpendicular distances from each rod center to the molecular axis for a given chain configuration. Clearly, for stiff systems there will always be a high degree of internal alignment along the principal axis, and we anticipate that the internal order parameter will be a much more sensitive function of length and stiffness than its external counterpart measured with respect to the director field.

RESULTS AND DISCUSSION

In Fig. 7 we show the nematic order parameter S determined relative to the external director field as a function of reduced reciprocal temperature A^* for

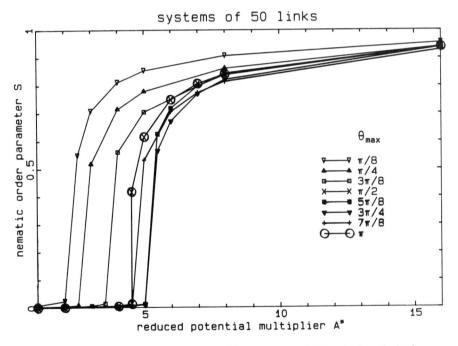

Figure 7 Nematic order–disorder transition parameter S determined on the basis of Monte Carlo simulations as a function of reduced potential multiplier A^* for $N = 10$ and $N = 100$. The first-order transition appears totally independent of length and chain stiffness and coincides closely with the Maier–Saupe result for a single rod $A^* = 4.541$. The iterative convolution results for $N = 10$ and $\theta_{max} \leq \pi/2$ are shown for comparison.

$N = 100$. To within statistical accuracy the curve shows a total independence upon chain length (for $N \gtrsim 10$) and stiffness regardless of whether rectangular or harmonic bend potentials are used. These conclusions were anticipated above where the iterative convolution results were presented. For comparison we show the corresponding IC result for the rectangular bend potential ($\theta_{max} = \pi/2, N = 10$), and the agreement with the MC data is seen to be good. Certainly, the IC predictions do appear to show a weak dependence of the location of the transition on stiffness, but this is not systematic and we believe this to be an artifact of the IC technique.

A striking feature of the order parameter curve is the first-order transition at $A_{NI}^* \sim 4.5$, which coincides very closely with the Maier–Saupe result ($A_{NI}^* = 4.541$) for a perfectly stiff rod and a perfectly flexible chain. Within these two limiting stiffnesses it is hard to see how any substantial stiffness dependence can possibly develop. Neither the IC nor the MC determinations appear to confirm the highly sensitive dependence of A_{NI}^* on flexibility predicted by Warner et al. [5,14], although their theoretical estimates were based on a "worm" model in which chain flexibility was homogeneously distributed along the contour length rather than the localized flexibility at rod junctions adopted here and in our IC analysis. Even so, it seems unlikely that such fundamentally distinct behavior can be accounted for purely on the basis of the distribution of chain flexibility. However, orientational confinement of the principal molecular axis does appear to introduce length and stiffness dependences, and we discuss these aspects below.

In Fig. 8 we show the internal order parameter S_{int} versus A^*, defined similarly to its external counterpart, but with respect to the principal molecular axis of the system. It is immediately apparent that this function is highly sensitive to chain length and stiffness, and although there is evidence of a transition in S_{int}, this is undoubtedly a projection of S along the principal molecular axis. Indeed, like S, the location of the transition in S_{int} along the A^* axis appears quite independent of both length and stiffness.

At low A^* (high T) $< A_{NI}^*$ the only internal ordering influence is the stiffness, and S_{int} is obviously nonzero for all $k > 0$. The degree of internal ordering is obviously higher for large k than for small k, and more so in small N than for large. For $A^* > A_{NI}^*$ orientational effects also contribute, but the principal molecular axis need not align along the director in the nematic phase, particularly for small k (e.g., Fig. 9a). Again, we have the same N dependence as for low A^*.

For stiffer systems, however, there will be some association of S_{int} with S at $A^* > A_{NI}^*$, and although the alignment of the molecular axis may still be only weakly coupled to the director field, a component of the individual rod alignments producing high S will make some contribution to the development of S_{int}. The very strong N, k dependence of S_{int} may well be confused with and attributed to the behavior of S, particularly if it is believed that geometric packing of the system implies the close registration and identification of

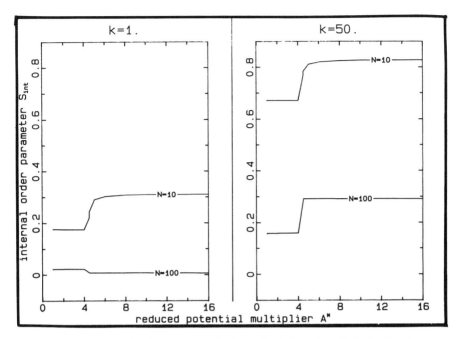

Figure 8 Internal order parameter S_{int} determined along the principal molecular axis as a function of the reduced potential multiplier A^* for $N = 10$ and $N = 100$ corresponding to highly flexible chains (a), $k = 1.0$, and stiff chains (b), $k = 50.0$.

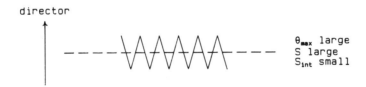

Figure 9 Coupling of S and S_{int} via the principal molecular axis for (a) flexible chains and (b) stiff chains.

molecular and director axes. A principal molecular axis will exist for all systems having $k > 0$ or $\theta_{max} < \pi$ for all $A^* > 0$; however, in the first set of simulations, no constraint is placed on its orientation. If, for whatever reason, the principal molecular axis *is* orientationally constrained, it follows that the internal ordering $S_{int}(k, N)$ will project a length and stiffness dependence into S, and this may well account for the dependences reported by Warner et al.

To illustrate the effects of orientational constraint on the principal molecular axis, the Monte Carlo runs were repeated, recording data only for those configurations whose principal axis fell within a prescribed angular range Ω measured with respect to the director. The individual rods constituting the molecule were not subjected to any orientational confinement, and continued to experience only the bend and orientational forces already described for the unconfined systems.

The same equilibration and batching procedures were followed as for the unconfined sequences; however, the equilibrium values of the external order parameter differed from the unconfined systems and showed both a length and stiffness dependence notably absent from the rotationally free chains. Moreover, there was a systematic decrease in A_{NI}^* with increasing orientational confinement of the principal molecular axis. This suggests that if the principal molecular axes of the molecules are believed to be in close registry with the director, and that the order–disorder transition is fully accounted for in terms of the meanderings of the chain backbone about the principal axis, we must anticipate substantially different behavior to that of a chain whose principal axis is only partially confined, or indeed, totally unconfined (Fig. 10).

Returning to the orientationally unconfined system, the mean-square end-to-end length $\langle R_N^2 \rangle$ and the mean-square radius of gyration $\langle S_N^2 \rangle$, unlike the order parameter S, both show a highly sensitive dependence on both length and stiffness of the sequence (Figs. 11 and 12). However, the qualitative form of the curves appears insensitive to the detailed nature of the bend potential. In all cases, except for perfectly flexible systems, there is an expansion in both $\langle R_N^2 \rangle$ and $\langle S_N^2 \rangle$ at A_{NI}^* regardless of stiffness; the mean-square length and radius of gyration show a progressively weaker transition with increasing flexibility. The distinction between the order–disorder parameter and the rod–coil transition is at its most marked for perfectly flexible systems when the latter retains its random walk values throughout, regardless of the pronounced transition in S. We do not agree that the rod–coil and order–disorder transitions can be separated simply by making the chains long enough, as suggested by Wang and Warner [14] in the context of the worm model. We note that in the nematic phase ($A^* > 4.5$) $\langle R_N^2 \rangle$ and $\langle S_N^2 \rangle$ grow very slowly with A^*, if at all, and appear to show no tendency toward their rod limits.

Basically, for both the "rectangular" and harmonic bend potentials, both the configurational means $\langle R_N^2 \rangle$ and $\langle S_N^2 \rangle$ assume plateau values, independent of A^*, in the isotropic ($A^* < A_{NI}^*$) and nematic ($A^* > A_{NI}^*$) regions, with a sharp transition between the two at $A^* \sim A_{NI}^*$. This behavior is contrary to a

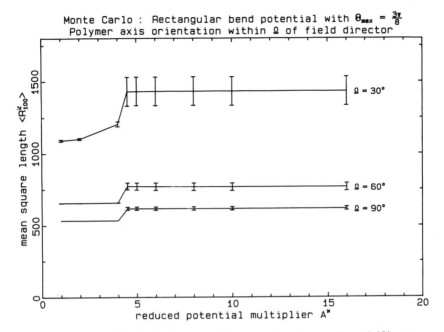

Figure 10 (a) Shift in the location of the order–disorder transition $A_{NI}^*(\Omega)$ as a function of the orientational confinement Ω of the principal molecular axis. (b) Effect of chain length on the order–disorder transition for orientationally confined sequences.

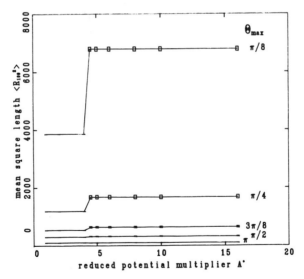

Figure 11 Mean-square end-to-end length $\langle R_N^2 \rangle$ for chains of length $N = 100$ as a function of reduced potential multiplier A^* for various stiffnesses $(0.125\pi \leq \theta_{max} \leq \pi)$.

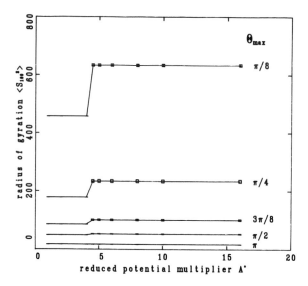

Figure 12 Mean-square radius of gyration $\langle S_N^2 \rangle$ for chains of length $N = 100$ as a function of reduced potential multiplier A^* for various stiffnesses $(0.125\pi \leq \theta_{max} \leq \pi)$.

priori expectations; however, we attribute this to the entrapment of hairpins, as we discuss in more detail below.

Given that the mean-square length for perfectly flexible (random walk) systems shows no variation with A^*, the ratio $\langle R_N^2 \rangle/$(random walk value) is of interest, and we show this quantity as a function of A^* and chain flexibility in Fig. 13. The stiffer chains are expanded with respect to their more flexible counterparts as expected, and chains of a given flexibility are expanded in the nematic phase $A^* > A_{NI}$ with respect to their isotropic counterpart ($A^* < A_{NI}$). Of some interest, however, is the *separation* of the two curves of given flexibility, representing the size of the elongation at A_{NI}^* with increasing chain length. The data clearly suggest that the size of the expansion progressively decreases with increasing N at a rate that depends quite sensitively on stiffness. A similar conclusion holds for the radius of gyration data. Moreover, it appears from the tendency of the ratio to become horizontal with increasing N that the asymptotic value of the exponent $\gamma \to 1$ in the representation

$$\langle R_N^2 \rangle = CN^\gamma$$

where $C(k, A^*)$ is a constant of proportionality. Similar observations again hold for $\langle S_N^2 \rangle$.

The inevitable conclusion appears to be that although the order–disorder transition remains unaffected by the chain length and stiffness, at least for orientationally unrestricted molecules, the geometrical parameters $\langle R_N^2 \rangle$, $\langle S_N^2 \rangle$ show a discontinuous *elongation* of the sequence at A_{NI}^*. This elongation,

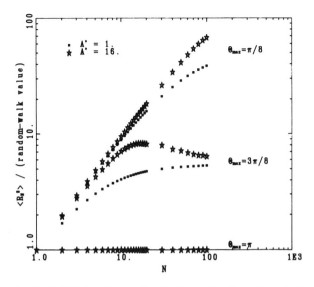

Figure 13 Ratio $\langle R_N^2 \rangle$/(random walk value) as a function of length for various stiffnesses at $A^* = 1$ and $A^* = 16$.

however, appears to be progressively "accommodated" within the internal structure of the molecule with increasing N, although the rate at which this accommodation develops quite clearly depends on the stiffness of the system. The system appears to tend toward an expanded random walk dependence on chain length, where the expansion factor C clearly depends on the stiffness of the system.

Hairpin reversals along the director field are to be identified as $\cos \theta_{0i} \cos \theta_{0i+1} < 0$. Even for relatively stiff chains in which hairpins are necessarily spatially extended features of the sequence there will nevertheless be an adjacent pair of rods for which the cosine product will be less than zero. Generally, the energy of the hairpin E_H will comprise a bend component E_B and a nematic component E_S arising from the traversal of the director field:

$$E_H = E_B + E_S$$

Hairpins are high-energy occlusions within the system and may be expected to exhibit an Arrhenius type of behavior. Indeed, de Gennes [4] discusses the probability of their formation in terms of a Boltzmann factor

$$\langle n_H \rangle \sim C \exp(-A^* E_H) \tag{15}$$

where $\langle n_H \rangle$ represents the number of hairpins and C is a constant of proportionality within a chain of given length and stiffness. At first sight the MC data (Fig. 14) do not appear to support such a dependence: $\langle n_H \rangle$ shows two distinctly constant, temperature-independent values in the isotropic and nematic phases connected by a discontinuous drop at A_{NI}^*. However, in the isotropic

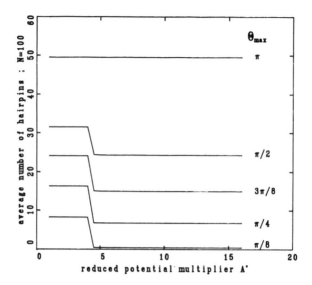

Figure 14 Number of hairpins $\langle n_N \rangle$ versus A^* for chains of length $N = 100$ of various stiffnesses.

phase $(A^* < 4.5)$ $E_S = 0$ since $S = 0$, and for the "rectangular" bend potential [Eq. (10a)] $E_B = 0$ within the allowed range of flexibility $\theta < \theta_{max}$, and accordingly we have

$$\langle n_H \rangle \sim C \qquad A^* < A^*_{NI} \tag{16a}$$

Similarly, in the case of the harmonic bend potential we have $E_B \sim k_B T = (A^*)^{-1}$, whereupon

$$\langle n_H \rangle \sim Ce^{-1} \qquad A^* < A^*_{NI} \tag{16b}$$

In other words, $\langle n_H \rangle$ should be constant throughout the isotropic phase $(A^* < A^*_{NI})$, and this behavior appears to be confirmed by the Monte Carlo data for both the rectangular and harmonic bend potentials.

At $A^* \sim A^*_{NI}$, S discontinuously increases from zero, as does E_S, with an associated discontinuous drop in the number of hairpins. We conclude that at lower temperatures (increasing A^*) the hairpins become "frozen in": The activation energy associated with the cooperative traversal of the nematic field by large sections of the macromolecule required to release the hairpin is too great and further thermal elimination of hairpins does not occur whereupon there remains a residuum of trapped hairpins. This accounts at once for the sustained high levels of the order parameter S with increasing A^*, while $\langle R_N^2 \rangle$ and $\langle S_N^2 \rangle$ remain substantially reduced by the occlusion of hairpins even for the highest values of A^*.

It is appropriate to point out that for perfectly flexible sequences $E_B = 0$ throughout and $E_S = 0$ in both the isotropic and nematic phases with the result that $\langle n_H \rangle = (N - 1)/2$ for all A^*. This result is clearly confirmed by the simulations. A number of workers [4,5] suggest that

$$E_H \sim \sqrt{\epsilon S} \tag{17}$$

where S is the order parameter appropriate to the particular value of A^* for a chain of given length and flexibility and ϵ is a constant, independent of S and A^* for a chain of given flexibility. We do not, however, find any support for this proposal on the basis of the present MC data.

The development of $\langle R_N^2 \rangle$ with A^* occurs largely (though not entirely) by thermal elimination of hairpins within the system, and given the double-plateau form for the dependence of $\langle n_H \rangle$ on reciprocal temperature, the behavior of $\langle R_N^2 \rangle$ and $\langle S_N^2 \rangle$ is readily understood. Alignment of the system along the director may proceed regardless of the thermal elimination of hairpins, and accordingly, we may reconcile the development of the order parameter S with that of $\langle R_N^2 \rangle$ and $\langle S_N^2 \rangle$, particularly for highly flexible systems that seem to contradict a priori expectations.

COMPARISON OF THE IC AND MC ESTIMATES

One of the primary objectives of these studies is the direct comparison of theoretical and simulated estimates on the basis of the same parametric description: Almost invariably small, but possibly significant differences between the various investigations reported in the literature prevent an unequivocal comparison of the results presented. For this reason all previously reported applications of the iterative convolution technique to polymeric systems have been complemented by Monte Carlo simulations on precisely identical parametric bases, and just such a direct comparison is made here. For computational reasons the IC determinations are necessarily restricted to relatively short sequences ($N \leq 20$), and we report here a comparison of the IC and MC results for $N = 10$. The presentation of the MC data for these shorter chains also provides an illustration in the shift in behavior from the longer MC sequences reported elsewhere in this chapter.

In Fig. 15 we compare the order parameters for $N = 10$, $\theta_{max} = \pi/2$ on the basis of the IC and MC treatments. The agreement is seen to be good over the entire range of A^* investigated. The MC curve has virtually achieved its asymptotic form by $N = 10$, after which it shows little further development with chain length. The location of the transition at $A_{NI}^* \sim 4.5$ coincides well for both the IC and MC analyses, and although the location of the IC transition showed a small, nonsystematic variation in its location with stiffness, this was believed to be an artifact of the numerical analysis. For the range of chain lengths investigated, the independence of the order–disorder transition curve on both length and stiffness appeared to be confirmed by the MC results.

The mean-square end-to-end length $\langle R_{10}^2 \rangle$ for IC and MC chains of length $N = 10$ and flexibility $\theta_{max} = \pi/2$ are compared in Fig. 16, from which we see

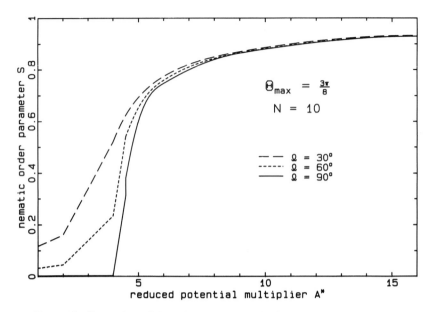

Figure 15 Comparison of the order parameters $S(A^*)$ on the basis of the iterative convolution (IC) and Monte Carlo (MC) analyses for $N = 10$, $\theta_{max} = \pi/2$.

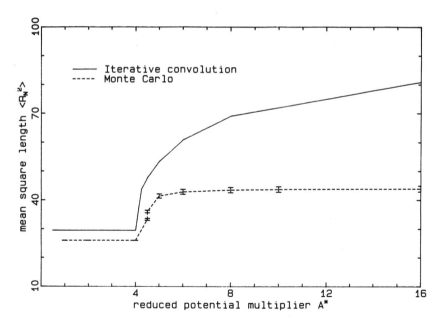

Figure 16 Comparison of the iterative convolution (IC) and Monte Carlo (MC) estimates of the mean-square end-to-end length $\langle R_{10}^2 \rangle$ for $N = 10$, $\theta_{max} = \pi/2$.

that both analyses predict a discontinuous elongation at the order–disorder transition. The amplitude of this sudden elongation is a function of both chain length and stiffness, and both the IC and MC analyses concur that for perfectly flexible sequences ($\theta_{max} = \pi$) there is *no* elongation, either systematic or discontinuous, with increasing A^* notwithstanding the pronounced transition in the order parameter. We note that for $A^* < A_{NI}^*$ the IC and MC estimates of $\langle R_{10}^2 \rangle$ are in good agreement, but that with increasing A^* the curves diverge somewhat, the MC data assuming a plateau value while the IC results show a continuing elongation toward a rodlike state ($\langle R_{10}^2 \rangle_{rod} = 100$). Similar behavior is observed for the mean-square radius of gyration $\langle S_{10}^2 \rangle$. Unfortunately, the iterative convolution technique does not permit an estimate of the number of hairpins which on the basis of the Monte Carlo investigations is believed to govern the behavior of the geometrical properties of the sequence, and so no immediate conclusions regarding their differing behavior can yet be drawn. We do note, however, that with increasing flexibility the IC estimates of both $\langle R_{10}^2 \rangle$ and $\langle S_{10}^2 \rangle$ reveal a progressive tendency to plateau for $A^* > A_{NI}^*$, culminating, of course, in the totally horizontal curve for perfectly flexible sequences mentioned above.

REFERENCES

1. F. Jähnig, *J. Chem. Phys.* **70**, 3279 (1979).
2. P. Flory, *Macromolecules* **11**, 1141 (1978).
3. A. ten Bosch, P. Maissa, and P. Sixon, *J. Phys. (Paris) Lett.* **44**, L105 (1983).
4. P.-G. de Gennes, in *Polymer Liquid Crystals,* ed. A. Ciferri, W. R. Krigbaum, and R. B. Meyer, Academic Press, New York, 1982, Chap. 5.
5. M. A. Warner, J. M. F. Gunn, and A. B. Baumgärtner, *J. Phys.* **A18**, 3007 (1985).
6. G. J. Vroege and T. Odijk, *Macromolecules* **21**, 2848 (1988).
7. A. R. Khokhlov and A. N. Semenov, *J. Phys.* **A15**, 1361 (1982).
8. R. B. Meyer, in *Polymer Liquid Crystals,* ed. A. Ciferri,, W. R. Krigbaum, and R. B. Meyer, Academic Press, New York, 1982, Chap. 6.
9. W. Maier and A. Saupe, *Z. Naturforsch.* **13a**, 564 (1985); **14a**, 882 (1950); **15a**, 287 (1960). For an introduction to intermolecular forces, see J. O. Hirschfelder, C. F. Curtiss, and R. B. Bird, *Molecular Theory of Gases and Liquids,* Part 3, Wiley, New York; Chapman & Hall, London, 1954.
10. S. Chandrasekhar, *Liquid Crystals,* Cambridge University Press, Cambridge, 1977.
11. C. A. Croxton, *J. Phys.* **A17**, 2129 (1984).
12. See, for example, C. A. Croxton, *Macromolecules* **21**, 3023 (1988), and references contained therein.
13. P. G. Khalatur, Y. G. Papulov, and S. G. Pletneva, *Mol. Cryst. Liq. Cryst.* **130**, 195 (1985).
14. X-J Wang and M. A. Warner, *J. Phys.* **A19**, 2215 (1986).
15. C. A. Croxton, *Macromolecules* **23**, 2270 (1990).
16. G. M. Luckhurst and C. Zannoni, *Nature* **267**, 412 (1977).

17. S. Bluestone and M. J. Vold, *J. Chem. Phys.* **42,** 4175 (1965).

18. E. de Vos and A. Bellemans, *Macromolecules* **7,** 809 (1974).

19. A. Baumgartner and D. Y. Yoon, *J. Chem. Phys.* **79,** 521 (1983).

20. P. G. Khalatur, Y. G. Papulov, and S. G. Pletneva, *Mol. Cryst. Liq. Cryst.* **130,** 195 (1985).

21. C. A. Croxton, *Macromolecules* (in press) (1989).

22. P. R. Bevington, *Data Reduction and Error Analysis for the Physical Sciences,* McGraw-Hill, New York, 1969.

22

Computer Simulations of Structural and Dynamical Properties of Polymers in the Solid State*

B. G. SUMPTER, D. W. NOID, and B. WUNDERLICH
Chemistry Division
Oak Ridge National Laboratory
Oak Ridge, Tennessee 37831
and Chemistry Department
The University of Tennessee
Knoxville, Tennessee 37996
and

S. Z. D. CHENG
Institute and Department of Polymer Science
University of Akron
Akron, Ohio 44325

*The submitted manuscript has been authored by a contractor of the U.S. Government under contract No. DE-AC05-840R21400. Accordingly, the U.S. Government retains a nonexclusive, royalty-free license to publish or reproduce the published form of this contribution, or allow others to do so, for U.S. Government purposes.

ABSTRACT

Molecular dynamics is used to study various structural and dynamical properties of polymers in the solid state. In particular, the details of structural changes of a polymer crystal going from the crystalline to the melted phase and the resulting dynamics are elucidated. Spectra and heat capacities are also calculated to reveal a number of interesting temperature-dependent phenomena.

INTRODUCTION

Details at the atomic or molecular level are important for a better understanding of polymer processes. A more detailed knowledge of both structure and dynamics is needed. The classification of the condensed states of matter can be summarized as shown in Fig. 1. In this diagram the large-amplitude mobility decreases from bottom to top. The extreme of order is reached for the crystalline solids, in which atoms assume periodic arrays. At the other extreme are the melt and the amorphous glass, in which the atoms or molecules are completely disordered and the systems are both orientationally and positionally isotropic. The transitions between the states that go with a change in order (entropy and thus heat of transition) are marked on the right side of the diagram. The transitions that go without change in order are glass transitions and are marked on the left. Between the two extremes (solids and fluids) are several other well-defined states, the mesophases. In the liquid-crystalline state, the molecules are distributed positionally at random, as in a fluid or glass, but the system is orientationally anisotropic on a macroscopic scale, as in a crystal. The LC glass is the corresponding solid to the liquid crystal, reached on cooling via a glass transition, as indicated by the arrow on the left of Fig. 1. The plastic crystal similarly shows orientational disorder, but positional order. A distinctly different type of mesophase with dynamic, conformational disorder is shown below the crystal state. This state is known as the condis crystal [1,2]. The condis crystal (crystal with dynamic, conformational disorder and orientational and positional order) does not fit into the early, standard classification of mesophases that required only plastic and liquid crystals [3].

It seems reasonable, and has now been illustrated in numerous examples, that molecules with dynamic, conformational disorder in the liquid state show such conformational disorder also in the liquid crystalline and plastic crystalline states [2]. The major need in distinguishing condis crystals from

ORNL DWG 89A-737

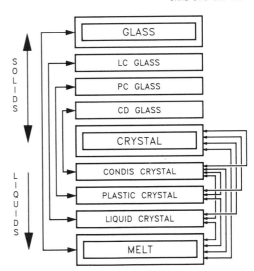

Figure 1 Schematic diagram of the relationship between condensed phases.

other mesophases is thus the identification of translational motion and positional disorder of the molecular centers of gravity in the case of liquid crystals, and of molecular rotation in the case of plastic crystals.

The major motion type for the crystalline and glassy solid states is vibration. For the linear macromolecules, one finds large deviations from the classical Einstein or Debye treatments of the heat capacity. Also, an early attempt to approximate the heat capacities with a combination of a one-dimensional and three-dimensional continuum to account for the anisotropy of the chain molecules was, by itself, not applicable over a wide temperature range. The method that was finally successful involves an approximation of the frequency spectrum by separation of the high-frequency group vibrations from the skeletal chain vibrations [4].

In our ATHAS laboratory, skeletal heat capacities of crystalline, linear macromolecules have been computed for many years using the Tarasov model. A three-dimensional Debye term with a quadratic distribution function, $g_3(\omega)$, is used to describe the lower skeletal frequencies, and a one-dimensional Debye term with a constant distribution function, $g_1(\omega)$, for the higher skeletal frequencies of the polymer crystal. The procedure involves a two-parameter fit of the Tarasov approximation to the experiment. The additional group vibrations are assigned values, computed from isolated polymer chains with parameters fitted to the experimental Raman and infrared spectra. The model appears to work well for both crystalline and amorphous solids (glasses) from a few kelvin to the glass transition temperature, and for crystals to the vicinity of the melting temperature.

As the precision of experiment and computation has increased, a number of questions about the limits of this approach have arisen [5]. One basic question to be addressed in this chapter is the influence of anharmonicity. Using MD, one is no longer restricted to the harmonic oscillator approximation used exclusively before. Any realistic force law can be assumed for the MD simulation. One can thus investigate the effect of anharmonicity on the behavior of the molecular vibrations (skeletal and group) in polymers.

In the course of analysis of motion types in crystals and mesophases, it has become clear that large-scale motion is also possible in molecular crystals. Conformational, rotation, and even diffusional motion can start considerably below the melting or disordering transition temperatures. The major tools for the determination of this motion, commonly connected only with the liquid or mesophase, are the various forms of solid-state nuclear magnetic resonance [6]. In order not to lead to dynamic disorder, these larger-scale motions must be jump motions between positions of identical symmetry within the crystal. Since the intermediate in a jump motion is present only during a negligible fraction of the total time, it can hardly be determined by most structure analysis techniques, such as x-ray diffraction and Raman or infrared spectroscopy. Similarly, there are hardly any effects thermodynamically as long as the order (entropy) sees little change. The changes in heat capacity when a torsional oscillator changes to a hindered rotator, for example, are gradual and usually occur over a temperature range of several hundred kelvin, so that its initial jumping from minimum to minimum in potential energy is difficult to distinguish from a pure oscillator. Ultimately, this motion leads always to disordering of the crystal to mesophases or the melt.

Melting of a polymer crystal becomes possible when the temperature is above the melting temperature, T_m. At this temperature, the free enthalpy between crystal and melt are equal. If the crystal is in equilibrium, the equilibrium melting temperature, T_m^0, signifies an equilibrium between crystal growth and melting. In case the crystal is not in equilibrium, as is frequent in polymer crystals due to chain folding and other defects, T_m signifies a zero-entropy-production point. Two-dimensional polymer crystal melting on an extended chain crystal surface was observed experimentally some years ago. The best known studies of polymer melting are for polyethylene [7], selenium [8,9], and poly(ethylene oxide) [10].

A method that is particularly well suited for investigating the various structural and dynamic properties of polymers is the molecular dynamics (MD) method. The method is well known [11], but its application to the solid state of polymers has not been extensive. Indeed, polymer applications present some unique difficulties, discussed in the next section.

In our detailed computer simulations of the dynamics of solid polymers, we have focused on the following three topics: (1) simulations of condis crystals and mesophase transitions [12], (2) calculation of vibrational spectra and heat capacities of solids [13], and (3) melting and crystallization of polymers [14].

In the next section we describe the methods we have used to study the structure and dynamics of polymers. Following that, we give the results, and then the conclusions are presented.

METHOD

Molecular Dynamics Method

Since recent reviews exist, we present only the equations necessary to set the notation. For the set of atoms of interest arranged into polymer chains, Hamilton's equations (a variation of Newton's equations [15]) are solved. These equations are

$$H = T + V_{2B} + V_{3B} + V_{4B} + V_{NB} \tag{1}$$

where H is the Hamiltonian or total energy operator, composed of kinetic energy terms [first term in Eq. (1)] and potential energy [remaining terms in Eq. (1)]. The tedious and time-consuming evaluations of the derivatives of the internal coordinates were drastically reduced using a new geometric statement function method [16].

Initial Conditions and Methods of Analysis

Initially, seven dynamic chains of 100 atoms each are set in an extended zigzag conformation on lattice sites of the orthorhombic phase of PE with the unit cell dimensions $a = 0.74$ nm, $b = 0.49$ nm, and $c = 0.26$ nm. A randomly chosen amount of kinetic energy is then placed in each atom of the dynamic chain. The energy quickly becomes distributed between kinetic and potential energy to generate an average temperature, computed from the standard equation

$$\frac{3}{2} NkT = \text{KE} = \sum_{i=1}^{N} \frac{p_{x_i}^2 + p_{y_i}^2 + p_{z_i}^2}{2m_i} \tag{2}$$

Trajectories are computed for a specific duration and a collective mode correlation function is extracted:

$$\rho(K, t) = \sum_{i=1}^{100} e^{iKZ_i(t)} \tag{3}$$

where Z_i is the Z-component of the Cartesian coordinates for the atoms and lies along the crystallographic c-axis and K is the wave vector, which varies from $0 \leq K \leq 2\pi N/L$, where L is the end-to-end distance of the polymer chain and N is an integer. This range of K ensures that we include all possible frequencies in the dispersion curves such that there is a unique relationship between the state of vibration of the lattice and the wave vector [17]. A local mode correlation function is calculated from

$$C_{LM}^i(t) = Z_{i+1}(t) - Z_i(t) \tag{4}$$

where i gives the atom number in the polymer chain. There are 99 different local-mode correlation functions for a 100-atom PE chain. For each of the correlation functions, a sampling frequency of 240 THz was used, which corresponds to 8000 cm^{-1}. To compute $g(\omega)$ for the first two types of correlation functions [Eqs. (3) and (4)], the correlation functions are appropriately histogrammed. For both the collective and local modes, the computed frequencies were histogrammed with a 10-cm^{-1} width.

The constant-volume heat capacity for a crystal can be determined by evaluation of the partition function [18]

$$Q = \prod_{j=1}^{3N} \frac{e^{-\hbar\omega_j/2kT}}{1 - e^{-\hbar\omega_j/kT}} \, e^{u(0,\rho)/kT} \tag{5}$$

Assuming that there are $3N$ normal frequencies that are essentially continuously distributed, with the introduction of the function $g(\omega)\,d\omega$, which gives the number of normal frequencies between ω and $\omega + d\omega$, the constant heat capacity can be written as

$$C_V = k \int_0^\infty \frac{(\hbar v/kT)^2 e^{-\hbar\omega/kT} g(\omega)\,d\omega}{(1 - e^{-\hbar\omega/kT})^2} \tag{6}$$

Equation (6) is essentially exact. It treats each of the $3N$ vibrations as a quantum-mechanical, harmonic oscillator, each mode having discrete energy levels, equally spaced by $\hbar\omega$. We include anharmonicity into this equation by using classical mechanics to calculate the density of states $g(\omega)$ and the frequencies ω of the modes. Thus we calculate the constant-volume heat capacity using a semiclassical approximation. As we show in the following section, this approximation is capable of determining the contribution of the optical and acoustical modes to the constant-volume heat capacity with the inclusion of anharmonicity.

Spectral Techniques

Standard fast Fourier transform (FFT) methods can be applied to compute spectra from molecular dynamics calculations; however, the resolution of this method is approximately proportional to 1/total time and thus it requires trajectories of 33.3 ps to give a reasonable resolution of 1 cm^{-1}. To study the dynamics of polymers, it is necessary to examine spectra on a subpicosecond time scale, due to the onset of classical chaos and macroconformational changes [19]. In large molecular systems such as polymers, there can be many phase changes over a 10-ps time scale. These changes in the structure and dynamics make the standard FFT of the motion appear very grassy and uninterpretable. We have recently found a new method which provides a solution to this formidable problem [13,20].

Of the methods proposed for solving such parametric identification problems, the class of algorithms known as high-resolution or signal-subspace methods have recently received a great deal of attention [21]. The term *signal subspace* refers to the fact that these algorithms rely on separating the space spanned by the received data into signal and noise subspaces to achieve their high-resolution capability. These methods exploit the underlying data model typically assumed for these problems, and consequently provide reliable parameter estimates when the assumed model is accurate.

Within this class of signal-subspace algorithms, the MUSIC (Multiple Signal Identification and Classification) algorithm [22] has received the most attention and has been studied widely. Its popularity stems primarily from its generality. For example, in direction-of-arrival estimation problems, it is applicable to arrays of arbitrary (but known) configuration and response. In time-series analysis and system identification problems such as the one addressed herein, nonuniform (arbitrary) sampling schemes can be employed. The price paid for this generality is the requirement that the array response must be known for all possible combinations of source parameters. The MUSIC method is based on an eigenanalysis of the pth-order Toeplitz autocorrelation matrix. Signal and noise matrices are computed, and the corresponding eigenvectors and eigenvalues are used to calculate the MUSIC frequency estimate. We have demonstrated the applicability of this method to MD calculations in a number of previous publications [13,20].

Recently we have investigated another, more accurate and computationally efficient method for spectral analysis [23]. This method, called ESPRIT (Estimation of Signal Parameters via Rotational Invariance Technique) [23], achieves resolution superiority by effectively performing a d-dimensional parameter search (compared to a one-dimensional search in the MUSIC method) in the complex plane for cisoid poles that best fit observations. For applications to molecular dynamics simulations, the ESPRIT method manifests significant performance and computational advantages [23].

RESULTS

Simulation of the Condis State of Polyethylene

Using the realistic model for polyethylene (PE) described earlier, the molecular dynamics technique is used to simulate atomic motion in a crystal. The model for the polyethylene crystal includes, in addition to the seven dynamic chains, a shell of 12 fixed chains as surroundings. A diagram of this 19-chain model is shown in Ref. 12. The model is similar to the one used by Reneker and Mazur [24] to study the motion of defects, except that our chain is longer and the end groups are not constrained.

With this model we have computed trajectories for several initial momentum distributions of the polymer atoms, which serve to explore the dynamics of the system at several temperatures. To illustrate the MD method, two time sequences are shown in Figs. 2 and 3 for temperatures of about 80

ORNL DWG 89-12

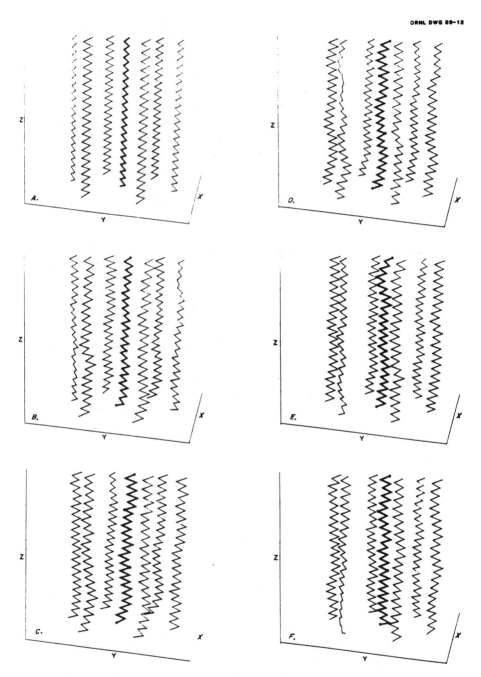

Figure 2 Movie sequence for values of (a) $t = 0$, (b) $t = 0.1$, (c) 0.3, (d) 0.5, (e) 0.7, and (f) 0.9 ps for low-temperature motion $T = 80$ K.

ORNL DWG 89-11

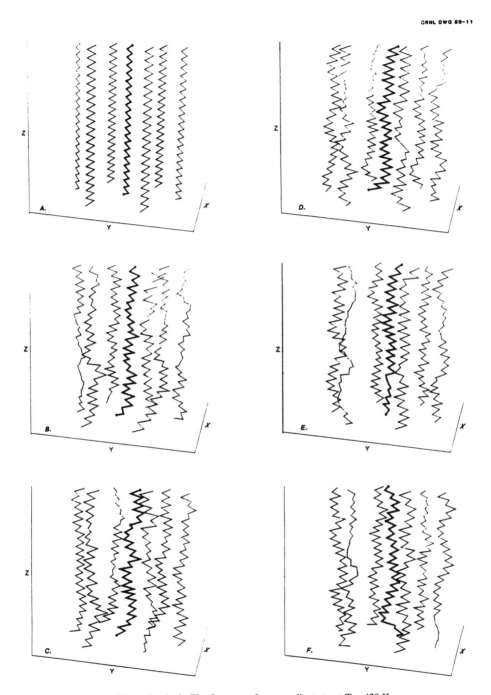

Figure 3   As in Fig. 2, except for a condis state at $T = 430$ K.

and 430 K, respectively, in the interval from 0 to 1 ps. The sequences of pictures show the movement of the lower 40 atoms of the seven dynamic chains in the crystal. A complete movie of the motion in the crystal has been produced. The low-temperature motion of Fig. 2 is much more stable than the highly chaotic motion [12] of Fig. 3. For a general discussion of chaotic motion, see Ref. 12c.

Further analysis of the trajectory can reveal how fast the system equilibrates for rotational isomerization. The distribution functions of $\rho(\cos \tau)$ versus $\cos \tau$ reveal that after a few picoseconds, the distribution function reaches a stable value. The distribution function $\rho(\cos \tau)$ at time zero contains only trans conformations; however, a very short time later some of the gauche conformational states and all intermediates become populated, with the majority of the conformations remaining trans at $\cos \tau$ close to -1. As the temperature is increased, the dihedral angle is found to assume all possible positions, with only a small preponderance for the gauche conformation (Figs. 4 and 5 of Ref. 12a).

Other interesting results that were extracted from the simulations involve shrinking of the PE crystal (change in fiber period) as indicated by the size of the central dynamic chain. The end-to-end distance R is reduced by approximately 0.3% upon heating to 324 K (about room temperature). This agrees well with the experimental shrinking of the fiber period for PE [25].

Spectra and Heat Capacities

One-dimensional studies.   The spectral properties of several linear chain models have been studied. In particular, the polymers +SE+_n, and +CO+_n are examined for various conditions. In the simulation of the dynamics of these polymers, we have constrained the vibrations to one dimension.

In each of the polymer chains we see a decrease in the heat capacity above a critical temperature [26]. To understand the significance of this observation in more detail, we have performed MUSIC analysis of the vibrational motion.

Examination of the time-dependent frequencies of the optical and acoustical modes of +CO+_n [anharmonic model] gives some interesting insight into the behavior of these modes. Figure 4 is a plot of the optical (high frequency) and acoustical (low frequency) modes as a function of time at different temperatures. Comparisons of Fig. 4a and b show the effect of increasing temperature. As the temperature is increased from 10 K (Fig. 4a) to 4000 K (Fig. 4b), the behavior of the optical mode becomes increasingly chaotic. A broad range of frequencies is shown in Fig. 4b compared to Fig. 4a. One point of interest is that the optical frequency decreases very rapidly with temperature, while the acoustical mode is relatively stable. It is also interesting to note that the temperature which causes the optical mode to become chaotic in frequency space is the temperature at which we have observed a beginning decrease in heat capacity for polyethylene. For the vibrational models studied in these examples, it appears that the reason for this decrease is that the group vibrations which make up the optical mode no longer exist. This is because the energy in the molecule is distributed such that some bonds have very large displacements. These irregularities are of *local nature*, and thus any group motion consisting of a combination of these local-type modes will become chaotic. This constitutes a change in the description of the motion of the atoms. There are still the same number of vibrational modes, but the group vibrations that described the optical mode at low temperature no longer strictly exist. In the harmonic model, the group vibrations always exist and there can never be chaotic motion. This conclusion is readily seen by comparing Fig. 4c, which is the time-dependent frequencies of the optical and acoustical modes of the harmonic model for a temperature of 4000 K or to Fig. 4b and a. As can be seen, there is little difference between Fig. 4a and c, but there is a large difference between Fig. 4b and c. Again, this shows that the effect of anharmonicity at large temperatures (Fig. 4b) is significant, whereas there are no effects at low temperatures (Fig. 4a). For anharmonic, nonseparable Hamiltonians, chaotic motion destroys the coherence of the collective modes.

Three-dimensional studies.   A more detailed study of the constant-volume heat capacity of polyethylene is presented next. In the first series of calculations, a comparison of $g_L(\omega)$ (local modes) versus $g_C(\omega)$ (collective modes) was made for five temperatures, ranging from about 20 to 600 K [27]. In each case, 10 samples of 64 points each ($\Delta t = 0.004166$ ps) were frequency-analyzed by MUSIC. From the 10 samples, a collection of the frequency peaks was histogrammed. The example in Fig. 5 shows a comparison between $g_C(\omega)$ calculated from the collective-mode correlation function [Eq. (3)] with the $g_L(\omega)$ calculated from the local-mode correlation function [Eq. (4)]. Both $g(\omega)$ spectra show good agreement. There are major peaks centered around the frequencies expected for the optical (1100 cm^{-1}) and acoustical (0 to 600 cm^{-1}) modes. Both approximations have the general form of a three-dimensional Debye term for the low frequencies and a one-dimensional Debye term for the higher frequencies. The structure of the local mode $g_L(\omega)$ more closely matches that of experiment [28]. The important point that we wish to stress here is that our MD calculations include anharmonicity and allow us to calculate $g(\omega)$ using any description of the molecular vibrations for any temperature. This method should also be applicable to other liquid phases and Fermi resonances.

Our model of crystalline polyethylene includes anharmonicity in all the interactions given in Eq. (1) (see Ref. 12). Anharmonicity can become very important in dynamics at higher temperatures, as shown in a previous paper [26]. One effect that we have observed is the shifting of the frequencies for the optical and acoustical modes. Figure 6 shows the effect of increasing the

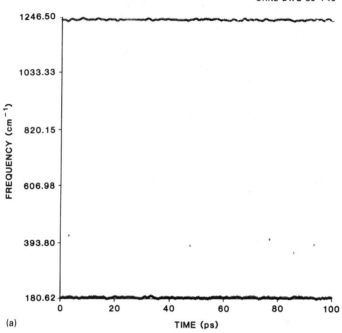

(a)

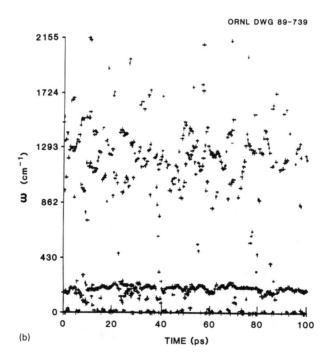

(b)

ORNL DWG 89-738

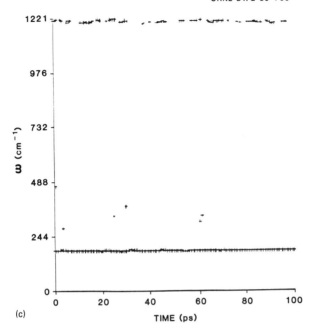

(c)

Figure 4 (a) Time-dependent frequencies of $S(K, \omega)$ for $K = 2\pi n/L = 0.116756a_0^{-1}$ at 10 K. (b) Same as Fig. 1a except for 4000 K. (c) Same as Fig. 1c except for the harmonic model of --(CO)--.

temperature on the $g(\omega)$ spectra in both the local (Fig. 6a) and collective (Fig. 6b) mode approximations. As can be seen, the frequency of the optical mode decreases and the frequency of the acoustical mode increases. This effect is more pronounced for the local-mode representation $g_L(\omega)$.

Calculation of C_V using the partition function [Eq. (6)] reveals that C_V values calculated in the collective- and local-mode approximations are in good agreement (see Ref. 27). Furthermore, the contribution of the optical modes to C_V shows that at low temperature, the low-frequency (acoustical) modes predominate, and as the temperature increases, the high-frequency (optical) modes are dominant. To understand better the effects of anharmonicity, we have calculated C_V for a linear model of a diatomic chain in which the forces are described by (1) N harmonic oscillators and (2) N anharmonic (Morse) oscillators (see Ref. 27 for details of the model). The results indicate that there are both negative and positive differences between the two models.

Melting and Crystallization

A more detailed picture regarding the melting on a molecular level has been developed using molecular dynamics techniques. In these simulations

ORNL DWG 89-332

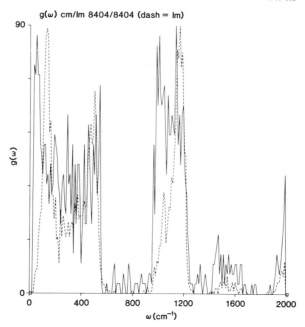

g(ω) cm/lm 8404/8404 (dash = lm)

Figure 5   Local and collective $g(\omega)$ at 347 K. The solid line is $g(\omega)$ for the collective modes, and the dashed line is $g(\omega)$ for the local modes.

the model for a polyethylene (PE) crystal surface consists of several static chains arranged in an extended zigzag of the orthorhombic phase with $a = 0.74$ nm, $b = 0.49$ nm, and $c = 0.26$ nm. As part of the crystal, one chain with n atoms and m folds lies on the crystal surface. Such a chain provides a new interface layer between the crystal and the melt. The total surface area of the crystal, however, does not change by addition of such a chain (see Fig. 1 of Ref. 14b).

We have carried out molecular dynamics simulations of both single and multiple folded chains [14b]. Up to 1000 atoms have been simulated. The effects of bending and torsional modes on the melting have also been studied. From the trajectories generated, we computed the end-to-end distance (EED) and radius of gyration (Rg) of the molecule as a function of time. When melting starts under isothermal conditions, EED and Rg of the chain decrease with increasing time. Simultaneously, the chain gradually leaves the crystal surface to become a more or less coiled structure. It approaches the "self-avoiding walk" in a limited, three-dimensional space (with a wall formed by the crystal surface). Furthermore, such a melting process is highly temperature and molecular-length (n) dependent. We found that a faster-melting process occurs at a shorter molecular length and higher melting temperature.

The effects of the torsion and number of folds have been studied. Mak-

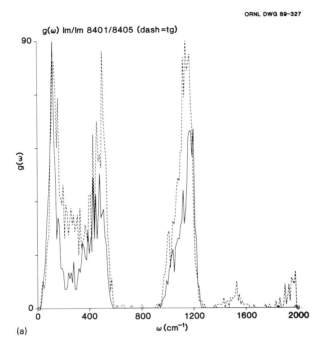

(a)

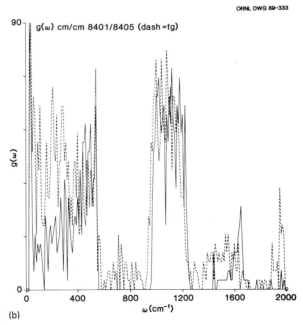

(b)

Figure 6 Comparison of $g(\omega)$ for temperature changes. (a) Local mode $g_L(\omega)$. The solid line is for a temperature of 23 K and the dashed line is for 588 K. (b) Same as for part (a) except that $g_C(\omega)$ is for the collective modes.

ing the torsional mode into a free rotor tends to decrease substantially the temperature required to melt a PE chain. The effect of folds in a chain with constant lamella thickness is to decrease the rate of melting with increasing number of folds. However, this observation holds only for temperatures very near the equilibrium melting temperature. For large ΔT ($\Delta T = T - T_m^\circ$), the rate tends to increase with increasing folding. Two different pathways for the melting process are suggested to explain this observed behavior for multiply folded chains. One of these pathways involves a $1:1$ competition between melting and crystallization. The initial step of this process is the peeling off of the ends of the chain, as illustrated for the single chain in Fig. 7 (see Fig. 7a). The ends of the chains then form "loops," which increase in size over time (see Fig. 7b). Some of the atoms in the loops find lattice sites and "sit down" into an ordered (crystallographic axis) arrangement. The rate of formation of the disordered loops compared to the formation of new N-atom crystal segments is approximately the same. This $1:1$ competition continues until the center of the chain begins to diffuse (see Fig. 7c). At this time, the equilibrium is somewhat shifted toward melting and the polymer chain reaches a randomly disordered macroconformation (random coil). In this disordered state, the EED and Rg do not change significantly with time. Contrary to the pathway just discussed, a second pathway involving extreme melting is found for high ΔT values. In this case the total energy available to the system is sufficient to cause a rapid propagation of the "peeling off process" to the center of the polymer chain (as, again, illustrated on the melting of a single chain in Fig. 8). While the initial events of the melting process along this pathway are similar to those of the first pathway, the time required to shift the melting–crystallization equilibrium is much smaller. In short, the polymer chain goes almost directly to the melted (fluid) state, bypassing the competing nucleation-controlled crystallization.

For a single extended PE chain, the temperature dependence of either melting mechanism is qualitatively the same. We have observed a subtle difference [14a], whereby for high ΔT, the logarithms of the rate of melting deviate from a linear relation to $1/T$. For folded polymer chains, the added instability of the folds leads to a greater shift in the melting–crystallization equilibrium and leads to a *kinetic* effect of increasing the rate of melting as the number of folds increases.

CONCLUSIONS

The conclusions from our simulation of polyethylene and its condis state [12a,b] seem to be that (1) conformational disorder exists in the crystal before melting, (2) the simple rotational isomer model may not be a suitable model for the description of crystal defects, and (3) anharmonic vibrations cause measurable deviations from the normal-mode description of motion in crys-

ORNL DWG 89-741

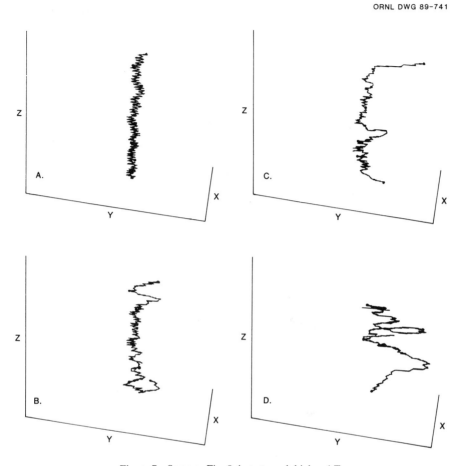

Figure 7 Same as Fig. 8, but at much higher ΔT.

tals. Whereas anharmonicity tends to destroy, via chaotic motion, the coherence of the collective modes, in particular the optical modes, the collective mode description still remains useful at low and moderate temperatures.

In the process of investigating these properties, we have developed a new method for calculating the density of states $g(\omega)$ for macromolecules in the solid phase [27]. The method allows for the calculation of $g(\omega)$ using various definitions of the molecular vibrations (i.e., collective- versus local-mode description).

Analysis of the temperature-dependent $g(\omega)$ spectra shows that anharmonicity tends to cause the optical modes to decrease in frequency and the acoustical modes to increase in frequency as a function of increasing temperature. Using a semiclassical approximation to calculate constant-volume

ORNL DWG 89-742

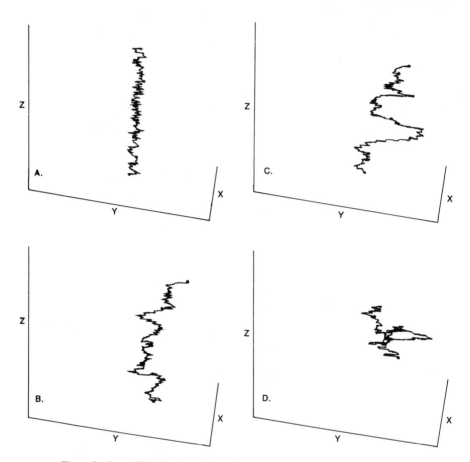

Figure 8   Low ΔT ($\Delta T = T - T_m^\circ$) melting time sequence. Time steps 5 ps.

heat capacity [Eq. (6)], it was shown that the optical modes do not significantly contribute to the heat capacity at low temperatures, but become the predominant factor at high temperatures. Finally, we have found that anharmonicity tends to lead to both negative and positive deviations from a harmonic model.

The transition from the crystalline state to the fluid state reveals a number of interesting phenomena. The conclusions from our simulation results are [14]:

1. The rate of melting increases with increasing temperature.
2. The rate of melting of a single extended PE chain decreases with increasing molecular length at a given superheating ΔT.

3. The diffusion coefficients for 10-atom segments increases as the segment gets closer to the ends or folds of the PE chain and as the temperature increases.

4. Melting occurs first by peeling off at the ends or at folds layer after layer of molecular chains in a direction perpendicular to the chain directions.

5. For a constant lamella thickness and small ΔT, increasing the number of folds in a chain will decrease the rate of melting, while for large ΔT, increasing the number of folds will increase the rate of melting.

6. The rate of recrystallization appears to depend on the type of phase from which it starts (i.e., the rate is faster from a mesophase than from a melt).

Most of these conclusions on fusion have found their counterparts in experimental observations of chain molecules. A typical example is the behavior of low-molecular-mass poly(ethylene oxide) fractions [29].

In addition to resolving the questions of melting of macromolecules, it is planned to study the nucleation process in more detail, to obtain better insight into the melting–recrystallization equilibrium.

ACKNOWLEDGMENTS

This work was supported in part by the Polymer Program of the National Science Foundation, present Grant DMR-8818412, and by the Division of Materials Sciences, Office of Basic Energy Sciences, U.S. Department of Energy, under contract DE-AC05-840R21400 with Martin Marietta Energy Systems, Inc. Some of the work reported here was carried out on the Cray X-MP/48, Grant TRA-890046N, Projects YPO, XCK, National Center for Supercomputing Applications (NCSA) at the University of Illinois at Urbana–Champaign. The major part of the substantial computation time was supported by the University of Tennessee. Program development and testing was done on the Cray X/MP computer of ORNL.

REFERENCES

1. B. Wunderlich and J. Grebowicz, *Adv. Polym. Sci.* **60/61,** 1 (1984).
2. B. Wunderlich, M. Moeller, J. Grebowicz, and H. Baur, *Adv. Polym. Sci.* **87,** 1 (1988).
3. G. W. Smith, *Advances in Liquid Crystals,* Vol. 1, ed. G. H. Brown, Academic Press, New York, 1975.
4. B. Wunderlich and S. Z. D. Cheng, *Gazz. Chim. Ital.* **226,** 348 (1986).
5. H. S. Bu, S. Z. D. Cheng, and B. Wunderlich, *J. Phys. Chem.* **91,** 4179 (1987).
6. B. Wunderlich and H. Baur, *Adv. Polym. Sci.* **7,** 151 (1970).
7. G. Czornyj and B. Wunderlich, *J. Polym. Sci., Polym. Phys. Ed.* **15,** 1905 (1977); see also B. Wunderlich, *Faraday Discuss. Chem. Soc.* **68,** 237 (1979).

8. P. H-C. Shu and B. Wunderlich, *J. Cryst. Growth* **48,** 337 (1980).

9. A. H. Crystal, *J. Polym. Sci., Polym. Phys. Ed.* **8,** 2153 (1970).

10. A. H. Kovacs, A. Gonthier, and C. Straupe, *J. Polym. Sci., Polym. Symp.* **50,** 283 (1975).

11. M. Klein, *Annu. Rev. Phys. Chem.* **36,** 525 (1985); W. G. Hoover, *Annu. Rev. Phys. Chem.* **34,** 103 (1983).

12. (a) D. W. Noid, B. G. Sumpter, and B. Wunderlich, *Macromolecules* **23,** 664 (1990); (b) D. W. Noid, B. G. Sumpter, M. Varma, and B. Wunderlich, *Makromol. Chem.* **10,** 377 (1989); (c) D. W. Noid, M. L. Koszykowski, and R. A. Marcus, *Annu. Rev. Phys. Chem.* **32,** 261 (1981).

13. (a) D. W. Noid, B. T. Broocks, S. K. Gray, and S. L. Marple, *J. Phys. Chem.* **92,** 3386 (1988); (b) D. W. Noid and G. A. Pfeffer, *J. Polym. Sci., Polym. Phys. Ed.* **27,** 2321 (1989).

14. (a) D. W. Noid, G. A. Pfeffer, S. Z. D. Cheng, and B. Wunderlich, *Macromolecules* **21,** 3482 (1988); (b) B. G. Sumpter, D. W. Noid, B. Wunderlich, and S. Z. D. Cheng, *Macromolecules* (in press).

15. H. Goldstein, *Classical Mechanics,* Addison-Wesley, Reading, Mass., 1950, Chap. 5.

16. D. W. Noid, B. G. Sumpter, G. A. Pfeffer, and B. Wunderlich, *J. Comput. Chem.* **11,** 236 (1990).

17. P. C. Painter, M. N. Coleman, and J. L. Koenig, *The Theory of Vibrational Spectroscopy and Its Application to Polymeric Materials,* Wiley, New York, 1982.

18. McQuarrie, *Statistical Mechanics,* Harper & Row, New York, 1976.

19. A. J. Kovacs and A. Gonthier, *Kolloid Z. Z. Polym.* **250,** 530 (1972).

20. D. W. Noid, G. A. Pfeffer, B. G. Sumpter, and S. K. Gray, *Inf. Q. Comput. Simul. Condens. Phases* **31,** Oct. (1989).

21. S. L. Marple, *Digital Spectral Analysis with Applications,* Prentice-Hall, Englewood Cliffs, N.J., 1987.

22. R. O. Schmidt, Ph.D. thesis, Stanford University, Stanford, Calif., 1987.

23. R. Roy, B. G. Sumpter, D. W. Noid, and B. Wunderlich, *J. Phys. Chem.* (in press); R. Roy and T. Kailath, *IEEE Trans. Acoust. Speech Signal Process.* **37,** 984 (1989).

24. D. H. Reneker and J. Mazur, *Polymer* **29,** 3 (1988).

25. G. T. Davis, R. K. Eby, and J. P. Colson, *J. Appl. Phys.* **41,** 4316 (1970).

26. B. G. Sumpter, D. W. Noid, and B. Wunderlich, *Polymer* (in press).

27. D. W. Noid, B. G. Sumpter, and B. Wunderlich, *Anal. Chem. Acta* (in press).

28. B. Wunderlich and H. Baur, *Adv. Polym. Sci.* **7,** 151 (1970).

29. S. Z. D. Cheng, A. Zhang, J. Chen, and D. P. Heheson, *J. Polym. Sci., Polym. Phys. Ed.* (in press).

23

Crystallographic Defects in Polymers and What They Can Do

DARRELL H. RENEKER
Polymers Division
National Institute of Standards and Technology
Gaithersburg, Maryland 20899

JACOB MAZUR
Program Resource Inc.
National Cancer Institute Research Facility
Frederick, Maryland 21701

ABSTRACT

A family of five crystallographic defects, characterized by a defect loop that encircles one chain, is described. The present description is restricted to defects in polyethylene, but it can be generalized to other crystalline polymers. The interactions of the defects with themselves and with each other are also described. Use of these crystallographic defects to examine the nature of an entanglement in polyethylene consisting of two chains hooked together like two hairpins is described.

INTRODUCTION

A family of five crystallographic defects, characterized by a defect loop that encircles one chain, was described for polyethylene [1]. The conformation of each defect was adjusted to minimize its energy (i.e., the sum of the interatomic interactions) in a polyethylene crystal that contained 19 chains and was 60 repeat units long.

These defects can interact with each other, or with larger defects of all sorts, including vacancies, entanglements, folds, side branches, and larger crystallographic defects, such as edge and screw dislocations. The defects and their interactions provide an improved model for the description of larger-scale phenomena in polyethylene. The concept of crystallographic defects can be generalized to other crystalline polymers. The five defects are:

1. Dispiration loop, interstitial-like. As it diffuses along a chain, or moves in response to strain fields, this defect carries with it one extra CH_2 group and a 180° twist around the chain axis.

2. Dispiration loop, vacancy-like. This defect carries a vacancy amounting to a missing CH_2 group and 180° of twist. The sense of the twist may be either the same as or opposite to the twist in a nearby defect in the same chain.

3. Disclination loop. This defect carries a twist of 360°. The twist may be of either sense.

4. Dislocation loop, interstitial-like. This defect carries two extra CH_2 groups (one crystallographic repeat unit along the chain axis). There is no net twist associated with this defect.

5. Dislocation loop, vacancy-like. This defect carries a vacancy amounting to two CH_2 groups (one crystallographic repeat along the chain axis). There is no net twist associated with this defect.

Both 4 and 5 are edge dislocation loops, in which the Burgers vector is perpendicular to the dislocation line. The connection of these defects to the established terminology for defects in crystals of low molecular weight or atomic substances is discussed elsewhere [2]. A screw dislocation loop that encircled one chain of polyethylene would require shear displacements within the molecule, which would distort the chemical nature of the molecule. Small screw dislocations may be interesting for molecules with two or more identifiable backbone chains, or for cases in which the loop encircles two or more chains. For brevity, in the remainder of this chapter, shorter names are used for some of the defects. No distinction is intended in this shortening of the names.

INTERACTION OF TWO DEFECTS OF THE SAME TYPE

In this section we summarize the interaction of two defects of the same kind. For interactions that are inherently repulsive—for example, those that would increase the number of residues in the interaction region or would produce more distortion of the molecular geometry—external forces must be applied to the molecule to force the two defects together. A strain field is the most likely possibility for supplying that energy, although electromagnetic fields and other possibilities exist. For defects that carry a twist around the molecular axis, a pair of such defects can have their twist in the same sense or in the opposite sense. These two possibilities affect the nature of the interaction between the defects.

It is a straightforward matter to calculate the energy associated with these defect combinations, but this has not been done except for a few cases. The energy of the combination is approximately equal to the sum of the energy associated with each of the defects, except that the energy decreases when twist in one sense cancels twist in the other, when twist disappears by a 360° rotation around a bond, or when the extra CH_2 associated with an interstitial-like defect is canceled by the missing CH_2 in a vacancy-like defect. The energy may increase significantly if the polyethylene chain is stretched or if extra CH_2 groups are crowded into a small volume.

1. Two dispirations (interstitial-like). If the two dispirations are twisted in the same sense, the defect resulting from their combination contains 360° of twist and two extra CH_2 groups. The combination therefore contains a disclination and an interstitial-like dislocation. If the two dispirations are twisted in the opposite sense, the combined defect contains zero twist and two extra CH_2 groups. The combination is therefore an interstitial-like dislocation.

It is possible to convert the 360° of twist that is distributed over five to eight CH_2 groups into twist around a single bond if the defect region is loosely constrained, as it might be in the neighborhood of folds, crystal surfaces, or other defects. In such an environment, disclinations in polyethylene can disappear in a complicated process that concentrates all the twist at a single backbone bond. This possibility provides a connection between the two rather different states that result from the sense of the twist in the two interacting dispirations. The energy barriers that must be crossed in this process appear to be high, but probably not so high that molecular scission is likely.

2. Two dispirations (vacancy-like). If the two dispirations are twisted in the same sense, the defect resulting from their combination is two CH_2 groups deficient and contains 360° degrees of twist. It is the same as the combination of a disclination and a vacancy-like dislocation. If the two dispirations are twisted in the opposite sense, the defect resulting from their combination is also two CH_2 groups deficient but contains no twist. It is a vacancy-like dislocation. These are defects in which the chain is highly stretched because of the two missing CH_2 groups. The stretched and twisted region resulting from the dispirations with the twist in the same sense is some tens of repeat units long. As a result, the conversion of twist distributed over many bonds into twist around a single bond is much less likely to occur than in the case of two interstitial-like dispirations.

3. Two disclinations. Two disclinations with twist in the same sense combine to form a larger defect with 720° of twist. The conversion of half of this twist to one 360° rotation around a single bond can occur. Two disclinations with twist of opposite sense simply annihilate each other and disappear, leaving the chain in its proper crystallographic conformation.

4. Two dislocations (interstitial-like). Two interstitial-like dislocations in a single chain, if forced together, produce a region that contains four extra CH_2 groups. Many conformations may occur, depending on the way the dislocations are constrained. Transformation to two pairs of interstitial dispirations with the individual dispirations in each pair having a twist of opposite sense is geometrically possible. This means that dispirations (as an alternative to the original dislocations) could be emitted from such a region.

5. Two dislocations (vacancy-like). The forces that produce this combination would result from stretching the chain. The combination would need to be very long to avoid chain scission. Conversion to pairs of vacancy-like dispirations is possible. Vacancy-like dispirations could be emitted from such a region.

OTHER BINARY COMBINATIONS OF THESE FIVE DEFECTS

In this section the results of binary combinations of different defects are described. There are 10 such combinations.

6. Dispiration (vacancy-like) and dispiration (interstitial-like). When these defects combine, the missing CH_2 on the first cancels the extra CH_2 on the second. If the twist is in the same sense, a disclination with 360° of twist remains. If the twist is in the opposite sense, it cancels and the chain is left on its proper position in the crystal lattice.

7. Disclination and dispiration (interstitial-like). If the twist is in the same sense on each defect, the combined defect contains one extra CH_2 and 540° of twist. Rotation by 360° about a single bond could reduce this to 180°, the value that would be present if the twist of the disclination was in the opposite sense to that on the dispiration.

8. Disclination and dispiration (vacancy-like). If the twist is in the same sense on each defect, the combined defect contains 540° of twist and is deficient by one CH_2 group. Reduction of the twist by rotation around a single bond is unlikely, because of the stretch associated with the dispiration. If the sense of the twist is opposite in the defects, the twist is 180° in the combined defect.

9. Dislocation (interstitial-like) and dispiration (interstitial-like). This combined defect contains 180° of twist and three extra CH_2 groups. Relatively large external forces are required to hold these three groups in close proximity.

10. Dislocation (interstitial-like) and dispiration (vacancy-like). The missing CH_2 groups in the dispiration cancel one of the extra CH_2 groups in the dislocation. This leaves an interstitial-like dispiration.

11. Dislocation (interstitial-like) and disclination. This combination can be viewed as two interstitial-like dispirations with twist in the same sense, which is described in item 1 above.

12. Dislocation (vacancy-like) and dispiration (interstitial-like). The extra CH_2 group of the dispiration cancels one of the missing CH_2 groups in the dislocation. This leaves a vacancy-like dispiration.

13. Dislocation (vacancy-like) and dispiration (vacancy-like). This combination is a highly stretched zigzag with 180° of twist. It needs to be extraordinarily long to avoid chain scission.

14. Dislocation (vacancy-like) and disclination. This combination can be viewed as two vacancy-like dispirations with twist in the same sense, which is described in item 2 above.

15. Dislocation (vacancy-like) and dislocation (interstitial-like). These defects cancel, leaving the chain on the crystal lattice sites.

INTERACTION OF CRYSTALLOGRAPHIC DEFECTS WITH LARGER DEFECTS

In addition to interactions with themselves, these defects interact with other morphological structures that occur in polyethylene. The production, motion, structural evolution, and even disappearance of these morphological structures, such as vacancies, folds, branches, larger edge of screw dislocations, and entanglements, can be mediated by the motion of crystallographic defects along single chains.

Entanglements that are trapped in solid polyethylene provide an example. Entanglements play an essential role in the formation of high-modulus fibers [3]. Some of the entanglements remain trapped in the fibers and therefore can affect the mechanical properties, for example, by controlling the rate at which a tensile stress causes a fiber to become longer.

The role of entanglements in the determination of structural properties rests, in part, on the motion of one entangled chain through a loop in the other chain. This motion can be modeled at the molecular level as a process in which defects that bear extra repeat units migrate along the chain, arrive at the entanglement, pass into and out of the entanglement, and allow the entanglement to move.

Such processes can be modeled in detail if the entanglement is embedded in a crystal. Figure 1 shows the trajectory of the chains in an entanglement that consists of two U-shaped chains that are hooked together (i.e., entangled) and embedded in the lattice of polyethylene. The U-shaped chains connect lattice site A to site B at the top, and site C to D at the bottom. Two additional chains are required to avoid rows of vacant lattice sites. These chains each spiral a quarter turn around the hooked region of the U-shaped chains. One of the additional chains connects lattice site C at the top to site A at the bottom, and the other connects site D at the top to B at the bottom.

The result is an arrangement that has four chains passing through four lattice sites at the top and also through the four corresponding lattice sites at the bottom of the bottom. In between is the entanglement, which, since it contains extra CH_2 groups, forces the surrounding chains to bow out slightly, into a barrel-like shape, but does not cause their dihedral angles to depart from the trans potential well.

In a crystal, lattice sites A, B, C, and D at the top are connected to the

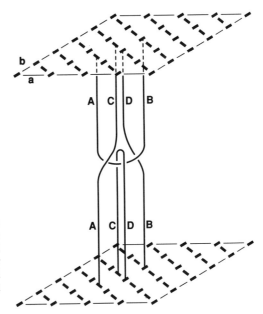

Figure 1 Trajectory of the chains in an entanglement that is inside a crystal. The crystallographic *a* and *b* axes are labeled on the top plane of the model. The four lattice sites at which the chains enter and leave the model are labeled A, B, C, and D at both the top and the bottom plane.

corresponding sites at the bottom by planar zigzag (all-trans) segments that contain 21 CH_2 groups, which corresponds to the separation, in the crystal, of the top and bottom planes of the model. This separation was chosen as a balance between the need to provide a long segment to accommodate the defect, and the need to keep the number of atoms in the model within the range that contemporary computers can handle in the determination of minimum-energy conformations.

In this model the four chains described above, which are involved in the entanglement, are surrounded by a shell of 10 chains (Fig. 2). It is these 10 chains which are bowed outward like a barrel to accommodate the entanglement.

The second shell contains 16 chains that were held at their lattice positions in the initial calculations, although it may prove desirable to deform or move them radially outward from the center of the entanglement in order to achieve a conformation with the lowest energy.

The perfect crystal model contains lattice sites for 30 chain segments, with each segment containing 21 CH_2 groups, for a total of 630 CH_2 groups. The model of the entanglement has 12 more CH_2 groups than the perfect crystal. Each of the U-shaped chains contain 26 CH_2 groups, and each of the spiraling chains contains 22 CH_2 groups. It is these extra groups that cause the barrel-like expansion into the surrounding region, against the elastic strain of the surrounding crystal. The resulting increase in the mass of the region is about 2%.

There are several possible ways to arrange an entanglement inside a

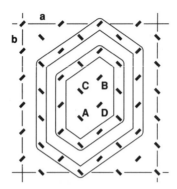

Figure 2 The chains involved in the calculation are the four labeled A, B, C, and D; the 10 in the first shell around them; and the 16 in the surrounding shell.

polyethylene crystal. This one was chosen because it is compact and symmetrical. The starting conformation was created by building a physical model of the defect and measuring the position of the carbon atoms. Then a combination of Newton–Raphson and conjugate gradient methods were used to adjust the entanglement toward a low-energy conformation. The physical model was helpful in the determination of the number of extra CH_2 groups that were required in both the U-shaped and in spiraling chains. The coordinates of the carbon atoms in the four chains in the entanglement are given in the appendix.

An interstitial-like dispiration, which carries an extra CH_2 group, was placed in a leg of one of the U-shaped chains, as far from the hooked region as possible. The dispiration was driven [4] into the entanglement in a calculation in which the constraints of the two shells were removed and only the ends of the U-shaped chains were held at lattice sites. The larger calculation for the model, which included the first shell of surrounding molecules, encountered difficulties because the changes in the energy of the system produced by the incremental steps involved in driving the defect were larger than the barriers the defect encounters as it is moved along the chain. Simplification of the model and improvement in the algorithm for finding minimum energy conformations are expected to overcome this difficulty. The coordinates of the carbon atoms in the leg of a. U-shaped chain that contains a dispiration are available from the authors.

SUMMARY

The behavior of binary combinations of small crystal defects in polyethylene is described. Pairs of defects combine in specific ways to produce predictable consequences when the defects are either forced together or attracted together. The conformation of an entanglement of polyethylene molecules in a crystal was approximated with a physical model and improved by adjustment of the atom positions to minimize the total calculated energy of the system. A systematic procedure for investigating the interaction of small crystallographic defects with entanglements was described and demonstrated in a simplified model consisting of entangled chains constrained only at their ends.

APPENDIX: Coordinates of Carbon Atoms in the Four Central Chains of an Entanglement

Refer to Fig. 1 for a diagram of the sites. Coordinates are in angstrom units.

	x	y	z		x	y	z
			U-Shaped Chains				
Site A, top	3.85	2.20	25.45	Site D, bottom	0.41	0.43	0.10
	4.39	1.59	23.96		1.24	1.03	1.24
	4.09	2.45	22.70		0.59	0.72	2.61
	4.61	1.72	21.45		1.43	1.23	3.78
	4.37	2.47	20.14		0.68	1.12	5.11
	4.98	1.69	18.96		1.46	1.62	6.35
	4.65	2.31	17.62		0.53	1.95	7.52
	5.45	1.63	16.45		1.22	1.99	8.90
	5.01	2.19	15.10		0.31	2.67	9.95
	5.59	1.41	13.85		0.28	1.95	11.34
	4.76	1.56	12.54		1.34	2.47	12.35
	4.40	0.27	11.56		1.55	1.60	13.69
	2.80	0.42	11.59		2.81	0.52	14.01
	2.09	1.34	11.92		2.65	0.93	14.23
	0.57	1.28	12.06		3.45	2.49	13.97
	0.06	2.43	12.90		3.34	3.72	13.00
	0.33	2.58	14.41		4.03	3.53	11.59
	0.34	3.62	15.36		3.18	4.21	10.41
	0.46	3.67	16.67		3.54	3.53	9.03
	0.13	4.61	17.77		2.66	3.93	7.81
	0.19	4.02	19.14		3.27	3.28	6.55
	0.15	4.91	20.35		2.57	3.50	5.18
	0.37	4.29	21.65		3.81	3.39	4.09
	0.09	5.01	22.92		3.28	2.48	2.65
	0.67	4.51	24.17		4.15	3.51	0.52
Site B, top	0.00	5.02	25.52	Site C, bottom	3.54	3.12	0.15
			Spiraling Chains				
Site C, top	4.09	2.70	25.53	Site D, top	0.19	0.29	25.45
	4.50	3.24	24.05		0.57	0.77	24.00
	4.24	2.30	22.84		0.28	0.06	22.93
	4.59	3.01	21.60		0.06	0.52	21.49
	4.49	2.15	20.35		0.96	0.02	20.48
	4.83	2.94	19.11		0.76	0.59	19.08
	4.90	2.00	17.88		1.57	0.15	18.00
	5.85	2.63	16.81		1.18	0.34	16.60
	6.04	1.68	15.59		2.11	0.24	15.50
	6.77	2.44	14.44		1.30	0.00	14.08
	7.06	1.41	13.25		2.55	0.63	13.16
	7.25	2.13	11.91		2.36	0.35	11.63
	7.35	1.10	10.72		2.19	1.63	10.82
	6.36	1.51	9.57		2.18	1.33	9.29
	6.14	0.58	8.49		1.20	2.24	8.53
	5.14	0.91	7.32		1.38	2.07	7.00
	5.49	0.01	6.08		0.26	2.78	6.22
	4.33	0.24	5.06		0.40	2.61	4.70
	4.81	0.93	3.81		0.63	3.45	3.90
	3.70	1.24	2.73		0.07	3.53	2.44
	4.44	1.60	1.37		0.86	4.46	1.52
Site A, bottom	3.68	2.24	0.23	Site B, bottom	0.24	4.67	0.14

REFERENCES

1. D. H. Reneker and J. Mazur, *Polymer* **29,** 3–13 (1988).
2. D. H. Reneker and J. Mazur, *J. Phys. (Paris) Colloq.,* Suppl. 12, **46,** 499–505 (1985).
3. Y. Termonia, P. Meakin, and P. Smith, *Macromolecules* **19,** 154–159 (1986).
4. D. H. Reneker and J. Mazur, *Polymer* **24,** 1387 (1983).

24

Behavior of the Correlation Coefficient for r^2 and s^2 in Rotational Isomeric State Chains of Finite n

NEAL A. NEUBURGER and WAYNE L. MATTICE
Institute of Polymer Science
The University of Akron
Akron, Ohio 44325

ABSTRACT

The dimensionless ratios $\langle r^4 \rangle_0 / \langle r^2 \rangle_0^2$, $\langle s^4 \rangle_0 / \langle s^2 \rangle_0^2$, and $\langle r^2 s^2 \rangle_0 / \langle r^2 \rangle_0 \langle s^2 \rangle_0$ have been computed as a function of n for chains subject to symmetric threefold rotation potentials. The squared end-to-end distance and radius of gyration for a particular conformation are denoted by r^2 and s^2, respectively, and angle brackets denote the average of the enclosed property. The correlation coefficient, ρ, between r^2 and s^2 was calculated from these dimensionless ratios. In the limit where $n \to \infty$, $\rho \to (\frac{5}{8})^{1/2}$, as expected from an earlier rigorous analytical treatment of ρ for freely jointed chains. The approach to this limit, $[\partial \rho / \partial (1/n)]_\infty$, is determined by properties that include the asymptotic limit for the characteristic ratio, $(\langle r^2 \rangle_0 / n l^2)_\infty$, and the nature of the local conformation that is responsible for the extension of the chain. Local maxima and minima in ρ versus n are exhibited by some of the chains, in contrast to the monotonic decrease in ρ with increasing n in random flight chains. A minimum in ρ will be observed at n_{min} if chains with $n < n_{min}$ exhibit one type of approximately linear dependence of s^2 on r^2, chains with $n > n_{min}$ exhibit a *different* type of approximately linear dependence of s^2 on r^2, and both types of behavior coexist in the chain with $n = n_{min}$.

Keywords: Correlation coefficient, End-to-end distance, Freely jointed chain, Generator matrix, Radius of gyration, Rotational isomeric state theory

INTRODUCTION

Ratios of different moments of conformation-dependent physical properties play an important role in the physical chemistry of macromolecules. An example is the expression of the Fourier transform, $G(\mathbf{q})$, of the distribution function for the end-to-end vector, $W(\mathbf{r})$, by an equation that utilizes dimensionless ratios of the various moments of the end-to-end distance r [1,2].

$$G(\mathbf{q}) = \exp(-v)[1 + g_2(2v) + g_4(2v)^2 + \cdots] \tag{1}$$

$$v = \frac{q^2 \langle r^2 \rangle_0}{6} \tag{2}$$

$$g_{2p} = 2^{-p}\left[-\frac{\Delta_4}{2!\,(p-2)!} + \frac{\Delta_6}{3!\,(p-3)!} - \cdots \frac{(-1)^{p-1}\Delta_{2p}}{p!}\right] \tag{3}$$

$$\Delta_{2p} = 1 - \frac{3p}{3 \cdot 5 \cdot 7 \cdots (2p+1)} \frac{\langle r^{2p} \rangle_0}{\langle r^2 \rangle_0^p} \tag{4}$$

This expression reduces to the familiar Gaussian equation for $W(\mathbf{r})$ when the dimensionless ratios, $\langle r^{2p}\rangle_0/\langle r^2\rangle_0^p$, reach their asymptotic limits at $n \to \infty$. Chains with a finite number of bonds, n, may have distribution functions that are non-Gaussian. The number of terms required for convergence of $G(\mathbf{q})$ and the pertinent g_{2p}, and hence the upper limit for p in the Δ_{2p} of Eq. (4), will depend on n and the local structure of that chain. An elaboration of this approach has been used to treat macrocyclization equilibrium constants [3] and the closely related constants that describe the intramolecular formation of excimers in end-labeled polymers [4].

The dimensionless ratios required in Eq. (4) yield the special case of the even moments of r. This is a subset of the more general ratios, that can be written as $\langle x^a y^b\rangle_0/\langle x^a\rangle_0\langle y^b\rangle_0$. These ratios allow for the investigation of the correlation between two different conformation-dependent properties. The dimensionless ratio $\langle xy\rangle_0/\langle x\rangle_0\langle y\rangle_0$ provides vital information for this inquiry. This ratio must have a value of 1 if the two properties are uncorrelated. When a correlation exists, the strength of the correlation is measured by the customary correlation coefficient,

$$\rho = \left(\frac{\langle xy\rangle_0}{\langle x\rangle_0\langle y\rangle_0} - 1\right)\left[\left(\frac{\langle x^2\rangle_0}{\langle x\rangle_0^2} - 1\right)\left(\frac{\langle y^2\rangle_0}{\langle y\rangle_0^2} - 1\right)\right]^{-1/2} \tag{5}$$

which has a range of -1 to 1.

When $x = r^2$ and $y = s^2$, casual reflection quickly shows that $0 < \rho < 1$ if n is large and the chain is flexible. Here s^2 denotes the squared radius of gyration. The expectation that $0 < \rho$ comes from the fact that conformations with large end-to-end distances tend to have large radii of gyration. The expectation that $\rho < 1$ is easily justified by consideration of those conformations that generate cyclic chains, for which r^2 is zero. Since cyclic chains of large n can adopt a variety of conformations with different values of s^2, it is clear that s^2 cannot be a unique linear function of r^2. In fact, the correlation coefficient for r^2 and s^2 for a freely jointed chain of 2 or more bonds is given exactly by [5]

$$\rho = \left[\frac{5(n + 1)(n + 2)}{4(2n^2 + 2n + 3)}\right]^{1/2} \tag{6}$$

The value of ρ is 1 at $n = 2$, and the value of ρ decreases as n increases, approaching $(\frac{5}{8})^{1/2}$ as $n \to \infty$. The derivative $\partial\rho/\partial(1/n)$ is positive at all n for which ρ is defined, and it approaches the value of $(\frac{5}{8})^{1/2}$ as $n \to \infty$.

It is well known [1,2] that the real conformational characteristics of chains with finite n can modify the form of $W(\mathbf{r})$, as shown by Eqs. (1) to (4). These same conformational characteristics must cause the dependence of ρ on n to be different from that of the freely jointed chain, Eq. (6). The only aspect of Eq. (6) that will necessarily carry over to real flexible chains is $\rho \to (\frac{5}{8})^{1/2}$ as $n \to \infty$. The objectives here are to define some of the factors that determine (1) how the conformational characteristics of real chains affect the value of

$\partial\rho/\partial(1/n)$ as $n \to \infty$, and (2) the circumstances under which local maxima and minima can occur in ρ versus n.

METHODS

The generator matrix techniques described by Flory [6] were used to calculate $\langle r^2\rangle_0$, $\langle r^4\rangle_0$, $\langle s^2\rangle_0$, and $\langle s^4\rangle_0$. A generalization [7] of the procedure for the calculation of $\langle r^4\rangle_0$ and $\langle s^4\rangle_0$ was used for the computation of $\langle r^2 s^2\rangle_0$. The correlation coefficient is calculated from these five properties as

$$\rho = \left(\frac{\langle r^2 s^2\rangle_0}{\langle r^2\rangle_0\langle s^2\rangle_0} - 1\right)\left[\left(\frac{\langle r^4\rangle_0}{\langle r^2\rangle_0^2} - 1\right)\left(\frac{\langle s^4\rangle_0}{\langle s^2\rangle_0^2} - 1\right)\right]^{-1/2} \tag{7}$$

All of the chains that are considered here have a conformation partition function, Z, that can be written as a serial product of n statistical weight matrices, U_i.

$$Z = U_1 U_2 \cdots U_n \tag{8}$$

For the simple chains there are three rotational states, t, g^+, g^-, separated by rotations about bonds of 120°, and weighted according to the statistical weight matrix, which is most concisely written in condensed form [6] as

$$U_1 = [1 \quad 0] \tag{9}$$

$$U_i = \begin{bmatrix} 1 & 2\sigma \\ 1 & \sigma(1+\omega) \end{bmatrix}, \qquad 1 < i < n \tag{10}$$

$$U_n = [1 \quad 1]^T \tag{11}$$

The value of ω is one for the chains subject to independent rotations and $\omega \neq 1$ for chains subject to interdependent rotations.

Rotational isomeric state models described previously in the literature were used for the real chains. Specifically, the rotational isomeric state models used are those of Flory et al. [8] for poly(dimethyl siloxane), Abe et al. [9] for polyethylene, Semlyen and Flory [10] for poly(phosphate), Tonelli [11] for poly(vinylidene fluoride), Abe and Mark [12] for poly(oxymethylene), and Welsh et al. [13] for poly(silane) and poly(dimethylsilane). We also examine models for simpler chains, which are chosen to illustrate the origin of the interesting features that can be observed in ρ versus n for several real chains.

SIMPLE CHAINS WITH BONDS SUBJECT TO INDEPENDENT ROTATIONS

Figure 1 depicts a stack plot of ρ versus $1/n$ for 19 chains in which the bond angle is 110°, $\omega = 1$, and σ is assigned values ranging from 0.1 to 10. The limit is $(\frac{5}{8})^{1/2}$ as $n \to \infty$, as expected from Eq. (6). The values of $[\partial\rho/\partial(1/n)]_\infty$ are positive, but they depend on the value for σ. The largest values of the deriva-

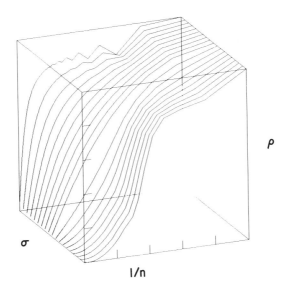

Figure 1   ρ versus $1/n$ for 19 simple chains with independent symmetric threefold rotation potentials. Reading from the rear to the front of the stack, the values of σ are 0.1 to 1.0, at intervals of 0.1, followed by values of 1 to 10, at intervals of 1. The vertical axis is ρ, ranging from 0.79 to 1.00, and the horizontal axis is $1/n$, ranging from 0 to $\frac{1}{3}$.

tive are obtained when σ is small. In these chains with independent bonds, the probability for a trans placement is $1/(1 + 2\sigma)$. This probability approaches 1 as σ decreases. In the limit where $\sigma \to 0$, chains are rigid rods even at large n, and ρ will never fall to the value of $\left(\frac{5}{8}\right)^{1/2}$ that is characteristic of a very long, flexible chain. In the presence of a very slight degree of flexibility, which is obtained when σ has a value only slightly above zero, a large value of n is required in order for ρ to approach $\left(\frac{5}{8}\right)^{1/2}$. For this reason, $[\partial\rho/\partial(1/n)]_\infty$ is largest for the curve at the rear of the stack in Fig. 1. At the opposite limit of Fig. 1, where σ is large, the chain contains predominantly g^+ and g^- placements, each with probability $\sigma/(1 + 2\sigma)$, that are randomly distributed along the chain. Short chains are highly disordered, and extended conformations are improbable, causing $[\partial\rho/\partial(1/n)]_\infty$ to be small.

The qualitative rationalization presented in the preceding paragraph suggests that there may be a correlation between $[\partial\rho/\partial(1/n)]_\infty$ and the limiting value of the characteristic ratio, $C_\infty = (\langle r^2 \rangle_0/nl^2)_\infty$, since the characteristic ratio is large for chains that prefer locally extended conformations. The validity of this suggestion for the chains with independent bonds is shown by the relationship between $[\partial\rho/\partial(1/n)]_\infty$ and C_∞ that is depicted by the circles in Fig. 2. The data for these circles is provided by the 19 chains described in Fig. 1.

Local maxima and minima are apparent in the plots at the rear of the stack in Fig. 1. They arise from the twofold nature of the planar zigzag that

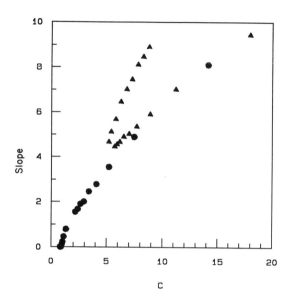

Figure 2   Relationship between $[\partial\rho/\partial(1/n)]_\infty$ and C_∞ for 38 simple chains with $\sigma \neq 1$ and $\omega = 1$ (circles) or $\omega = 0.01$ (triangles).

becomes the dominant conformation of short chains as the trans state displaces gauche states. A more interesting occurrence of a local minimum is suggested by the curves near the front of the stack in Fig. 1. Close inspection of Fig. 2 shows that $[\partial\rho/\partial(1/n)]_\infty$ will become negative at sufficiently small C_∞ (i.e., at sufficiently large σ) and there will then be a local minimum in ρ versus n. Furthermore, the depth of this local minimum will signify that the correlation between r^2 and s^2 is *worse* at some finite value of n than at infinite n. The origin of this local minimum will be investigated further in the next section.

SIMPLE CHAINS WITH BONDS SUBJECT TO INTERDEPENDENT ROTATIONS

Figure 3 depicts ρ versus $1/n$ for 19 chains in which $\omega = 0.01$ and σ is assigned values ranging from 0.1 to 10. These chains differ from those considered in Fig. 1 in that the bonds are now interdependent. A severe penalty is accessed to each contiguous pair of bonds that adopt gauche placements of opposite sign. There is a strong similarity between the curves at the rear of the stacks in Figs. 1 and 3, because the suppression of contiguous pairs of gauche placements of opposite sign is of little consequence when trans placements are dominant.

Marked differences are apparent in the curves at the front of the stacks in these two figures. For the curve in the front of Fig. 3, the dominant local conformation is a string of gauche placements that are all of the same sign, but in the curve in the front of Fig. 1 it is a string of gauche placements with both signs mixed randomly. The preference for gauche helices in the polymer described by the curve at the front of Fig. 3 produces an extended chain, with a large value for C_∞. Hence $[\partial\rho/\partial(1/n)]_\infty$ is strongly positive.

The triangles in Fig. 2 depict the relationship between $[\partial\rho/\partial(1/n)]_\infty$ and C_∞ for the chains described in Fig. 3. These triangles describe a pattern that is reminiscent of the letter V. The triangles that form the right-hand side of the V fall close to the line described by the circles. These triangles are provided by the curves toward the rear of the stack in Fig. 3. The circles in the same region of Fig. 2 are from the curves in the rear of the stack in Fig. 1. In these curves, $\sigma < 1$ and trans placements are dominant.

The triangles that form the left-hand side of the V are provided by the curves toward the front of the stack in Fig. 3. Here $\sigma > 1$, and the dominant local conformations are the right- and left-handed gauche helices. The existence of the V shows that the nature of the locally extended structures also plays a role in determining $[\partial\rho/\partial(1/n)]_\infty$. The value of C_∞ alone is a useful

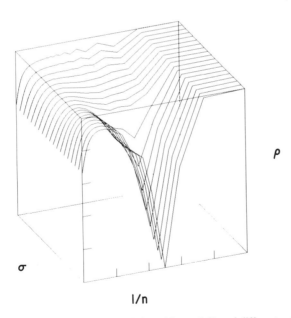

Figure 3 ρ versus $1/n$ for 19 simple chains with $\omega = 0.01$ and different values of σ, which range from 0.1 for the plot at the rear of the stack to $\sigma = 10$ for the plot at the front of the stack. The intervals for σ are 0.1 for 0.1 to 1.0 and 1 for 1 to 10. The vertical axis is ρ, ranging from 0.4 to 1.0, and the horizontal axis is $1/n$, ranging from 0 to $\frac{1}{3}$.

TABLE 1 Dominant Conformations when $\sigma = 10$, $\omega = 0.01$, and the Bond Angle is 109.5° [a]

n	Conformation	Probability	r^2/l^2	s^2/l^2
5	$g^+g^+g^+$	0.688	9.0	1.28
	tg^+g^+	0.138	9.0	1.44
	g^+tg^+	0.069	9.0	1.51
	g^+tg^-	0.069	11.7	1.58
	ttg^+	0.014	11.7	1.73
	tg^+t	0.007	14.3	1.73
6	$g^+g^+g^+g^+$	0.577	13.3	1.66
	$tg^+g^+g^+$	0.115	13.3	1.77
	$g^+tg^-g^-$	0.115	13.3	1.89
	$g^+tg^+g^+$	0.115	8.0	1.72
	ttg^+g^+	0.012	13.3	1.99
	tg^+tg^+	0.012	16.0	2.10
	tg^+tg^-	0.012	18.7	2.21
	g^+ttg^-	0.006	10.7	2.04
	g^+ttg^+	0.006	13.3	2.10
	tg^+g^+t	0.006	13.3	1.88
7	$g^+g^+g^+g^+g^+$	0.484	17.0	2.08
	$g^+tg^+g^+g^+$	0.097	11.7	2.04
	$g^+tg^-g^-g^-$	0.097	17.0	2.25
	$tg^+g^+g^+g^+$	0.097	19.7	2.21
	$g^+g^+tg^+g^+$	0.048	6.3	1.87
	$g^+g^+tg^-g^-$	0.048	17.0	2.25
	$tg^+g^+tg^+$	0.010	11.7	2.12
	$g^+ttg^-g^-$	0.010	11.7	2.29
	$tg^+tg^+g^+$	0.010	14.3	2.29
	$g^+ttg^+g^+$	0.010	17.0	2.41
	$ttg^+g^+g^+$	0.010	19.7	2.38
	$tg^+g^+tg^-$	0.010	19.7	2.38
	$tg^+tg^-g^-$	0.010	19.7	2.54
	$g^+tg^+tg^-$	0.010	19.7	2.63
	$tg^+g^+g^+t$	0.005	17.0	2.25
	$g^+tg^-tg^+$	0.005	22.3	2.75

[a] The probability listed includes degenerate conformations; that is, $g^-g^-g^-$ is included with $g^+g^+g^+$, and tg^-g^-, g^+g^+t, and g^-g^-t are included with tg^+g^+. The statistical weights of the conformations listed account for 98% of the value of Z at $n = 5$ and 6, and 96% of the value of Z at $n = 7$. The square of the length of a bond is denoted by l^2.

predictor of $[\partial\rho/\partial(1/n)]_\infty$ when one is examining a family of chains that have differing preferences for the *same* type of locally ordered structure.

A second dramatic difference between the curves at the front of the stacks in Figs. 1 and 3 is the occurrence of a local minimum at $n = 6$ in the latter figure. The value of ρ at this minimum is much smaller than ρ_∞. An explanation for the occurrence of this pronounced local minimum must consider those chains with statistical weights that make the major contribution to Z when $n = 5$, 6, and 7. A useful feature of the local minimum is that it occurs at a value of n that is sufficiently small so that all important conformations can be enumerated and analyzed individually. The dominant conformations and their probabilities, r^2 and s^2, are listed in Table 1.

At each value of n, the most important conformation is the gauche helix. When $n = 5$, two of the three next-most-important conformations, tg^+g^+, and also g^+tg^+, have the same value of r^2 as the gauche helix, but different values of s^2. These three conformations, which together have statistical weights that account for 90% of the value of Z, have values of r^2 and s^2 that describe a straight line with a slope that is infinite. The most important conformation that does not fall on this line is g^+tg^-, which has a statistical weight that is only 8% of that of the preceding three entries in Table 1. It is this conformation, assisted by several others with even lower probabilities, that causes ρ to be less than 1.

At $n = 6$, 92% of the value of Z is provided by the gauche helix and by $tg^+g^+g^+$, $g^+tg^-g^-$, and $g^+tg^+g^+$. The first three of these conformations all have the same value of r^2 and lie on the line $r^2/l^2 = 13.3$, but the last conformation has a value of r^2 that places it off this line. The major outlier has a statistical weight that is 14% of that of the three points on the line with infinite slope. In contrast, when n was 5, the major outlier had a statistical weight that was only 8% of that of the three conformations on the line $r^2/l^2 = 9$. Consequently, ρ is smaller at $n = 6$ than at $n = 5$.

At $n = 7$, the first six conformations listed in Table 1 account for 87% of the value of Z. The third and sixth conformations have the same values for r^2 and s^2. The five distinguishable points provide a good approximation to a straight line, as shown in Fig. 4. Consequently, ρ is rather large at $n = 7$.

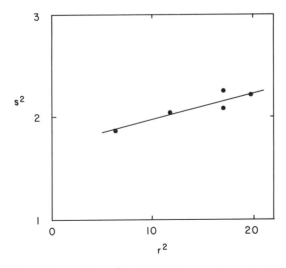

Figure 4   Dependence of s^2 on r^2 for the dominant conformations in a chain with $n = 6$, bonds of unit length, a bond angle of 109.5°, $\sigma = 10$, and $\omega = 0.01$. The points plotted are for conformations whose summed statistical weights account for 87% of the value of Z.

It is noteworthy that the best straight line through the points in Fig. 4 has a much smaller slope (about 0.02) than the straight line through the three points for the dominant conformations at $n = 5$. (The slope of the latter line is infinite.) At $n = 6$ one can see aspects of both types of behavior, and hence ρ is very small. The first three conformations listed for $n = 6$ in Table 1 would describe a line with infinite slope (reminiscent of the behavior at $n = 5$), but pairing the fourth conformation with any of the first three would give a line with a slope less than 0.04 (reminiscent of the behavior at $n = 7$).

In more general terms, the local minimum for ρ is seen at a value of n that marks the transition between one type of approximately linear dependence of s^2 on r^2 at smaller n, and a different type of approximately linear dependence of s^2 on r^2, at larger n, with both types of behavior coexisting at the value of n at which the transition occurs. Hence a means of detecting transitions from one type of linear dependence of x on y to another, where x and y may be any two conformation-dependent scalar physical properties that can be calculated by generator matrix methods (i.e., r^2, s^2, μ^2, γ^2, $\mathbf{r}^T \alpha \mathbf{r}$, $\mu^T \alpha \mu$, etc.) is to search for local minima in the behavior of the appropriate correlation coefficient as a function of n.

More than one local maximum and minimum can occur in ρ versus n, as shown in Fig. 5. Here the value of σ is 1, and the value of ω ranges from 0.1 at the rear of the stack to 10 at the front. Multiple maxima and minima are observed when $\omega > 1$. The behavior depicted in Fig. 5 is primarily a mathematical curiosity because of the paucity of real chains with $\omega > 1$.

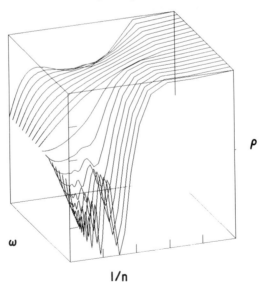

Figure 5 ρ versus $1/n$ for 19 simple chains with $\sigma = 1$ and values of ω that range from 0.1 for the plot at the rear of the stack to 10 at the front, at intervals of 0.1 for 0.1 to 1.0 and 1 for 1 to 10. The vertical axis is ρ, ranging from 0.68 to 1.00, and the horizontal axis is $1/n$, ranging from 0 to $\frac{1}{3}$.

REAL CHAINS WITH BONDS SUBJECT TO THREEFOLD ROTATION POTENTIALS

Figure 6 depicts ρ versus $1/n$, in the region where n is large, for rotational isomeric state models of seven real chains. Each line extrapolates to an intercept of $(\frac{5}{8})^{1/2}$. All of the $[\partial\rho/\partial(1/n)]_\infty$ are positive. The values of $[\partial\rho/\partial(1/n)]_\infty$ from Fig. 6 are depicted as a function of C_∞ in Fig. 7. The relationship between $[\partial\rho/\partial(1/n)]_\infty$ and C_∞ is stronger than might have been expected, in view of the behavior of the simpler models in Fig. 2 and the fact that Fig. 7 contains data for chains that prefer trans states (e.g., polyethylene) or gauche helices (e.g., polyoxymethylene).

Several of the rotational isomeric state models show local maxima and minima in ρ at finite n that can be attributed to an odd–even effect that persists for ρ at $n > 100$. More dramatic local minima can be observed at smaller n, as shown in Fig. 8 for polydimethylsilane, polydimethylsiloxane, polyphosphate, and polyoxymethylene. A broad minimum near $n = 8$ is superimposed on the odd–even effect. In these chains, s^2 is more strongly correlated with r^2 at $n = 20$ than at $n = 10$. The local minimum for polyoxymethylene in Fig. 8 occurs at $n = 7$, which is very close to the location ($n = 6$) of the local minimum for the curves at the front of the stack in Fig. 3. This result is expected because polyoxymethylene and the chains described by the curves at the front of the stack in Fig. 3 have $\sigma \gg 1$ and $\omega \ll 1$.

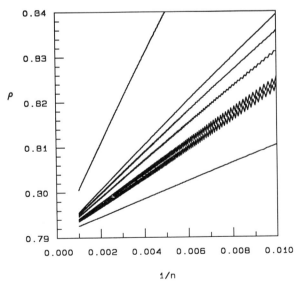

Figure 6 ρ versus $1/n$, for $100 \leq n \leq 1000$, for rotational isomeric state models for seven chains with symmetric threefold rotation potentials. In order of increasing slope, the chains are polysilane, polydimethylsiloxane, polyvinylidene fluoride, polyoxymethylene, polyphosphate, polyethylene, and polydimethylsilane.

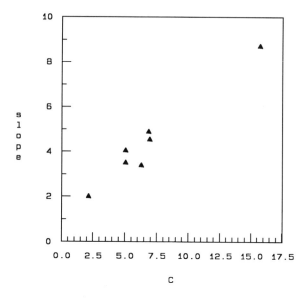

Figure 7   Relationship between $[\partial\rho/\partial(1/n)]_\infty$ and C_∞ for the seven chains in Fig. 6.

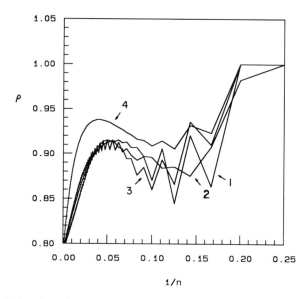

Figure 8   Local maxima and minima in ρ versus $1/n$ for rotational isomeric state models for (1) polyphosphate, (2) polyoxymethylene, (3) polydimethylsiloxane, and (4) polydimethylsilane.

CONCLUSIONS

The value of ρ for r^2 and s^2 for rotational isomeric state chains approaches $(\frac{5}{8})^{1/2}$ as n becomes infinite, as expected from the closed-form, analytical result for freely jointed chains. The approach to this limit is delayed for stiffer chains. Local maxima and minima can be seen in ρ as a function of n in rotational isomeric state chains, but they are absent in freely jointed chains. The local minima provide interesting information because they can be produced by the transition from one type of approximately linear dependence of s^2 on r^2, at smaller n, to another type of approximately linear dependence of s^2 on r^2, at larger n. This interpretation of the local minima in ρ versus n can be extended to studies of the values of ρ for other pairs of conformation-dependent scalar physical properties selected from those susceptible to calculation via generator matrix methods.

ACKNOWLEDGMENT

This research was supported by National Science Foundation Grant DMR 87-06166.

REFERENCES

1. K. Nagai, *J. Chem. Phys.* **38**, 924 (1963).
2. P. J. Flory, *Statistical Mechanics of Chain Molecules,* Wiley, New York, 1969, pp. 309–311.
3. P. J. Flory, U. W. Suter, and M. Mutter, *J. Am. Chem. Soc.* **98**, 5733 (1976).
4. A. Pannikottu and W. L. Mattice, *Macromolecules* **23**, 867 (1990).
5. W. L. Mattice and K. Sienicki, *J. Chem. Phys.* **90**, 1956 (1989).
6. P. J. Flory, *Macromolecules* **7**, 381 (1974).
7. W. L. Mattice, *J. Chem. Phys.* **87**, 5512 (1988).
8. P. J. Flory, V. Crescenzi, and J. E. Mark, *J. Am. Chem. Soc.* **86**, 146 (1964).
9. A. Abe, R. L. Jernigan, and P. J. Flory, *J. Am. Chem. Soc.* **88**, 631 (1966).
10. J. A. Semlyen and P. J. Flory, *Trans. Faraday Soc.* **62**, 2622 (1966).
11. A. E. Tonelli, *Macromolecules* **9**, 547 (1976).
12. A. Abe and J. E. Mark, *J. Am. Chem. Soc.* **98**, 6468 (1976).
13. W. J. Welsh, L. DeBolt, and J. E. Mark, *Macromolecules* **19**, 2978 (1986).

25

Computer Simulation Study of the Collapse Transition of Polymers in Two and Three Dimensions

HAGAI MEIROVITCH
Supercomputer Computations Research Institute
Florida State University
Tallahassee, Florida 32306

ABSTRACT

We have applied the scanning simulation method to tricritical self-avoiding walks (SAWs) and trails on a square and a simple cubic lattice at the θ-point. The method has been found to be highly efficient, where the values of the tricritical temperature T_t and the various exponents have been obtained with higher accuracy than in previous studies. In two dimensions (2D) we find that for the chains in the bulk, the results for the partition function exponent γ_t and the shape exponent ν_t agree with the exact results obtained by Duplantier and Saleur (DS) for a particular SAW model on the hexagonal lattice. However, the results of γ_{11t} and γ_{1t}, for trails that are terminally attached to an impenetrable linear boundary on the square lattice, are significantly different from the DS values. We also find that the functions which lead to the crossover exponent ϕ_t are affected, for all the models studied, by strong corrections to scaling and we suggest ways to take them into account. We show that in 2D the values of ϕ_t for trails, SAWs and the DS model are significantly different, which suggests that the three models belong to different universality classes. In 3D some differences between the exponents of trails and SAWs are also observed, but they are not significant enough to lead to the conclusion that the two models are not in the same universality class.

INTRODUCTION

The behavior of a polymer under various solvent conditions has been modeled by self-avoiding walks (SAWs) of N steps ($N + 1$ monomers) on a lattice, where an attraction energy ϵ ($\epsilon < 0$) is defined between a pair of nonbonded nearest-neighbor (nn) monomers. At high temperature T the chain is swollen; that is, its end-to-end distance R scales as N^ν, where $\nu \simeq 0.59$ in three dimensions (3D) [1]. This corresponds to the good solvent regime. As T is decreased (i.e., solvent conditions worsen) the attractions become more effective and at some temperature θ (the Flory θ temperature) they cancel the excluded volume repulsions and the chain behaves in many respects like a random walk [i.e., it is characterized by $\nu = 0.5$ and $\gamma = 1$ (γ is the exponent of the partition function Z)]. At $T < \theta$ the attractions prevail and the chain collapses (i.e., $\nu = 1/D$). This picture, which was first suggested by Flory [2], is supported by experiment [3–6]; however, it takes into account only the two-body term in the expansion of the free energy [2,7,8]. Later de Gennes also considered the three-body interactions and identified the θ transition as a tricritical point with an upper critical dimension three [1,9,10] (we therefore will denote θ by T_t, where the subscript t stands for tricritical); in 3D one still obtains the random walk exponents but R and Z become dependent on logarithmic corrections to scaling [9–17]. Even though the de Gennes theory in 3D is well accepted, not enough numerical work has been carried out to examine its validity [15,18–28]. In two dimensions the situation is less clear than in 3D since the number of relevant terms in the free-energy expansion is unknown [15,29,30].

    An important development in 2D has been a recent work of Duplantier and Saleur (DS) [31], where the exact tricritical exponents of a polymer in 2D are proposed. They find the polymer size exponent $v_t = \frac{4}{7} \approx 0.571$, the partition function exponent $\gamma_t = \frac{8}{7} \approx 1.143$, and the crossover exponent $\phi_t = \frac{3}{7} \approx 0.43$. They also calculate the partition function surface exponents (i.e., for a polymer that is terminally attached to an impenetrable boundary) and obtain $\gamma_{1t} = \gamma_t$ and $\gamma_{11t} = v_t$, where γ_{11t} is related to the subgroup of chains that also end on the surface. These values have been derived for a special model of SAWs on a hexagonal lattice with randomly forbidden hexagons using Coulomb gas techniques and conformal invariance with a conformal anomaly $C = 0$. However, it has been pointed out [32] that this model consists, in addition to the nn attractions, of a special subset of the next-nearest-neighbor (nnn) attractions, and therefore, instead of describing the usual θ-point, it might describe a multicritical θ′-point. Also, Vanderzande [33] has classified the θ-point on the basis of the $n \rightarrow 0$ limit of the $O(n)$ model with annealed vacancies by a superconformal field theory with a conformal anomaly $C = \frac{1}{2}$. He has found $v_t = \frac{4}{7}, \phi_t = \frac{3}{7}, \gamma_t = \frac{15}{14} \approx 1.07, \gamma_{1t} = \frac{15}{28} \approx 0.53, \gamma_{11t} = -\frac{4}{7}$, which means that the θ and θ′ points belong to different universality classes (see also the discussion in Ref. 34). Indeed, most recent numerical studies of SAWs with nn attractions have found exponents that differ from the DS values [34–36]. Seno and Stella [34,35] have obtained (from exact enumeration and simulation of SAWs of $N \leq 40$ on the square lattice) a result for v_t that agrees with the DS value, but their estimate $\gamma_t = 1.075 \pm 0.04$ is smaller than $\gamma_t(DS) = 1.143$, and their result for ϕ_t is slightly larger than the DS value. However, they have also calculated the surface exponents and have found values that are dramatically smaller than those of DS, $\gamma_{1t} = 0.57 \pm 0.09$ and $\gamma_{11t} = -0.53 \pm 0.10$. (Thus their results support the Vanderzande [33–35] theory.) Poole et al. find a marginal crossover from the θ-point to the θ′-point on the hexagonal lattice as the special nnn interactions are introduced and suggest that a continuous change of the universality class occurs between the two points [36]. One should also mention the exact results obtained by Bradley [37] for SAWs with nn attractions on the Manhattan lattice, $v_t = \frac{4}{7}$ (the DS value) but $\gamma_t = \frac{6}{7}$. Therefore, the factors that determine the universal behavior of a polymer in 2D at the θ-point are as yet not clear, and further work needs to be done.

    A related question is whether trails and SAWs share the same exponents at tricriticality. A trail is a walk on a lattice that may intersect itself but two bonds are not allowed to overlap. Also, an interaction energy ϵ ($\epsilon < 0$) is associated with each self-intersection [38,39]. Shapir and Oono have argued that these models may belong to different universality classes [40] contrary to SAWs and trails without attractions (i.e., at $T = \infty$) which are in the same universality class [38–42]. Therefore, Shapir and co-workers have carried out an extensive study of tricritical trails in 2D and 3D by exact enumeration [43–45]; however, their chains (of length $N < 21$) steps appeared to be too short to provide unequivocal results.

We have therefore decided to carry out a systematic study of the tricritical behavior of relatively long SAWs and trails ($N \leq 300$) in 2D and 3D by computer simulation, using the scanning method [46,47]. With this method a chain is generated step by step with the help of transition probabilities, obtained by scanning the so-called "future chains"; they are the possible continuations of the chain in future steps up to some cutoff. Therefore, the probability of construction of a chain is known and hence the entropy and the free energy are also known, and the exponent γ can be calculated. The method is applicable to a wide range of models, and the statistical error can be estimated reliably. Another advantage of the scanning method is the fact that results at *many* temperatures can be obtained from a *single* sample generated at a given temperature. This greatly facilitates the search for the tricritical temperature T_t. Thus we have simulated tricritical trails and SAWs on the square [48–51] and the simple cubic lattice [52,53] and trails that are terminally attached to an impenetrable boundary on a square lattice [51].

THEORY

Tricritical Temperature and v_t

The tricritical temperature T_t can be determined from the results of the radius of gyration G or the end-to-end distance, R close to T_t. One can adopt the generalized scaling behavior for SAWs or trails at tricriticality [1],

$$G_N \equiv \langle G^2 \rangle^{1/2} \sim N^{v_t} f_{\pm}(N^{\phi_t}\tau) \qquad (1)$$

where ϕ_t is a crossover exponent and τ is $|(T - T_t)/T_t|$. At high T ($T > T_t$) the chains are swollen and are characterized by $[v(T = \infty) = v_\infty]$. At low T ($T < T_t$) they are expected to collapse ($v = 1/D$). Therefore, for small τ and large x (i.e., very large N) $f_{\pm}(x)$ must have the form

$$f_{\pm}(x) \sim x^{\mu} \begin{cases} \mu = \mu^+ = \dfrac{v_\infty - v_t}{\phi_t} & T > T_t \\[2ex] \mu = 0 & T = T_t \\[2ex] \mu = \mu^- = \dfrac{1/D - v_t}{\phi_t} & T < T_t \end{cases} \qquad (2)$$

However, short chains at $T > T_t$ will expand with $v < v_\infty$, which is expected to increase monotonically with increasing N and to approach v_∞. An opposite trend is expected at $T < T_t$, where the shape of the shorter chains grows with $v > 1/D$, which decreases asymptotically to $1/D$; thus at $T = T_t$, $v_t(N)$ is expected to become flat. It should be noted that Eqs. (1) and (2) also hold for the end-to-end distance, where R and R_N replace G and G_N, respectively; we shall denote the results for v obtained from the radius of gyration and the end-to-end distance by $v(G)$ and $v(R)$, respectively. Thus, calculating the v values of several subchains at different temperatures (assuming that $G_N \sim N^v$) enables one to locate T_t on the basis of the foregoing flatness criterion. This criterion has indeed been employed to determine T_t in several previous studies [54,55]. It should be pointed out, however, that $f_\pm$ [Eq. (1)] becomes a function of $x = N^{\phi_t}\tau$ only for large enough N. For small N correction terms, which depend on N but are independent of τ, are expected to contribute significantly, in particular at T_t; therefore, the foregoing flatness criterion should be used with caution (i.e., it should be applied to the longer chains where the corrections are small). Indeed, for the continuum SAW model in the bulk, several terms of the expansion of $f_\pm(x)$ (for small x) have been calculated [16,17] and in 3D a logarithmic correction leading term $[\sim(\log N)^{-1}]$ (independent of τ) has been obtained (see also Refs. 1 and 9–15). A strong correction to scaling for G and R (which is consistent with a logarithmic correction) has also been observed in our simulation of SAWs on a simple cubic lattice. For random walks and SAWs adsorbed to a surface in 3D [55,56], a correction to scaling leading term $(\sim 1/N^{1/2})$ has been found for the component of R that is perpendicular to the surface. This component is also affected by strong corrections for trails terminally attached to an impenetrable linear boundary on a square lattice [51]. These corrections reflect the fact that the repulsion of the chain by the surface increases rather slowly with increasing N, which means that the shorter chains are relatively more compact than the longer ones and are therefore expected to lead to an overestimation of T_t.

Crossover Exponent

The crossover exponent ϕ_t can be calculated from the results of the specific heat $C(T, N)$, which close to T_t are expected to scale like

$$C(T, N) = N^{y_t} g(N^{\phi_t}\tau) \qquad (3)$$

where $y_t = \alpha_t \phi_t$ and α_t is the tricritical exponent of the specific heat. For large $x = N^{\phi_t}\tau$ (small τ) one expects that

$$g(x) \sim \begin{cases} A^+ x^{-\alpha_t} & T > T_t \\ \text{constant} & T = T_t \\ A^- x^{-\alpha_t} & T < T_t \end{cases} \qquad (4)$$

where A^+ and A^- are prefactors. Thus one can calculate the exponent y_t from the results of C at T_t and ϕ_t can be obtained from the relation [45,57,58]

$$\alpha_t = 2 - \frac{1}{\phi_t} \tag{5}$$

which leads to

$$\phi_t = \frac{1 + y_t}{2} \tag{6}$$

The crossover exponent ϕ_t can also be estimated from the results for $C_{max}[N, T_{max}(N)]$, which will be denoted by $C_{max}(N)$. $C_{max}(N)$ is the maximal value of C for a given N obtained at $T_{max}(N)$ (see Chang et al. in Ref. 45). Thus the specific heat scales like

$$C_{max}(N) \sim \tau^{-\alpha_t} f(N^{\phi_t}\tau) \sim N^{\alpha_t \phi_t} = N^{y_t} \tag{7}$$

where $\tau = |(T_{max}(N) - T_t)/T_t|$. Therefore, one can also estimate y_t from the slope of the results for $\log[C_{max}(N)]$ versus $\log N$ and obtain ϕ_t from Eqs. (5) and (6).

An additional way to obtain ϕ_t is from the derivative of G_N^2 [Eq. (1)] with respect to $K = -\epsilon/k_B T$. One obtains [59]

$$G' = \frac{\partial G_N^2}{\partial K}/G_N^2 = \frac{\langle G^2 E \rangle - \langle G^2 \rangle \langle E \rangle}{\langle G^2 \rangle} \sim N^{\phi_t} \tag{8}$$

where the derivative is evaluated at K_t. Correspondingly, ϕ_t can be obtained from the derivative of the end-to-end distance R_N^2, which will be denoted by R'. One can log-log plot the results at K_t for G', R', and C versus N and estimate ϕ_t from the related slopes. We have found that for all the models studied, the plots of G', R', and C are strongly concave (i.e., they are affected by strong corrections to scaling). The slopes have therefore been calculated from the results for the longer chains, where the lines have seemingly become straight. However, our results for ϕ_t have always been found to satisfy the relation

$$\phi_t(R) < \phi_t(G) < \phi_t(C)$$

which means that finite size effects are still significant for the chain lengths studied. To take these corrections into account in 2D we have best-fitted the results for G' (and R') at K_t to the function

$$G' = AN^{\phi_t}\left(1 + \frac{B}{N^\delta}\right) \tag{9}$$

where A and B are constants, and have searched for the value of δ that leads to $\phi_t(G) = \phi_t(R)$.

It should be pointed out that the values of ϕ_t obtained from the results of $C_{max}(N)$ have led approximately to the values obtained from Eq. (7). On the other hand, the values of ϕ_t, which are based on the results of C at K_t itself, are always larger than the other estimates. This suggests that the latter method is the least reliable among the four.

Partition Function

As we have already pointed out, the scanning method, in contrast to other simulation techniques, leads to the partition function Z. For $T > T_t$ and large N, one expects the results for Z to behave as $N^{\gamma-1}\mu^N(T)$, where the growth parameter $\mu(T)$ is a function of the temperature, while γ obtains its value at $T = \infty$. However, at $T = T_t$, these quantities are expected to attain their tricritical values μ_t and γ_t. To estimate them from finite chains, one has to verify that corrections to scaling are negligible. This can be done by best-fitting the results of Z at K_t for various chain lengths $(N_{\min}, N_{\max})$ to the function

$$Z(T_t) = BN^{\gamma_t - 1}\mu_t^N \tag{10}$$

where B is a constant. If corrections to scaling are negligible, the values of $\gamma_t(N_{\min}, N_{\max})$ or $\mu_t(N_{\min}, N_{\max})$ should be constant (i.e., independent of $N_{\min}$ and $N_{\max}$). Our calculations have revealed that for trails and SAWs on the simple cubic lattice, finite size effects in Z are negligible; however, for chains on the square lattice the results have been found to be more N-dependent.

RESULTS AND DISCUSSION

Results in 2D

Using the flatness criterion described earlier, we have obtained for SAWs and trails on the square lattice the following estimates for K_t (with 95% confidence limits) [48,49,51]:

$$K_t(\text{SAWs}) = 0.658 \pm 0.004 \qquad K_t(\text{trails}) = 1.086 \pm 0.002$$

It should be pointed out that, as expected, the above estimate for $K_t(\text{trails})$ has been obtained for both trails in the bulk [48] and for trails in the presence of an impenetrable linear boundary [51]. These estimates are significantly more accurate than those obtained in previous studies [35,43 and references therein].

The exponent ν_t for both SAWs and trails has been found to be close to the DS exact value $\nu_t = 0.571\ldots$. However there are small differences between the results for $\nu_t(G)$ and $\nu_t(R)$ due to finite size effects. We obtain (with 95% confidence limits)

$$\nu_t(G, \text{SAWs}) = 0.5795 \pm 0.0030 \qquad \nu_t(R, \text{SAWs}) = 0.574 \pm 0.006$$

$$\nu_t(G, \text{trails}) = 0.563 \pm 0.003 \qquad \nu_t(R, \text{trails}) = 0.574 \pm 0.005$$

The results above are for chains in the bulk. For trails on a surface the results are comparable. These results are in accord with estimates for ν_t obtained using other techniques [34–37].

Our results for γ_t of SAWs and trails in the bulk are (with 95% confidence limits)

$$\gamma_t(\text{SAWs}) = 1.110 \pm 0.022 \qquad \gamma_t(\text{trails}) = 1.133 \pm 0.024$$

The value for trails above is equal within the error bars to the DS prediction $\gamma_t = 1.143\ldots$, whereas the result for SAWs is slightly smaller than the DS value. Seno and Stella [34,35] have obtained for SAWs a smaller estimate (1.075) than ours and Bradley's exact result for the Manhattan lattice [37] is $\frac{6}{7} = 0.857\ldots$. The latter result suggests that γ_t might not be universal but lattice dependent. For μ_t we obtain values that are larger and smaller than 3 (the μ value of a nonreversal random walk) for SAWs and trails, respectively,

$$\mu_t(\text{SAWs}) = 3.213 \pm 0.013 \qquad \mu_t(\text{trails}) = 2.9901 \pm 0.0020$$

While our results for $\nu_t(\text{trails})$ and $\gamma_t(\text{trails})$ are in accord with the DS values, a significant difference is found between the DS results and ours for the surface exponents γ_{1t} and γ_{11t}. We obtain (with 95% confidence limits) [51]

$$\gamma_{1t} = 0.634 \pm 0.025 \qquad \gamma_{11t} = -0.44 \pm 0.02$$

compared to the DS values ≈ 1.143 and ≈ 0.571, respectively. Our results are equal to the Seno and Stella results within their relatively large error bars [34,35].

To obtain ϕ_t we have analyzed the results for G' and R' [Eq. (8)] using Eq. (9), which takes into account corrections to scaling. For both SAWs and surface trails we have obtained that for $\delta \approx 0.60$, $\phi_t(G) \approx \phi_t(R)$ and the results are [51,60]

$$\phi_t(\text{SAWs}) = 0.50 \pm 0.02 \qquad \phi_t(\text{trails}) = 0.70 \pm 0.02$$

The value for trails above is in accord with the value obtained from $C_{\max}(N)$ [Eq. (7)], which is $\phi_t = 0.72 \pm 0.02$ [51]. These results suggest that SAWs and trails belong to different universality classes. Also, the value for SAWs above is significantly larger than the DS value $\phi_t \approx 0.43$, which suggests that the θ' point is in different universality class than the θ-point.

Results in 3D

We have studied tricritical SAWs and trails of $N \leq 250$ on the simple cubic lattice [52,53]. It should be pointed out that even though the samples generated for these models are larger than those of the 2D models, the statistical errors of the results for G and R are larger for the 3D models. Also, for 3D SAWs strong corrections to scaling have been observed; therefore, it is more difficult to apply the flatness criterion in 3D than in 2D. The results for K_t (with one standard deviation) are

$$K_t(\text{SAWs}) = 0.274 \pm 0.006 \qquad K_t(\text{trails}) = 0.550 \pm 0.004$$

For SAWs we have found that the estimate of K_t above is consistent with the theoretical predictions $v_t = 0.5$ and a logarithmic correction to scaling leading term for both G and R. Thus we have best-fitted the results for G^2 (R^2) at each temperature to a function similar to Eq. (9) (using $2v_t = 1$) but with $\ln N$ instead of N^8 in the denominator and have found the optimal fit to occur always within the error bars of the estimated K_t. However, this is not the case for trails where the optimal fit has not been found close to the estimated $K_t = 0.550$, but in the much colder region, $K \geq 0.585$. At K_t we have obtained for the two models $v_t = 0.515 \pm 0.03$, which is larger than 0.5, the theoretical value for SAWs. The results for γ_t are also slightly different,

$$\gamma_t(\text{SAWs}) = 1.005 \pm 0.017, \qquad \gamma_t(\text{trails}) = 1.040 \pm 0.005$$

$$\mu_t(\text{SAWs}) = 5.058 \pm 0.014 \qquad \mu_t(\text{trails}) = 5.003 \pm 0.002$$

The result for $\gamma_t(\text{SAWs})$ is equal to the theoretical value $\gamma_t = 1$, whereas the value for trails is larger than 1. For both models the correction to scaling in the results for Z appears to be very small and at K_t it is not consistent with a logarithmic correction.

To obtain ϕ_t we have used Eq. (8), where the results for the longer chains are taken into account ($140 \leq N \leq 220$ for SAWs and $170 \leq N \leq 240$ for trails). For SAWs we obtain

$$\phi_t(G) \simeq 0.581 \qquad \phi_t(R) \simeq 0.551$$

and for trails

$$\phi_t(G) \simeq 0.510 \qquad \phi_t(R) \simeq 0.479$$

The results for trails are close to the theoretical value for SAWs, $\phi_t = 0.5$, whereas the values of SAWs are larger. As expected, the results for $\phi_t(C)$, which are based on the results of the specific heat at K_t, are much larger than the estimates above; we obtained $\simeq 0.672$ and $\simeq 0.624$ for SAWs and trails, respectively. On the other hand, for trails the results of $C_{\max}(N)$ led to $\phi_t = 0.50 \pm 0.02$, which is equal to the estimates of $\phi_t(G)$ and $\phi_t(R)$ above.

We have also tried to best-fit the data for G' and R' at each temperature to the theoretical function for SAWs, which is based on a multiplicative logarithmic correction to scaling [16,17]. However, the optimal fit has been found to occur at much hotter temperatures than the values of K_t for both SAWs and trails.

SUMMARY

We have applied the scanning method to tricritical SAWs and trails on a square and a simple cubic lattice. The method has been found to be highly efficient, where the values of K_t and the various exponents have been obtained with higher accuracy than in previous studies. In 2D we find that for the chains in the bulk, the results for γ_t and v_t agree with the DS exact values. However, the results of γ_{11t} and γ_{1t} for the surface trails are significantly different from the DS values. We also find that the functions which lead to ϕ_t are affected, for all the models studied, by strong corrections to scaling and we suggest ways to take them into account. We show that in 2D the values of ϕ_t for trails, SAWs, and the DS model are significantly different, which suggests that the three models belong to different universality classes. In 3D some differences between the exponents of trails and SAWs are also found, but they are not significant enough to lead to the conclusion that the two models are in different universality classes.

ACKNOWLEDGMENTS

The author acknowledges support from the Florida State University Supercomputer Computations Research Institute, which is partially funded by the U.S. Department of Energy under Contract DE-FC05-85ER250000.

REFERENCES

1. P.-G. de Gennes, *Scaling Concepts in Polymer Physics,* Cornell University Press, Ithaca, N.Y., 1985.
2. P. J. Flory, *Principles of Polymer Chemistry,* Cornell University Press, Ithaca, N.Y., 1953.
3. J. Mazur and D. McIntyre, *Macromolecules* **8,** 464 (1975).
4. M. Nierlich, J. P. Cotton, and B. Farnoux, *J. Chem. Phys.* **69,** 1379 (1978).
5. G. Swislow, S.-T. Sun, I. Nishio, and T. Tanaka, *Phys. Rev. Lett.* **44,** 796 (1980).
6. R. Perzynski, M. Adam, and M. Delsanti, *J. Phys. (Paris)* **43,** 129 (1982).
7. S. F. Edwards, in *Critical Phenomena,* ed. M. S. Green and J. V. Sengers, National Bureau of Standards Miscellaneous Publication No. **273,** U.S. Government Printing Office, Washington, D.C., 1966, p. 225.
8. I. M. Lifshitz, A. Y. Grosberg, and A. R. Khokhlov, *Rev. Mod. Phys.* **50,** 683 (1978).
9. P.-G. de Gennes, *J. Phys. (Paris) Lett.* **36,** L55 (1975).
10. P.-G. de Gennes, *J. Phys. (Paris) Lett.* **39,** L299 (1978).
11. M. J. Stephen and J. L. McCauley, *Phys. Lett.* **A44,** 89 (1973).
12. M. J. Stephen, *Phys. Lett.* **A53,** 363 (1975).
13. L. Kholodenko and K. F. Freed, *J. Chem. Phys.* **80,** 900 (1984).
14. J. F. Douglas and K. F. Freed, *Macromolecules* **18,** 2445 (1985).
15. B. J. Cherayil, J. F. Douglas, and K. F. Freed, *J. Chem. Phys.* **87,** 3089 (1987); J. F. Douglas, B. J. Cherayil, and K. F. Freed, *Macromolecules* **18,** 2455 (1985).
16. B. Duplantier, *Europhys. Lett.* **1,** 491 (1986).
17. B. Duplantier, *J. Chem. Phys.* **86,** 4233 (1986).
18. J. Mazur and F. L. McCrackin, *J. Chem. Phys.* **49,** 648 (1968).
19. F. L. McCrackin, J. Mazur, and C. L. Guttman, *Macromolecules* **6,** 859 (1973).
20. M. Janssens and A. Bellemans, *Macromolecules* **9,** 303 (1976).
21. R. Finsy, M. Janssens, and A. Bellemans, *J. Phys.* **A8,** L106 (1975).
22. M. Lal and D. Spencer, *Mol. Phys.* **22,** 649 (1971).

23. K. Kremer, A. Baumgärtner, and K. Binder, *J. Phys.* **A15,** 2879 (1982).
24. D. C. Rapaport, *Phys. Lett.* **A48,** 339 (1974).
25. D. C. Rapaport, *J. Phys.* **A10,** 637 (1977).
26. J. Webman, J. L. Lebowitz, and M. H. Kalos, *J. Phys. (Paris)* **41,** 579 (1980); *Macromolecules* **14,** 1495 (1981).
27. J. Tobochnik, I. Webman, J. L. Lebowitz, and M. H. Kalos, *Macromolecules* **15,** 549 (1982).
28. A. Baumgärtner, *J. Chem. Phys.* **72,** 871 (1980).
29. D. Thirumalai, *Phys. Rev.* **A37,** 269 (1988).
30. B. Duplantier, *Phys. Rev.* **A38,** 3647 (1988).
31. B. Duplantier and H. Saleur, *Phys. Rev. Lett.* **59,** 539 (1987).
32. P. H. Poole, A. Coniglio, N. Jan, and H. E. Stanley, *Phys. Rev. Lett.* **60,** 1203 (1988); B. Duplantier and H. Saleur, *Phys. Rev. Lett.* **60,** 1204.
33. C. Vanderzande, *Phys. Rev.* **B38,** 2865 (1988).
34. F. Seno, A. L. Stella, and C. Vanderzande, *Phys. Rev. Lett.* **61,** 1520 (1988); B. Duplantier and H. Saleur, *Phys. Rev. Lett.* **61,** 1521 (1988).
35. F. Seno and A. L. Stella, *J. Phys. (Paris)* **49,** 739 (1988); F. Seno and A. L. Stella, *Europhys. Lett.* **7,** 605 (1989).
36. P. H. Poole, A. Coniglio, N. Jan, and H. E. Stanley, *Phys. Rev.* **B39,** 495 (1989).
37. R. M. Bradley, *Phys. Rev.* **A39,** 3738 (1989).
38. A. Malakis, *J. Phys.* **A8,** 1885 (1975); **A9,** 1283 (1976).
39. A. R. Massih and M. A. Moore, *J. Phys.* **A8,** 237 (1975).
40. Y. Shapir and Y. Oono, *J. Phys.* **A17,** L39 (1984).
41. D. C. Rapaport, *J. Phys.* **A18,** L475 (1985); A. J. Guttmann, *J. Phys.* **A18,** 567, 575 (1985); A. J. Guttmann and T. R. Osborn, *J. Phys.* **A21,** 513 (1988).
42. H. Meirovitch and H. A. Lim, *Phys. Rev.* **A38,** 1670 (1988); H. A. Lim and H. Meirovitch, *Phys. Rev.* **A39,** 4176 (1989).
43. H. A. Lim, A. Guha, and Y. Shapir, *J. Phys.* **A21,** 773 (1988).
44. A. Guha, H. A. Lim, and Y. Shapir, *J. Phys.* **A21,** 1043 (1988).
45. I. S. Chang, A. Guha, H. A. Lim, and Y. Shapir, *J. Phys.* **A21,** L559 (1988).
46. H. Meirovitch, *J. Phys.* **A15,** L735 (1982).
47. H.Meirovitch, *J. Chem. Phys.* **89,** 2514 (1988).
48. H. Meirovitch and H. A. Lim, *Phys. Rev.* **A39,** 4186 (1989).
49. H. Meirovitch and H. A. Lim, *J. Chem. Phys.* **91,** 2544 (1989).
50. H. Meirovitch and H. A. Lim, *Phys. Rev. Lett.* **62,** 2640 (1989); B. Duplantier and H. Saleur, *Phys. Rev. Lett.* **62,** 2641 (1989).
51. H. Meirovitch, I. S. Chang, and Y. Shapir, *Phys. Rev.* **A40,** 2879 (1989); I. S. Chang, H. Meirovitch, and Y. Shapir, *Phys. Rev.* **A41,** 1808 (1990).
52. H. Meirovitch and H. A. Lim, *J. Chem. Phys.* **92,** 5144 (1990).
53. H. Meirovitch and H. A. Lim, *J. Chem. Phys.* **92,** 5155 (1990).
54. K. Kremer, A. Baumgärtner, and K. Binder, *J. Phys.* **A15,** 2879 (1982).
55. E. Eisenriegler, K. Kremer, and K. Binder, *J. Chem. Phys.* **77,** 6296 (1982).
56. S. Livne and H. Meirovitch, *J. Chem. Phys.* **88,** 4498 (1988); H. Meirovitch and H. Livne, *J. Chem. Phys.* **88,** 4507 (1988).
57. B. Derrida and H. J. Herrmann, *J. Phys. (Paris)* **44,** 1365 (1983).
58. P. M. Lam, *Phys. Rev.* **B36,** 6988 (1987).
59. V. Privman, *J. Phys.* **A19,** 3287 (1986).
60. H. Meirovitch, unpublished results.

Data Structures and Algorithms for Computation of Topological Invariants of Entanglements: Link, Twist, and Writhe

R. C. LACHER
Department of Computer Science
Florida State University
Tallahassee, Florida 32306

D. W. SUMNERS
Department of Mathematics
Florida State University
Tallahassee, Florida 32306

ABSTRACT

Data structures and algorithms for calculation of linking numbers in lattice models are presented and analyzed. The twist of an embedded annulus, and the writhe of a single curve, can be calculated with these methods. Especially simple methods for computations of writhe are found. As an application, it is proved that four times the writhe of any curve on the simple cubic lattice is an integer.

INTRODUCTION

Knots and links are the mathematical atoms for notions of entanglement of molecular chains, whether dynamic or static. Dynamic entanglements depend on rates of diffusion or reptation and can be defined for linear chains, while static entanglements (which may be thought of as dynamic entanglements with infinite time allowed for disentanglement) exist only for closed loops of some sort [1–5]. In any case, the idea of molecular entanglement ultimately rests on linking and knotting of closed curves in space. There has therefore been interest in simulations and calculations of knots and links both to investigate the statistics of random knotting and to differentiate molecular types using knot or link invariants.

Although a knot or link consists a priori of a cube of information, it is often the case that much of that information, including all of the topologically invariant information, is redundant or reconstructible from significantly less than a cube of data. For example, it is common practice to represent a knot by inscribing an image on a piece of paper, indicating orientation and crossing data, in a manner that completely specifies a three-dimensional knot type. This is a reduction of the dimension of stored information by at least one. The inscribed image is essentially an annotated planar projection of the knot, the annotations indicating orientation and giving a crossover type at each self-intersection point in the projection of the curve. It is the creation, storage, and use of annotated planar projections of knots and links that is the major concern of this chapter.

REGULAR PROJECTIONS

First let us fix a setting in which to discuss knot and link invariants. We confine this discussion to the category of piecewise linear objects and mappings; an equivalent discussion can be made in the category of smooth objects and mappings. We will also confine much of the discussion to closed curves in space; analogous results also hold for proper embeddings of 1-manifolds in a cube, ball, or half-space. Consider a closed oriented one-dimensional submanifold K of real three-dimensional space with coordinates x, y, and z. Then K consists of a finite number of components each of which is an oriented simple closed polygonal curve. If K has only one component, it is called a knot; otherwise, it is called a link. Two such submanifolds are *topologically equivalent* if and only if there is a (piecewise linear, orientation-preserving) homeomorphism of space taking one submanifold to the other. Knot theory can be (pedantically and cryptically) described as the search for invariants that distinguish these topological equivalence classes. The *knot type* of a submanifold is its topological equivalence class. It is far beyond the scope of this chapter to give a general treatise on knot theory [6–9].

A *regular projection* of a knot or link K is an orthographic projection into a plane such that (1) no vertex of K projects to the same image point as another point of K, and (2) no three points of K project to the same image point. A *crossing point* of the projection is a point in the plane that is the image of two points of K. A *crossing* of the projection consists of the image of two edges of K that have no point in common in K but that project to two edges in the plane that intersect in a common interior point. Regularity ensures that each crossing point is represented by a crossing. An *annotated projection* (called a *knot diagram* in Ref. 9) of K consists of a regular projection together with the inherited orientation of each component and type information for each crossing. Type information for a crossing indicates which of the two edges goes under the other. After an affine transformation of space, a piecewise linear knot or link K may be assumed to be situated so that orthogonal projection into the plane $z = 0$ is regular.

LINKING NUMBERS

Since K can be uniquely reconstructed from an annotated projection, it follows that any knot invariant is in principle obtainable from the information captured in an annotated projection. Usually, the first step in evaluating an invariant uses the crossing data. For example, suppose that $K = \alpha \cup \beta$ is a link with two components. The linking number of the two curves α and β may be defined by

$$\mathrm{lk}(\alpha, \beta) = \tfrac{1}{2} \sum_{p \in \alpha \sqcap \beta} \epsilon(p) \tag{1}$$

where the sum is over all crossing points p, and $\epsilon(p)$ denotes the sign of the crossing at p (Ref. 9, p. 14). The sign of a crossing is determined by the right-hand rule: With thumb pointing in the direction of α, $\epsilon = 1$ if the fingers agree with the direction of β and $\epsilon = -1$ if the fingers disagree. A variant is the *signed undercrossing method,* expressing $lk(\alpha,\beta)$ as the sum of $\epsilon(p)$ where p ranges over the subset of $\alpha\ \square\ \beta$, corresponding to places where α crosses under β (Ref. 8, p. 132).

A variation of the discussion above pertains to nonclosed curves with "clamped" ends (i.e., arcs properly embedded in 3-cells). (The calculation in the clamped case must be done relative to a coordinate system specified in the 3-cell.) Similarly, in all of the following discussion, the curves considered may be taken to be either closed or clamped; usually, one or the other case is used for specificity.

LINKING IN LATTICE MODELS

Data Structure

When working with curves confined to lattices it is convenient to use projections to planes defined by standard sublattices. At least two advantages accrue by use of these projections: (1) there is a discrete set of points in the projection plane at which all crossings must occur, and (2) there is a finite set of directions that represent all possibilities for a projected edge. Consider, for example, the simple cubic lattice L of points whose vertices have integer coordinates and the shifted lattice $L_\sigma = L + \sigma$ obtained by adding $\sigma = (s_1, s_2, s_3)$ to each point of L. We assume that no coordinate of σ is zero and that σ has length less than 1. If α is a curve in L and β is a curve in L_σ, projection into the plane $z = 0$ guarantees that crossings must occur at points of the form $(x + s_1, y)$ or $(x, y + s_2)$, where x and y are integers, and that each edge in the projection has one of four directions. Unfortunately, this projection is usually not regular: Vertical edges will collapse to points, and several horizontal edges differing only in their z-coordinate will all project to the same edge in the image plane. These problems can be overcome by modifying the notion of annotated projection to include the possibility of many edges projecting to the same edge in the image plane [10].

To annotate a crossing one needs the height and orientation of each projected edge. The orientation of a crossing can be reconstructed by knowing only the sign of the direction of the projected edge. Define the data type **edgerec** to be a record containing three fields: *curvename, height,* and *rsign,* where *height* and *rsign* are of type integer ($rsign = \pm 1$). Then define the data type **projdata** to be an array of lists of records of type **edgerec** indexed by the values (x, y) where crossing points may occur. Each list **projdata**$[x, y]$ is ordered by the *height* field.

Algorithm

Let PD be of type **projdata**. Assume that β is given in L_σ and that we wish to store annotated projection information on β for reference in computing lk(α, β) for some future α. Then projection of β is accomplished by traversing β once, inserting records into PD one edge at a time. If e is an edge of β from (x_1, y_1, z_1) to (x_2, y_2, z_2), the insertion proceeds as follows: If $z_1 \neq z_2$, go to the next edge (store nothing if e is vertical); else define a record of type **edgerec** for β with $rsign = (x_2 - x_1) + (y_2 - y_1)$ and insert this record in the list $PD[(1 - s_1)x_1 + s_1 x_2, (1 - s_2)y_1 + s_2 y_2]$. Note that either $x_1 = x_2$ or $y_1 = y_2$ and the unequal coordinates differ by 1, so $rsign$ and the array indices have permissible values. This projection process completely disassembles the curve β, reordering information and throwing away the vertical edges altogether. A great deal of information has been captured, but not quite enough to reconstruct the original curve. Information sufficient to reconstruct the curve can be maintained by adding a field to **edgerec,** giving the step number of the traversal. Note that there is no upper bound on the number of curves that can be projected into structure PD.

Now let α be a curve in the lattice L and let β_i be one of the curves projected into PD. We can compute lk(α, β_i) by traversing α once and comparing edge information with that stored for β_i in PD. Accumulate the signed undercrossings in lk_i, initialized to zero. If e is an edge of α from (x_1, y_1, z_1) to (x_2, y_2, z_2), update as follows: If e is vertical, go to the next edge; else define $lsign = (x_1 - x_2) + (y_2 - y_1)$ and traverse the list $PD[(1 - s_1)x_1 + s_1 x_2, (1 - s_2)y_1 + s_2 y_2]$ from the highest *height* until z_1 is passed. For each record encountered with *height* $> z_1$ and *name* $= \beta_i$, do $lk_i \leftarrow lk_i + lsign * rsign$. The sign conventions are chosen to implement the right-hand rule. At the end of the traversal of α, lk_i is the sum of signed crossings of α under β_i.

The problem of nonregularity of the projection must be addressed. Note that β_i can be moved to a curve γ so that the projection applied to γ is regular by applying a rotation that takes vertical segments of β_i to slightly nonvertical segments of γ and moves vertically aligned horizontal segments of β_i into nonvertically aligned segments of γ. (There are only finitely many segments in β_i.) If the rotation is small enough, no new crossings are introduced, so the signed undercrossing method applied to α and γ yields lk(α, γ) $= lk_i$. Making the rotation smaller if necessary, we can ensure that β_i does not move through α during the move, so lk(α, β_i) $=$ lk(α, γ) by topological invariance. Therefore, $lk_i =$ lk(α, β_i).

Efficiency

Two implementation details are of practical significance. (1) To obtain space efficiency, the ordered lists **projdata**[x, y] should use dynamically allocated memory. This can be done implicitly using pointers in a language such as Pascal or C. An alternative method, often more efficient in terms of run time, uses explicit memory management implemented with a cursor system and declared arrays. Our CROSSWALK simulation used Pascal pointers in its prototype versions and array-implemented alternatives in the data production stages [4,10]. (2) In a situation where curves β_i are generated in a way that makes control of their overall dimensions inappropriate, as was the case in CROSSWALK, the range of the array indices x, y determining where the lists **projdata**$[x, y]$ are stored may become inordinately large and hence wasteful, since for outlying index values the lists may have a high probability of being empty. A hashing technique eliminates this problem. For example, suppose that the indices for which there are likely to be nonempty lists are in the range $0 \le x, y < l$ but that other nonempty lists occur sparsely and randomly throughout a much larger range. The index set for the array **projdata** $[x, y]$ can be restricted to lie in the smaller range by adding X *offset* and Y *offset* fields to **edgerec**. The **edgerec** of a horizontal edge of β_i is stored in the list $PD[\eta, \zeta]$, where $\eta = (1 - s_1)x_1 + s_1 x_2 \bmod l$, $\zeta = (1 - s_2)y_1 + s_2 y_2 \bmod l$, X *offset* $= (1 - s_1)x_1 + s_1 x_2$ div l, and Y *offset* $= (1 - s_2)y_1 + s_2 y_2$ div l. The algorithm must be modified to compute the actual location of each encountered projected edge as α is traversed in order to decide whether to update lk_i for that edge.

With these implementation assumptions, the efficiency of the algorithm can be calculated. The average length of the lists $PD[x, y]$ is given by

$$\langle |PD[x, y]| \rangle = \frac{2 \sum |\beta_i|}{3 l^2} \qquad (2)$$

where $|\beta_i|$ is the number of edges in β_i and l is the length of a side of a square upon which most of the edges project. Thus the storage requirement is $O(\sum |\beta_i|)$. Since each curve is traversed exactly once in the algorithm, the run time to calculate one linking number $lk(\alpha, \beta)$ is $O(|\alpha| + |\beta|)$ and the run time to calculate all linking numbers $lk(\alpha, \beta_i)$ is $O(|\alpha| + \sum |\beta_i|)$.

Remarks

Simulations involving calculations of invariants of knots or links K can be roughly categorized using the parameters l, the length of the side of a cube containing the simulation; c, the number of components of K; and $n = (1/c)|K|$, the average number of edges in a component of K. When $O(n) \le l$ and $O(c) \ge l^2$, that is, a large number of relatively short curves, methods similar to the ones discussed above are efficient since the simplest invariants usually suffice to measure entanglements among short curves. This was the situation in our simulation of entanglement in semicrystalline polymers [4,10]. In contrast, when $O(n) \ge l^2$ and $O(c) \le l$, we have relatively long curves, and more complex invariants may be required to differentiate entanglement types. In this case structures of type **projdata** still give handy access to information needed for invariant calculations, but the algorithms are more complex and far more time consuming.

WRITHE AND TWIST

For self-avoiding polygons (SAPs) on cubical lattices, the number of edges must be rather large before significant knotting is observed. For example, calculations of van Rensberg and Whittington have found that about 1% of the SAPs of length 1600 on the body-centered cubic lattice are knotted [11]. Monte Carlo simulation work of Klenin et al. highlights the effect of volume exclusion on knotting, linking, and writhing for off-lattice SAPs in 3-space [12,13]. They found that knotting and linking are very sensitive to the volume exclusion parameter, a slowly increasing volume exclusion parameter resulting in rapidly declining rates of knot and link formation. For these reasons we concentrate on calculation of writhe.

Projections and structures of type **projdata** are useful in calculations of writhe of an energized curve. The *writhe* $wr(\alpha)$ of a curve α may be defined as the average value of $lk(\alpha, \alpha_{tu})$, where $\mathbf{u}$ is a unit vector, t is a small positive number, and $\alpha_{tu} = \alpha + t\mathbf{u}$ is a small "pushoff" of α in the $\mathbf{u}$ direction. If α is one of the boundary curves of an embedded annulus A, then *twist* $tw(\alpha, \beta)$ and writhe are related by

$$tw(\alpha, \beta) + wr(\alpha) = lk(\alpha, \beta) \tag{3}$$

where β is the other boundary curve of A. Thus the writhe of α, and the twist of α with respect to A, can be estimated by Monte Carlo methods that calculate $lk(\alpha, \alpha_{tu})$ for a random sample of unit vectors $\mathbf{u}$ [14–16].

It should be noted that whereas $lk(\alpha, \beta)$ is a topological invariant of the curves, neither $tw(\alpha, \beta)$ nor $wr(\alpha)$ is invariant under arbitrary topological transformation. Equation (3) captures a topologically invariant relationship among the three. Writhe and twist are *geometric* invariants whose variation may be interpreted in terms of the variation of configurational energy within a given topological equivalence class [16]. Writhe and twist are generally not integer quantities and in fact need not be rational numbers.

Writhe on a Lattice

In the case of a curve on a lattice, the Monte Carlo sampling of unit vectors can be replaced by calculation of a fixed number of linking numbers, thus shortening the calculation time and at once computing an exact value for writhe. Again, we concentrate on the simple cubic lattice L.

Consider the 2-sphere S of radius $\frac{1}{2}$ in $\mathbf{R}^3$, $S = \{\mathbf{u} = (u_1, u_2, u_3) | u_1^2 + u_2^2 + u_3^2 = \frac{1}{4}\}$. The coordinate planes in $\mathbf{R}^3$ separate S into eight connected regions ("octants") characterized by constancy of sign in each coordinate. Let α be a closed curve on the lattice. For $\mathbf{u} \in S$ define $\alpha_{\mathbf{u}} = \alpha + \mathbf{u}$. Note that $\alpha_{\mathbf{u}}$ lies on the shifted lattice $L_{\mathbf{u}}$, so that the algorithm described above may be used to calculate $lk(\alpha, \alpha_{\mathbf{u}})$. Since the set of all $\mathbf{u} \in S$ that do not belong to an octant (i.e., that lie in some coordinate plane) forms a set of measure zero in S, $\text{wr}(\alpha)$ is the average value of $\text{lk}(\alpha, \alpha_{tu})$, where $\mathbf{u}$ ranges over all octants. We now show that it suffices to take $t = 1$ and that $\text{lk}(\alpha, \alpha_{\mathbf{u}})$ can take on at most four distinct values.

Claim 1:   *If $\mathbf{u}$ lies in any octant, α and $\alpha_{\mathbf{u}}$ are disjoint.* We assume for simplicity that $\mathbf{u}$ lies in the octant with all coordinates positive. Then each coordinate of $\mathbf{u}$ satisfies $0 < u_i < \frac{1}{2}$. Points in α have the property that at least two of the coordinates are integers. The points in the pushoff $\alpha_{\mathbf{u}}$ are obtained by adding the vector $\mathbf{u} = (u_1, u_2, u_3)$ to all points of α. Now suppose that $(x, y, z) \in \alpha$ and $(x + u_1, y + u_2, z + u_3) \in \alpha \cap \alpha_{\mathbf{u}}$. By the pigeonhole principle, at least one of the following is true: both x and $x + u_1$ are integers; both y and $y + u_2$ are integers; both z and $z + u_3$ are integers. From this it follows that at least one of u_1, u_2, or u_3 is an integer, which is impossible.

Claim 2:   *If $\mathbf{u}$ and $\mathbf{v}$ lie in the same octant, $\text{lk}(\alpha, \alpha_{\mathbf{u}}) = \text{lk}(\alpha, \alpha_{\mathbf{v}})$.* Consider the great circle arc from $\mathbf{u}$ to $\mathbf{v}$. This arc lies completely within the single octant containing $\mathbf{u}$ and $\mathbf{v}$. The points along this arc define a one-parameter family of pushoffs, starting with $\mathbf{u}$ and ending with $\mathbf{v}$. Thus the curve $\alpha_{\mathbf{u}}$ can be moved to $\alpha_{\mathbf{v}}$ in the complement of α, and hence $\text{lk}(\alpha, \alpha_{\mathbf{u}}) = \text{lk}(\alpha, \alpha_{\mathbf{v}})$.

At this point we have shown that writhe is the average of eight linking numbers. These arguments generalize to any lattice in $\mathbf{R}^3$. We now use a symmetry property of the simple cubic lattice to reduce the number from eight to four.

Claim 3:   *If $\mathbf{u}$ is not on a coordinate plane, $\text{lk}(\alpha, \alpha_{\mathbf{u}}) = \text{lk}(\alpha, \alpha_{-\mathbf{u}})$.* For a parameter t, $-1 \le t \le 0$, let α_{tu} be obtained by adding $t\mathbf{u}$ to each point of α. As in Claim 1, for each value of t, $\alpha_{\mathbf{u}}$ and α_{tu} are disjoint. To see this, suppose that $(x + u_1, y + u_2, z + u_3) = (x' + tu_1, y' + tu_2, z' + tu_3)$ for some (x, y, z) and (x', y', z') in α. Suppose also that x and x' are integers. It follows that $(x - x') = (1 - t)u_1$. But $(1 - t)u_1$ cannot be an integer because $1 \le (1 - t) \le 2$ and $0 < u_1 < \frac{1}{2}$. Thus $\alpha_{\mathbf{u}}$ and α_{tu} are disjoint. Therefore we can move α to $\alpha_{-\mathbf{u}}$ in the complement of $\alpha_{\mathbf{u}}$, verifying that $\text{lk}(\alpha_{\mathbf{u}}, \alpha) = \text{lk}(\alpha_{\mathbf{u}}, \alpha_{-\mathbf{u}})$. Similarly, one can move $\alpha_{\mathbf{u}}$ to α in the complement of $\alpha_{-\mathbf{u}}$, showing that $\text{lk}(\alpha_{-\mathbf{u}}, \alpha_{\mathbf{u}}) = \text{lk}(\alpha_{-\mathbf{u}}, \alpha)$. The claim now follows from symmetry of linking numbers in $\mathbf{R}^3$.

From these claims we can see that the writhe of α is the average of four linking numbers:

$$\text{wr}(\alpha) = \tfrac{1}{4} \sum_{i=1}^{4} \text{lk}(\alpha, \alpha_i) \qquad (4)$$

where $\alpha_1, \ldots, \alpha_4$ are pushoffs of α a distance of $\frac{1}{2}$ or less into four mutually nonantipodal octants, for example, those with at least two signs positive or those with $z > 0$. In particular, it follows that four times the writhe of any curve on the simple cubic lattice is an integer.

CONCLUSIONS

Efficient data structures and algorithms exist for computation of linking numbers of curves on lattices. These may be applied to computation of the writhe of a curve on the same lattice. Further analysis reduces the computation of writhe to a finite number of linking number calculations, yielding an efficient computation of an exact value for writhe and showing that the writhe of a curve on the lattice must be rational.

ACKNOWLEDGMENTS

The authors thank John Bryant and Stewart Whittington for helpful discussions and E. J. J. van Rensburg and Whittington for providing unpublished data. The authors gratefully acknowledge support from the U.S. Office of Naval Research.

REFERENCES

1. P.-G. de Gennes, *Scaling Concepts in Polymer Physics,* Cornell University Press, Ithaca, N.Y., 1979, Chap. 8.
2. D. W. Sumners and S. G. Whittington, *J. Phys.* **A21,** 1689 (1988).
3. S. A. Wasserman and N. R. Cozzarelli, *Science* **232,** 951 (1986).
4. R. C. Lacher, J. L. Bryant, L. N. Howard, and D. W. Sumners, *Macromolecules* **19,** 2639 (1986).
5. R. C. Lacher, *Macromolecules* **20,** 3054 (1987).
6. K. Reidemeister, *Knottentheorie,* Ergebnisse der Mathematik und ihrer Grenzgebiete, Julius Springer, Berlin, 1932 (reprinted Chelsea, New York, 1948) (reprinted Springer Verlag, Heidelberg and New York, 1974).
7. R. H. Crowell and R. H. Fox, *Introduction to Knot Theory,* Ginn, Boston, 1963.
8. D. Rolfsen, *Knots and Links,* Publish or Perish Press, Berkeley, Calif., 1976.
9. L. H. Kauffman, *On Knots,* Princeton University Press, Princeton, N.J., 1987.
10. R. C. Lacher and C. R. Braswell, *Proceedings Supercomputing '87,* Vol. 2, International Supercomputing Institute, St. Petersburg, Fla., 1987, p. 229.
11. E. J. J. van Rensburg and S. G. Whittington, personal communication.
12. K. V. Klenin, A. V. Vologodskii, V. V. Anshelevich, A. M. Dykhne, and M. D. Frank-Kamenetskii, *J. Biomol. Struct. Dyn.* **5,** 1173 (1988).
13. K. V. Klenin, A. V. Vologodskii, V. V. Anshelevich, V. Yu. Klishko, A. M. Dykhne, and M. D. Frank-Kamenetskii, *J. Biomol. Struct. Dyn.* **6,** 707 (1989).
14. J. H. White, *Am. J. Math.* **91,** 693 (1969).
15. F. B. Fuller, *Proc. Nat. Acad. Sci. USA* **68,** 815 (1971).
16. W. R. Bauer, F. H. C. Crick, and J. H. White, *Sci. Am.* **243,** 118 (1980).

Spinodal Decomposition in Polymer–Polymer Systems: A Two-Dimensional Computer Simulation

M. V. ARIYAPADI and E. B. NAUMAN
Department of Chemical Engineering
Rensselaer Polytechnic Institute
Troy, New York 12180

J. W. HAUS
Department of Physics
Rensselaer Polytechnic Institute
Troy, New York 12180

ABSTRACT

Phase separation in polymer–polymer systems by spinodal decomposition has been studied using a continuum model. The NBCH equation, a modified form of the Cahn–Hilliard equation, was solved for two-dimensional concentration fluctuations to study the time evolution of phase structure during spinodal decomposition. The model was applied to predict the domain size in polymer blends formed by a novel manufacturing process known as compositional quenching. We also studied the late-stage domain size growth rate and compared the results with those obtained by solving the Cahn–Hilliard equation, with constant mobility.

INTRODUCTION

Phase separation by spinodal decomposition in binary systems has been an actively studied area since its introduction by Cahn [1]. The Cahn–Hilliard phenomenological theory has been applied to binary systems of glasses, metals, and polymers [2].

$$\frac{\partial \phi}{\partial t} = \nabla \left[M \nabla \left[\left(\frac{d \Delta g}{d \phi} - \kappa \nabla^2 \phi \right) \right] \right] \tag{1}$$

Here ϕ is the volume fraction of one of the polymers, M the mobility, Δg the free energy of mixing of the homogeneous system, κ the gradient energy parameter, and t the time.

Nauman and Balsara [3] have developed a diffusion equation to study the spinodal decomposition in binary systems. They started with the Landau–Ginzburg functional form for the free energy of mixing of an inhomogeneous system and used a gradient-dependent chemical potential [3] to obtain a modified form of the Cahn–Hilliard equation:

$$\frac{\partial \phi}{\partial t} = \nabla \left[D \phi (1 - \phi) \nabla \left(\frac{d \Delta g}{d \phi} - \kappa \nabla^2 \phi \right) \right] + \eta \tag{2}$$

Here D is the diffusion coefficient and η is the Cookian noise term [4], introduced to describe the system dynamics properly. Hereafter we denote Eq. (1) as the CH equation and Eq. (2) as the NBCH equation. Note that the NBCH equation provides a specific relationship between the mobility that appears in the CH equation and the diffusion coefficient.

$$M = D \phi (1 - \phi) \tag{3}$$

A similar relationship has been obtained by Kitahara and Imada [5] using a statistical approach and has also been noted by Langer et al. [6], Binder [7], and Oono and Puri [8]. The Nauman–Balsara approach has the advantage that it can easily be extended to spinodal decomposition in ternary systems [9].

Baumgartner and Heermann [10] were among the first researchers to use a Monte Carlo method to study spinodal decomposition in polymer films. Cifra et al. [11] have developed a Monte Carlo technique to study short-range fluctuations in the spinodal decomposition of polymer–polymer mixtures. Chakrabarti et al. [12] have used the Cahn–Hilliard equation to study spinodal decomposition in polymer–polymer systems in three spatial dimensions. Experiments [13,14] on polymer blends show an exponential increase in the scattering intensity for the early stages following a quench. Both the CH and the NBCH equations predict such behavior, indicating their applicability in the study of systems with long-range interactions, such as polymers.

SPINODAL DECOMPOSITION IN A QUENCHED POLYMER–POLYMER SYSTEM

The NBCH equation was applied to study the time evolution of phase structure in a quenched binary polymer–polymer system. We used the Flory–Huggins model for the free energy of mixing of the homogeneous polymer–polymer system.

$$\Delta g = \frac{1}{m}[\phi \ln(\phi) + (1 - \phi) \ln(1 - \phi)] + \chi_{12}\phi(1 - \phi) \tag{4}$$

Here χ_{12} is the Flory–Huggins interaction parameter between the two polymer molecules and m is the degree of polymerization (assumed to be the same for both polymers, $m = 1000$). The gradient energy parameter, κ, for a polymer–polymer system may be written as [7,15–17]

$$\kappa = \frac{1}{3}(R_{G1}^2 + R_{G2}^2)\chi_{12} + \frac{1}{3}m\left[\frac{R_{G1}^2}{\phi} + \frac{R_{G2}^2}{(1 - \phi)}\right] \tag{5}$$

where R_{G1}^2 and R_{G2}^2 are the radii of gyration of the two polymers. We have assumed that the radii of gyration of the two polymers are equal ($R_{G1} = R_{G2} = 80 \times 10^{-4}$ μm [18]). The interaction parameter, χ_{12}, between the two polymers is taken as 0.003. Here we treat the gradient energy parameter and the diffusivity as constants, although it is conceptually straightforward to incorporate the concentration dependency. The Cookian term has been found to accelerate only the initial stages of phase separation but was noted [19] to have little effect on the domain growth rate in the late stage. Hence in this study we have ignored the Cookian term. The spinodal decomposition in a quenched binary polymer–polymer system was studied by solving a scaled form of the NBCH equation, subject to two-dimensional concentration fluctuations:

$$\frac{\partial \phi}{\partial T} = \overline{\nabla} \left[\frac{\phi(1 - \phi)}{\overline{\phi}(1 - \overline{\phi})} \overline{\nabla} \left(m \frac{d \Delta g}{d \phi} - \overline{\nabla}^2 \phi \right) \right] \tag{6}$$

where

$$T = \frac{D \overline{\phi}(1 - \overline{\phi})}{\kappa m^2} t \tag{7}$$

and

$$\overline{\nabla} = (\kappa m)^{1/2} \left(\frac{\partial}{\partial x} + \frac{\partial}{\partial y} \right) = \left(\frac{\partial}{\partial X} + \frac{\partial}{\partial Y} \right) \tag{8}$$

Here $\overline{\phi}$ is the average volume fraction, and x, y and X, Y are the unscaled and scaled spatial coordinates, respectively. We solved Eq. (6) subject to random initial concentration fluctuations (about the average concentration) as the initial condition.

$$\phi\{X, Y, T = 0\} = \overline{\phi} + \epsilon \tag{9}$$

such that

$$\int_0^L \int_0^L \epsilon \, dX \, dY = 0 \tag{10}$$

Here ϵ is the white noise and L is the size of the periodic domain. A novel algorithm [20] that uses a discrete Fourier transform technique has been developed to solve Eq. (6). The periodic boundary conditions inherent in this approach help to simulate an infinite domain. A square domain with 64×64 grid points was used for the simulation.

RESULTS AND DISCUSSION

The extent of phase separation as predicted by the CH and the NBCH equations are illustrated in Fig. 1. Here we define the extent of phase separation, E, as

$$E = \frac{1}{N^2} \sum_{j=1}^{N} \sum_{i=1}^{N} \frac{(a_{i,j} - \overline{a})^2}{(\overline{a} - a_1)(a_u - \overline{a})} \tag{11}$$

where $a_{i,j}$ is the concentration at the (i, j)th grid point, a_1 and a_u are the lower and upper binodal concentrations, and N is the number of grid points in the X and Y directions. We find that for the initial stages the theories concur. The late-stage predictions of the CH equation show faster phase separation than those predicted by the NBCH theory. This is clearly due to the additional factor, $\phi(1 - \phi)/\overline{\phi}(1 - \overline{\phi})$, in the NBCH equation, which becomes important for the late stages of spinodal decomposition.

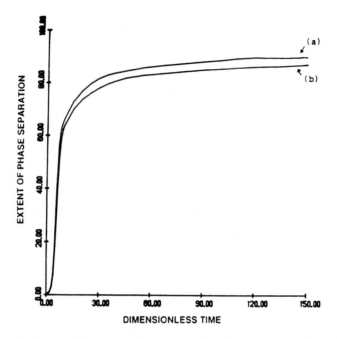

Figure 1 Extent of phase separation as a function of time for a two-dimensional simulation: (a) as predicted by the Cahn–Hilliard equation; (b) as predicted by the NBCH equation.

The time evolution of morphologies has been studied for systems with average concentrations of 0.35, 0.40, 0.45, and 0.50. We have used different gray levels to plot the concentration profiles. This helps to follow the evolution of a sharp interface from an initially diffuse interface. The black and white regions represent polymers 1 and 2, respectively. In all the cases, we observe coarsening of the initially formed phase structure as the time proceeds. This phenomenon is similar to that in Ostwald ripening. For the average concentration of 0.35, as shown in Fig. 2, we obtained circular regions of the minor component dispersed in the matrix of the major component. As we increased the concentration to 0.40, the mean particle size increased (Fig. 3) and some elliptically shaped particles were observed. The degree of connectivity of the minor phase increased (Fig. 4) as we increased the concentration to 0.45. Figures 5 to 7 illustrate the time evolution of concentration for a critical quench (equal volume fractions of both polymers). For this case we obtained a highly interconnected network in both phases.

The simulation results may be applied to predict the domain sizes obtained in polymer blends formed by compositional quenching. Details regarding this process may be found elsewhere [21]. We measured the domain sizes when the interfaces became sharp and their sizes became very small when compared to the domain sizes, as depicted in Figures 2d, 3d, 4d, and 7d. The

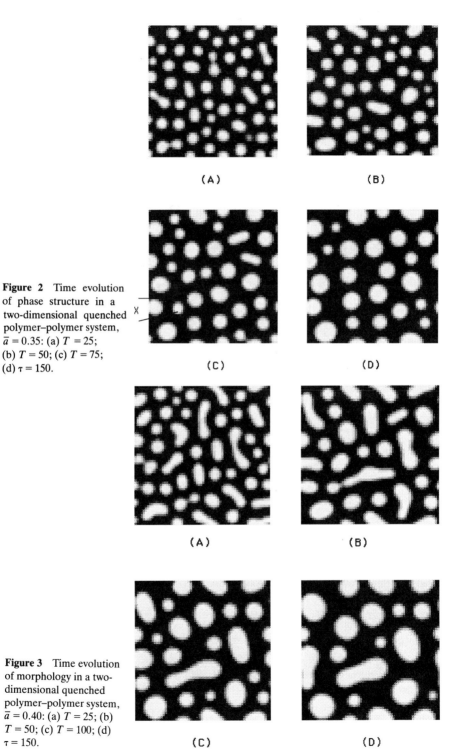

Figure 2 Time evolution of phase structure in a two-dimensional quenched polymer–polymer system, $\bar{a} = 0.35$: (a) $T = 25$; (b) $T = 50$; (c) $T = 75$; (d) $\tau = 150$.

(A)

(B)

(C)

(D)

(A)

(B)

Figure 3 Time evolution of morphology in a two-dimensional quenched polymer–polymer system, $\bar{a} = 0.40$: (a) $T = 25$; (b) $T = 50$; (c) $T = 100$; (d) $\tau = 150$.

(C)

(D)

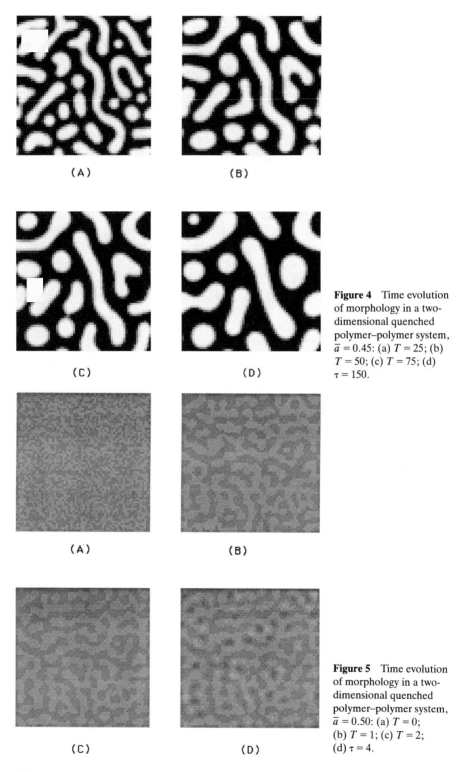

Figure 4 Time evolution of morphology in a two-dimensional quenched polymer–polymer system, $\bar{a} = 0.45$: (a) $T = 25$; (b) $T = 50$; (c) $T = 75$; (d) $\tau = 150$.

Figure 5 Time evolution of morphology in a two-dimensional quenched polymer–polymer system, $\bar{a} = 0.50$: (a) $T = 0$; (b) $T = 1$; (c) $T = 2$; (d) $\tau = 4$.

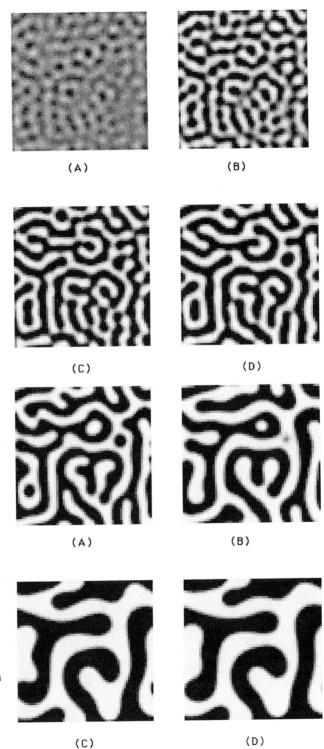

Figure 6 Time evolution of morphology in a two-dimensional quenched binary system, $\bar{a} = 0.5$: (a) $T = 5$; (b) $T = 7$; (c) $T = 10$; (d) $T = 15$.

(A)

(B)

(C)

(D)

(A)

(B)

Figure 7 Time evolution of morphology in a two-dimensional quenched binary system, $\bar{a} = 0.5$: (a) $T = 50$; (b) $T = 75$; (c) $T = 200$; (d) $T = 250$.

(C)

(D)

two-dimensional theory predicts submicron-sized particles for the particulate case. This is in qualitative agreement with the average particle size obtained by compositional quenching. Also, for the critical quench the interdomain distance predicted by the theory is in qualitative agreement with experiment [21]. However, it should be pointed out that the value of the interaction parameter we used in our simulation is much smaller than that for our experimental system of polystyrene–polybutadiene. This was essential to reduce the computational time, which otherwise became impossible to handle. Thus for a quantitative comparison of theory and experiment, knowledge of the effect of the interaction parameter on the domain size is essential. Also, facts such as the concentration dependencies of the diffusion coefficient and the gradient energy parameters, Brownian motion, and hydrodynamic effects should be incorporated in the theory to have good quantitative predictions.

The time evolution of the structure factor [22] was used to predict the late-stage scaling exponent. A modified form of the Lifshitz–Slyozov [23] model proposed by Huse [24] was used to study the domain scaling.

$$\lambda_m = A + BT^n$$

where λ is the dominating wavelength, n is the scaling exponent, and A and B are constants. Figures 8 and 9 illustrate the scaling behavior of the CH and NBCH theories. The CH theory predicts an exponent of about $\frac{1}{3}$, which is in agreement with the published results [19,25]. A similar analysis of the NBCH equation yields an exponent of $\frac{1}{4}$. A comparison of our result with the analyti-

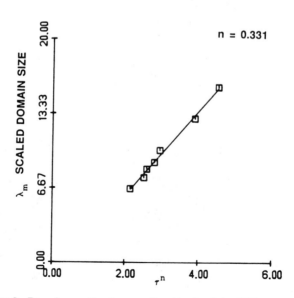

Figure 8   Domain growth rate as predicted by the Cahn–Hilliard equation.

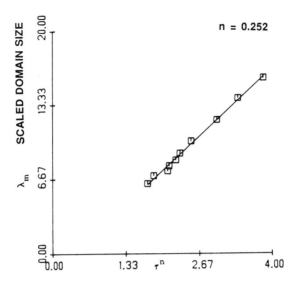

Figure 9 Domain growth rate as predicted by the NBCH equation.

cal treatment of Ohta [26] suggests that we are still in the intermediate stages. For this regime, Ohta predicts an exponent of $\frac{1}{4}$ that shifts to $\frac{1}{3}$ at longer times. Recent work of Rousar and Nauman [27] suggests that a shift in the scaling exponent does occur.

CONCLUSIONS

The time evolution of phase structure during spinodal decomposition in polymer–polymer systems was studied using the NBCH equation. The domain size predictions are in qualitative agreement with experiments. The predictions of both the Cahn–Hilliard and the NBCH equations concur for early stages of phase separation, while the concentration dependence of the mobility becomes important for the late stages. The scaling exponent predicted by the NBCH theory agrees with the analytical predictions of Ohta.

ACKNOWLEDGMENTS

The authors acknowledge George R. Clarkson for his help in preparing Figs. 2 to 7. Support for this research came from the General Electric Company and the New York State Science and Technology Foundation. All computations were performed using IBM 3090-600W at the Cornell National Supercomputer Facility, Ithaca, N.Y.

REFERENCES

1. J. W. Cahn, *J. Chem. Phys.* **113,** 93 (1965).
2. J. J. Van Aartsen, *Eur. Polym. J.* **6,** 919 (1976).
3. E. B. Nauman and N. P. Balsara, *Fluid Phase Equilibr.* **45,** 299 (1989).
4. H. E. Cook, *Acta Metall.* **18,** 297 (1970).
5. K. Kithahara and M. Imada, *Prog. Theor. Phys., Suppl.* **64,** 65 (1978).
6. J. S. Langer et al., *Phys. Rev.* **A11**(4), 1417 (1975).
7. K. Binder, *J. Chem. Phys.* **79,** 6387 (1983).
8. Y. Oono and S. Puri, *Phys. Rev.* **38**(1), 433 (1988).
9. M. V. Ariyapadi and E. B. Nauman, *Proc. Am. Chem. Soc., Div. PMSE* **61**(2), 89 (1989).
10. A. Baumgartner and D. W. Heermann, *Polymer* **27,** 1777 (1986).
11. P. Cifra, F. E. Karasz, and W. J. MacKnight, *Macromolecules* **22,** 3089 (1989).
12. A. Chakrabarti et al., *Phys. Rev. Latt.* **63** (19), 2072 (1989).
13. H. L. Snyder, P. Meakin, and S. Reich, *Macromolecules* **16,** 757 (1983).
14. F. S. Bates and P. Wiltzius, accepted for *J. Chem. Phys.* (1989).
15. L. P. McMaster, *Adv. Chem. Ser.* **142,** 43 (1975).
16. P.-G. de Gennes, *J. Chem. Phys.* **72,** 4756 (1980).
17. P. Pincus, *J. Chem. Phys.* **75,** 1996 (1983).
18. M. Daoud et al., *Macromolecules* **6,** 804 (1975).
19. K. R. Elder, T. M. Rogers, and R. C. Desai, *Phys. Rev.* **B38**(10), 4725 (1988).
20. M. V. Ariyapadi, J. W. Haus, and E. B. Nauman (in review).
21. E. B. Nauman et al., *Chem. Eng. Commun.* **66,** 29 (1988).
22. R. Petschek and H. Metiu, *J. Chem. Phys.* **79**(7), 3443 (1983).
23. I. M. Lifshitz and V. V. Slyozov, *J. Phys. Chem. Solids* **19,** 35 (1961).
24. D. A. Huse, *Phys. Rev.* **B34,** 7848 (1986).
25. T. Ohta, *J. Phys. Chem., Solid State Phys.* **21,** L361 (1988).
26. A. Chakrabarti and J. D. Gunton, *Phys. Rev. Lett.* **60**(22), 2311 (1988).
27. I. Rousar and E. B. Nauman (in review).

Phase Equilibria
of Ternary Polymer Solutions
by a Method
of Blob Rescaling

YONG-BYUNG BAN and JONG-DUK KIM
Department of Chemical Engineering
Korea Advanced Institute of Science and Technology
P.O. Box 131
Cheongryang, Seoul
Korea

ABSTRACT

A new method of calculating the liquid–liquid equilibria of polymeric ternary systems involving two polymers and solvent is proposed. The method incorporates the blob concept of polymer chains and a renormalization scheme from primary blobs to rescaled blobs, which we call pseudocomponents, the basic units responsible for intermolecular interactions. Following a conventional algorithm for the pseudocomponents, the phase equilibria are calculated in terms of the size and binary interaction parameters of local composition models. Phase diagrams of two distinct characteristics have been examined. The first is typified by polystyrenes of two different molecular weights and cyclohexane and the other by polyethyleneglycol–dextran–water. For both examples, the calculated results are in fair agreement with experiment.

Key Words: Phase equilibria, Blob rescaling, Polystyrenes-cyclohexane, and PEG-dextran-water.

INTRODUCTION

The phase behavior of polymer systems involving either polymer melts or polymer–solvent mixtures is of great importance. It is crucial in polymer processing with respect to the morphological structure and properties of polymer blends [1] and in biochemical processes involving phase separations for proteins [2]. A variety of phase diagrams have been explored experimentally: binary mixtures having upper or lower consolute solution points [3,4], or the very distinctive hourglass [5,6] and closed-loop [7,8] types, and multicomponent systems having multiphase behavior [2,9–12]. In general, P–T–X diagrams of binary or multicomponent mixtures provide concentrations of phases in equilibria, set for particular applications, and the microscopic states of polymer solution are important in determining the properties of materials through their microscopic adjustments.

Much effort has been devoted to exploring behavior and microstates through lattice-based models [3,13] and equations of state [14,15], but rigid lattice theories of polymer solutions have been the subject of much criticism [6,16–18]. Entropic consideration of polymer chains in the Flory–Huggins theory provides a good approximation for randomness in position and chain dimensions but cannot describe phase separation of liquid phases. The energetic contribution, more responsible for phase separations and microstates, was treated less seriously by approximating the mean-field theory in principle. Departure from Flory–Huggins theory starts semiempirically with Wilson's incorporation of local compositions in evaluating the volume fraction of athermal solutions [19], and the mean-field approximation was replaced by Guggenheim's quasi-chemical approximation. Later, Renon and Prausnitz [20] incorporated local compositions into the two-liquid theory of Scott [21], calculating the excess free energy, and Prausnitz and Abrams [22,23] incorporated both concepts into their model. Unfortunately, despite many successful applications to mixtures of relatively small molecules, most of these local-composition models failed in describing the phase separations of polymeric mixtures.

The microscopic states or molecular associations in the bulk phase have long been recognized as being responsible for the physicochemical properties of materials or mixtures. In polymers, de Gennes [24] recognized that the finite but persistent dimension of polymer, rather than the total dimension of a chain, is responsible for hydrodynamic interactions. Honnell and Hall [25] similarly applied the finite dimension of the polymer chain in their equation of state. Along with a similar observation, the propagation of momentum may relax only within a few segments of a blob, and instantaneous interaction between sites may be the most responsible part of the molecular interaction. In fact, these two concepts, local composition and finiteness, apparently have not been tested in describing the equilibrium behavior of polymer solutions.

The phase diagrams of binary solutions of polymer–solvent mixtures are of two types: those having the completely distinctive characteristics of hourglass [5,6] and closed-loop [7,8] diagrams. Typical examples of such diagrams are shown in Fig. 1. The general behavior of the former was explained as resulting from the free volume created by the thermal expansion of polymers and solvents [26], and for the latter, the strong interaction between polymer and solvents. Binary closed-loop diagrams of polyethylene glycol–water mixtures with strongly interacting groups, were calculated by Kim et al. [26]. Equilibrium concentrations of polymer solutions were evaluated by incorporating the blob configurations of polymers and strong interactions such as hydrogen bonding [28] toward one or more directions following the decorated lattice model [29], and it was shown that the local composition models are not only accessible to computer calculations but are also applicable for the equilibrium separation of polymers. However, the phase behavior of ternary mixtures of polymers and solvents has not been calculated despite many practical applications [1,2].

In this chapter, we report on a new method of calculating equilibrium concentrations and phase diagrams of multicomponent polymer solutions. The method developed approximates that of binary blobs for polymer solutions, and uses the successive renormalization of lattice and local-composition models. Example calculations for polystyrene–polystyrene–cyclohexane mix-

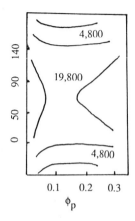

T-X diagram of PS-Acetone mixtures,
Patterson, *Macromolecules* **5**, 29, 1972

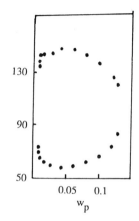

T-X Diagram of PVA-water Mixtures
Rehage, *Kunststoffe* **53**, 605,1963

Figure 1   Typical examples of binary-phase diagrams exhibiting two liquid phases. The behavior of polystyrenes–solvent [6] and polyethylene glycol–water [7] is different as temperature increases, showing hourglass and closed-loop behavior, respectively.

tures [10,11] and polyethylene glycol (PEG)–dextran–water mixtures [2,12] with significantly different types of phase behaviors and interactions, are also given, incorporating the size of polymer blobs and interaction parameters.

BLOB SOLUTION PRESENTATION

Let us solubilize a polymer chain in a solvent and focus on the interaction between solvent and polymer, between solvents, and between polymer chain segments. The orientational and configurational structures of a polymer chain in solution fluctuate near the minimum point of the free energy surface mainly by the thermal motion of the chain and its neighboring sites. A schematic picture of a polymer chain in solution is given in Fig. 2, representing the internal degree of chain dynamics or the local motion of the segments.

In the dynamic interaction of a polymer chain with a solvent or polymer, contact between the solvent and some finite dimensions of the polymer segments, called a *blob,* may be the most important part of the interaction. The effect of contact does not extend toward the entire dimension of a polymer chain. Within this picture, we observe first that the collision occurs within a finite contact area or cross section of polymer segments (i.e., blobs) relevant to solvent molecules. In interactions on or through the chain that does not belong to the target sites, the impact of collision is of a higher order. Second, the individual groups (i.e., blobs) are also assumed to have independent motions free from the neighboring groups in the chain and behave like inde-

Figure 2 Blob presentation of polymer chain and polymer–solvent mixtures. The circles represent the blobs, and the shaded circles designate small blobs internalized by polymer blobs.

pendent molecules—as subdivided pseudocomponents. Any linkages between such blobs are not allowed despite the connectivity of chains, but are relaxed as in freely rotating joints. Third, such a blob, having a finite number of segments, is assumed to undergo no significant structural change internally during the course of an instantaneous collision. In these assumptions, the polymer chains are treated fundamentally as chains of blobs (i.e., pseudocomponents), but the internal structure of blobs may depend only on the polymer itself and the solvents. The internal structure of a blob thus remains unchanged, as it is within the concentration range of calculation.

For a polymer chain in a solution, for example, let's take a random walk of N blobs on a periodic lattice and operate N successive steps from one end of a chain to the other. At each step of varying step size, the next jump of a blob may proceed toward any of its nearest-neighbor sites; the statistical weight is the same for all these possibilities. The length of the step in a chain is the size of a blob, whose distribution can be normalized by taking the proper distribution function. Here we face the problem of N/g independent variables, leading to Gaussian statistics provided that N/g is large and the mean-square end-to-end distance is linear in the number of blobs N:

$$\langle r^2 \rangle = \frac{N}{g}\langle c^2 \rangle = N\xi^2 \tag{1}$$

where ξ $[= (\langle c^2 \rangle/g)^{1/2}]$ is now an effective length per blob. Then we can break up the chain into a series of blobs each with an average size of ξ [24,29].

In addition to the size of blobs, the shape and molecular weight of blobs are also important factors accounting for the interactions between blobs and solvents. Therefore, another structural factor of blob q, representing the ratio of surface areas, is introduced. The shape of blobs and the blob weight are closely related to the step size g and are averaged along with ξ. For convenience, volume and surface area parameters of blobs in polymers can be obtained from the van der Waals volume and area of molecules relative to those of the reference segment. The size and area parameters r_i^* $(= V_{wi}/V_{ws})$ and q_i^* $(= A_{wi}/A_{ws})$ of species i can be determined by the standard method, where V_{wi} and A_{wi} are the van der Waals volumes and areas of the molecule given by Bondi [30] and the subscript s represents the reference segment of methylene group.

LATTICE AND GROUP RENORMALIZATION

It is also known that solvents adsorbed on or near the polymer chains may alter the chain properties and that the boundary layer structures of solvents near the chain would differ from their bulk structures. Therefore, our third assumption follows—that groups of polymer segments and solvents will form new groups of blobs, as illustrated in Fig. 2, where large circles represent renormalized blobs, consisting of polymers and solvents. Then we can break up the chain into a series of blobs, each having an average size of ξ, which can be evaluated from the primary blobs.

In fact, one of the simplest idealizations of these semiflexible polymer chains or groups in the chain consists of such a successive renormalization of lattices or groups. A successive lattice transformation of mixtures occurs from their sublattice onto the lattice, and the groups in a chain also are transformed into a wider range than that of their repeating unit scale. The transformations for both lattices and groups may be applied simultaneously or independently. Figure 3a shows a lattice of blobwise interaction and transformation from the sitewise interaction between polymer and solvent. The renormalized lattice of blobs, representing the binary solution blobs, was approximated by incorporating the two-step renormalization of lattices. First, the primary lattice ζ for polymer molecules is normalized to the blob lattice ξ with the effective interaction energy of the lattice bond U_{ij}. Second, the blob lattice with bond ij is renormalized under local compositions to yield the excess free energy of the system.

In Fig. 3b, blobs of A and B, consisting of polymer–solvent mixtures, were shown in a ternary diagram. The concentrations of polymers within blobs are selected, for example, as the concentrations of polymers on the intersection of the extension line of a tie line to the binary mixtures in which we are interested. For a phase, denoted by a prime, the overall material balance of these blob solutions must be satisfied and the total number of blobs N_{bt}', can be written

$$N_{bt}' = N_s' + N_A' + N_B' \qquad (2)$$

where N_A' and N_B' represent the numbers of blobs A and B in Fig. 3b. Then the results calculated by blob solutions will be constrained by the mole fractions of blobs defined by Eq. (2). The total number of molecules, N_t', is given by

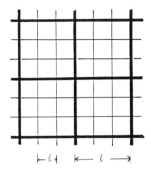

a) two-dimensional lattice representation of interactions and
renormalization of lattice ζ to ξ

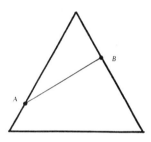

Figure 3 Lattice renormalization and pseudocomponents: (a) two-dimensional lattice representation of interactions and renormalization of lattice ζ to ξ; (b) blobs presented on a ternary-phase diagram equivalent to pseudocomponents in the calculations.

b) blobs presented on a ternary phase diagram are equivalent to pesudocomponents in the calculations

$$N_t' = N_s' + (N_s' + N_1')_A + (N_s' + N_2')_B \tag{3}$$

where the subscripts s, 1, and 2 stand for the solvent, polymer 1, and polymer 2, and N_1' and N_2' represent the numbers of polymers 1 and 2 in a phase. The relation between Eqs. (2) and (3) can be given by the material balance of solvent and polymer in one phase in terms of the numbers of moles of blobs. Following the third assumption, one binary blob contains one polymer blob and corresponding numbers of solvent molecules, and the total number of moles of solvent is the sum of free solvent and of solvent in both blobs:

$$N_{st}' = N_s' + N_A' R_{A1} + N_B' R_{B2} \tag{4}$$

where $R_{jk} = w_{js} M_k / w_{jk} M_s$, representing the number of solvent molecules per polymer blob k in blob j, w is the weight fraction of species, and M represents molecular weight. The subscript j represents blob j, A, or B, and k represents polymers 1 and 2. w_{js}/w_{jk} represents the ratio of weight fractions of solvent to polymer segments k of blob j. The number of polymers can be calculated from the number of blobs divided by the number of blobs per chain:

$$N_k' = \frac{N_j'}{D_k} \tag{5}$$

where D_k represents the numbers of pure polymer blobs per chain. Therefore, if we know N_j' and N_s', we can evaluate N_k' and N_{st}', the concentrations of polymers and solvent at a phase. The molecular weight of blob j containing polymer k can be calculated as

$$\frac{M_j}{\rho_j} = \frac{M_k w_{jk}}{\rho_k} + \frac{M_s w_{js}}{\rho_s} \tag{6}$$

where ρ_j are the densities of blobs j containing polymer k, equivalent to the densities of solution A or B, and w_{ji} represents the weight fraction of i species in a blob j.

From the material balances of molecules and blobs, the number of blobs can be converted to the number of molecules. Then the x_s' and x_j' can be converted to weight fractions in diagrams and the counter phase as well. Assuming that the polymer solution consists of the binary blob only, the scaling factor of the number of species, the ratio of the number of true molecules to the number of blobs, $f = N_t/N_b$, can be obtained from Eqs. (2), (3), (4), and (5) by

$$f = x_s + x_A\left(R_1 + \frac{1}{D_1}\right) + x_B\left(R_2 + \frac{1}{D_2}\right) \tag{7}$$

Here the number scaling factor is a strong function of D_1, D_2, and the ratio R. Then the concentrations of polymers 1 and 2 and solvent 3 can be deduced from Eq. (7) and the material balance by

$$x_1 = \frac{x_A}{fD_1} \tag{8a}$$

$$x_2 = \frac{x_B}{fD_2} \tag{8b}$$

$$x_3 = 1 - (x_1 + x_2) \tag{8c}$$

The concentrations of both phases in terms of blob concentrations can be calculated from the conditions of the equality of chemical potentials of each blob of s and A and B. Microscopic balance in phase equilibria can be established by the net exchange of chemicals having equal chemical potentials. To calculate the mixing free energies of blob solutions, the local composition models or any statistical models can be applied.

In evaluating the number of blobs, blob size and shape are important scales, and the counting of blobs in the polymer chains of given molecules is possible by introducing the size and number of blobs. Normalizing blobs in polymer, the shape of blobs need not be unvariant but assumed, for simplicity, as it is after this rescaling. Then the rescaled area parameter q_i of blobs of species i can be written

$$q_i = q_i^* \left(\frac{r_i}{r_i^*}\right)^l \tag{9}$$

where l is equal to 1 for the cylindrical blobs and to $\frac{2}{3}$ for the spherical blobs. Concerning the radius of gyration of polymer chains, blobs would be close to spherical, but for simplicity here, the case $l = 1$ was applied. This fact will give a somewhat larger blob size than they have and result in a decrease in blob molecular weight.

LOCAL COMPOSITIONS AND COMPUTER CALCULATION

Since each successive blob in a chain acts as an independent moiety, it is further assumed without difficulty that their interaction energies depend on the local compositions around the interacting sites. Assuming that the interactions between each blob and its neighbor are operated pairwise regardless of the sources of blobs, the probabilities of finding a molecule of species j near a molecule i may depend strongly on the net interaction energy between the site pair i, j. Since the interaction energy between sites is different from one site pair to another, any of the local-composition models [19,20,22,23] is appreciated. However, since all blobs may not be independent of their nearest neighbor on the same chain, some degrees of translational motion in the chain may be overestimated.

At equilibrium, the equality of chemical potentials of each blob of s and A and B between two phases must hold at constant temperature and pressure, given by

$$\mu_i' = \mu_i'' \qquad (10)$$

where μ_i' represents the chemical potential of blob i. If the reference fugacities at both phases are the same, Eq. (10) is equivalent to the equality of the activities of species i. Since the activities of species i can be calculated from the mole fraction and the activity coefficients of species i, any model for activity coefficient is appreciated conceptually. However, we tested with the NRTL model [20] and UNIQUAC model [23], given in the Appendix, although we report here only the results of NRTL. The interaction parameters in the models were evaluated based on the maximum likelihood principle by minimizing the sum of squares of the residuals of the calculated values from the experimental values [31–33]. Then using the textbook algorithm for calculating the liquid–liquid equilibria [22,32], the equilibrium concentration of blobs at each phase can be calculated. After calculating the mole fraction of blobs, the volume fraction or weight fraction of polymers was reevaluated.

RESULTS AND DISCUSSION

For PEG-6000, the monomer has a V_w value of 24.16 cm^3/mol and an A_w value of 3.3×10^9 cm^2/mol [30]. If g is 3, for example, the number of blobs is 45.5, and the r and q values of blobs are 4.78 and 3.96. From these, the counting of blobs in polymer chains is possible by introducing a renormalization of molecules: in fact, the size and number of blobs. In these calculations, the sizes of polymer blobs were fixed at 4 for PS-45,000, 7 for PS-102,000, 3 for dextran-17,000, and 4 for PEG-6000, giving the best results under our calculation program. Figures 4 and 5 show the phase envelopes of two-phase separations for two ternary examples of PS-45,000–PS-102,000–cyclohexane mixtures and PEG-6000–dextran-17,000–water mixtures. Two systems have the distinctive characteristics of phase diagrams: the former having a immiscible region near the cyclohexane-rich conditions where the polystyrene of lower molecular weight acts as a cosolvent [10,11], and the latter having one near water-rich conditions where water is the cosolvent [2,12]. The T–X diagrams of binary systems of PS–solvent and PEG–water mixtures have been known to show more different behaviors (Fig. 1), the former having a miscible region between UCST and LCST, of hourglass type [6,7], and the latter of the closed-loop type between these critical points [8,9].

Polystyrene–Polystyrene–Cyclohexane Mixtures

Liquid–liquid separation of ternary polymer solutions has long been studied by Flory to examine the theory with polystyrene–cyclohexane mixtures [10,11] because these combinations can minimize the specific attraction between species and also because of experimental convenience. Flory's calculation showed that although the relative trends agree with experimental values, the concentrations of components at both phases are far away from ex-

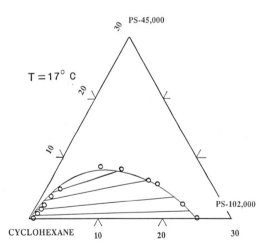

Figure 4 Ternary diagram of polystyrene ($M_w = 45{,}000$)–polystyrene ($M_w = 102{,}000$)–cyclohexane mixtures at 17°C. The solid line is that calculated with blobs of 30 wt % polystyrene by the NRTL model and the sizes of blobs 4 and 7, respectively. The circles represent experimental values [10], and the straight lines represent tie lines. Values are in weight percent.

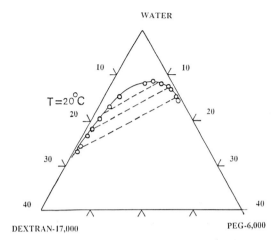

Figure 5 Ternary diagram of polyethylene glycol (PEG, $M_w = 6000$)–dextran ($M_w = 17,000$)–water mixtures at 20°C. The solid line is that calculated with blobs of 27.3 wt % dextran and 15.6 wt % polyethylene glycol by the NRTL model and the sizes of blob 3 for dextran and blob 4 for polyethylene glycol. The circles represent experimental values [2,12] and the dotted lines represent tie lines. Values are in weight percent.

periment [11]. Figure 4, a ternary diagram of polystyrene ($M_w = 45,000$)–polystyrene ($M_w = 102,000$)–cyclohexane mixtures at 17°C calculated at 30 wt % polystyrene blobs using the NRTL model, shows good agreement with experiment [11]. The solid line represents values calculated with blobs of 30 wt % polystyrene, the circles represent experimental values [11], and the straight lines represent calculated tie lines. For our calculations, both polystyrenes were counted blobwise. In principle, the blob size can be determined statistically, but it is difficult to determine the size in direct measurements. Here, for our first evaluations, we chose the sizes of blobs to reproduce the phase diagrams, giving the best fit. It is interesting that the low-molecular-weight polymer has the smaller blobs and keeps the solution miscible, but the larger has phase separation, although the interaction parameters remain the same.

PEG–Dextran–Water Mixtures

Binary mixtures of PEG–water mixtures have the closed-loop phase envelopes of two-phase separations in $T–X$ diagrams and were calculated by approximating blob size with the enhanced UNIQUAC [26]. Here the interaction energies of the directionals are fixed at −5 and −6 kcal/mol, close to the hydrogen bonding energy [28], while two nondirectionals are parameterized. Although the $T–X$ diagrams by the two-parameter UNIQUAC model are poor in the location of phase separation, the calculated results reproduced fairly closely the experiments of Saeki et al. [18], indicating that the blobs of polymers represent good statistical units for molecular interaction.

In a practical sense, ternary mixtures of PEG–dextran–water systems, sometime including salts, are useful for two-phase separation process of bio-chemicals [2,12]. Three cases—of dextran D17 ($M_w = 17,000$)–PEG ($M_w = 6000$), dextran T10 ($M_w = 10,500$)–PEG ($M_w = 3400$), and dextran T10–PEG ($M_w = 8000$)—have been calculated, showing excellent agreement with experiment [2,12], including tie lines. Figure 5 is a ternary diagram of PEG ($M_w = 6000$)–dextran ($M_w = 17,000$)–water mixtures at 20°C. The solid line represents values calculated with blobs of 27.3 wt % dextran and 15.6 wt % polyethylene glycol by the NRTL model. The circles represent experiment [2,12], and the straight lines represent tie lines. The concentration blobs were selected for convenience at the binary mixtures on the extension of a tie line.

In these calculations we used two different blobs: a temperature blob, representing increasing temperature, and a concentration blob, representing added solvent, which can be a method of maintaining a flexible polymer chain. If thermal energy or solvent is lacking at the interaction sites, the chain remains rigid and solidified in the liquid phase. In estimating the phase dia-gram, the blob size, which depends not only on the chain properties of the polymer but also on the solution state, was not described in detail, but the blob size increases for a rigid polymer chain and for a poor solvent. The size and number of blobs increase the immiscibility region just as the degree of polymerization does. Therefore, the r parameter would be an indication of solvency for a fixed chain or chain rigidity for a fixed solvent. In fact, the sizes of the blob in a solvent varies from solvent to solvent. For a polymer chain in a poor solvent, its size becomes larger than that of the chain in a good solvent, while the rigidness of a polymer chain increases the blob size. At present, we are attempting to elaborate on this matter. Further, the distribution function of blob size was not investigated, but averaging the size and weight of the blobs is required.

CONCLUSIONS

A new method for calculating liquid–liquid equilibria of ternary polymer–solvent mixtures was proposed, incorporating a scaled finite dimension of polymer chain with blobs of polymers in the liquid state. The blobs, having finite size and molecular weight, represent the effective dimension of polymer chains when interacting with solvent molecules, and are appropriate for cal-culating the equilibrium concentrations of mixtures by local-composition models.

Following a conventional algorithm, the phase separations of cyclohexane–polystyrenes of different molecular weights and polyethylene glycol–dextran–water mixtures were examined, having completely different phase behavior characteristics in binary or ternary mixtures. The equilibrium calculations were performed in terms of the size and binary interaction parameters of local composition models. The calculated concentrations and the size and shape of the region of phase separations showed good agreement with experiment.

ACKNOWLEDGMENTS

The authors wish to thank R. J. Roe for the invitation to participate in the symposium, and H. Yu and J. P. O'Connell for encouragement.

	LIST OF SYMBOLS
Symbol	Definition of symbol
A_{wi}	van der Waals area of species i
A_{ws}	van der Waals area of standard species, 2.5×10^9 cm^2/mol
$\langle c^2 \rangle$	average of the square of the step
D	number of blobs per polymer chain
G^E	molar excess Gibbs free energy, kcal/mol
f	number scaling factor, ratio of the number of species to the number of blobs
g	subunit of vector
l	geometric factor
M	molecular weight
N	number of molecules
q_i, q_i^*	surface area parameters of blob and linear polymer
$\langle r^2 \rangle$	mean-square end-to-end distance
r_i, r_i^*	size parameters of species i ($= V_{wi}/V_{ws}$) and linear polymer
R_{jk}	number of solvents per blob j containing single polymer blob k in the concentration blob
T	absolute temperature (K)
U_{ij}	interaction energy of ij pair blob
V_{wi}	van der Waals volume of species i
V_{ws}	van der Waals volume of standard species, 15.17 cm^3/mol
x_A, x_B	mole fractions of blobs A and B
X_i	mole fraction of species i
Greek	
α	nonrandomness parameter
ζ	primary lattice of polymer
μ_i	chemical potential of species i
Φ_i	volume fraction of blob i
θ_i	area fraction of blob i
θ_{ij}	area fraction of a blob j surrounded by blob i
τ_{ij}	interaction parameter
ξ	average blob size

APPENDIX: Binary NRTL and UNIQUAC Models

NRTL Model [20]

$$G^E/RT = x_1 x_2 \left(\frac{\tau_{21} G_{21}}{x_1 + x_2 G_{21}} + \frac{\tau_{12} G_{12}}{x_2 + x_1 G_{12}} \right) \tag{a}$$

$$\tau_{ij} = \frac{U_{ij} - U_{jj}}{RT}$$

$$G_{ij} = \exp(-\alpha_{ij} \tau_{ij})$$

where U_{ij} is the interaction energy between sites i and j and α_{ij} is the nonrandomness parameter.

UNIQUAC Model [23]

$$G^E = RT \left[x_1 \ln \frac{\Phi_1}{x_1} + x_2 \ln \frac{\Phi_2}{x_2} + \frac{z}{2} \left(q_1 x_1 \ln \frac{\theta_1}{\Phi_1} + q_2 x_2 \ln \frac{\theta_2}{\Phi_2} \right) \right.$$

$$\left. - q_1' x_1 \ln(\theta_1' + \theta_2 \tau_{21}) - q_2' x_2 \ln(\theta_2' + \theta_1 \tau_{12}) \right] \tag{b}$$

$$\tau_{ij} = \exp \left[-\left(\frac{U_{ij} - U_{jj}}{RT} \right) \right]$$

where x_i, θ_i, and Φ_i, respectively, represent the mole fraction, the area fraction, and the volume fraction of blob i, and z is the coordination number of the lattice.

REFERENCES

1. D. R. Paul and S. Newman, *Polymer Blends,* Academic Press, New York, 1978.
2. P. A. Albertson, *Partition of Cell Particles and Macromolecules,* 3rd ed., Wiley-Interscience, New York, 1986.
3. P. J. Flory, *Principles of Polymer Chemistry,* Cornell University Press, Ithaca, N.Y., 1953.
4. D. Patterson, *Macromolecules* **2,** 672 (1969).
5. H. O. Lee, Y.-B. Ban, and J.-D. Kim, *Korean J. Chem. Eng.* **5,** 147 (1988).
6. K. S. Siow, G. Delmas, and D. Patterson, *Macromolecules* **5,** 29 (1972).
7. G. Rehage, *Kunststoffe* **53,** 605 (1963).
8. G. N. Malcom and J. S. Rowlinson, *Trans. Faraday Soc.* **59,** 921 (1957).
9. S. Saeki, N. Kuwahara, M. Nakada, and M. Kaneko, *Polymer* **17,** 685 (1976).
10. J. Hashizume, A. Teramoto, and H. Fujita, *J. Polym. Sci., Polym. Phys.* **19,** 1405 (1981).
11. A. R. Shultz and P. J. Flory, *J. Am. Chem. Soc.* **74,** 4760 (1952).
12. B.-K. Kim, Master's thesis, KAIST, Seoul, Korea, 1989.

13. I. Prigogine, N. Trappeniers and V. Mathot, *J. Chem. Phys.* **21,** 559 (1953).

14. P. J. Flory, R. A. Orwell, and A. Vrij, *J. Am. Chem. Soc.* **86,** 3507 (1964).

15. D. Patterson, S. N. Bhattacharyya, and P. Picker, *Trans. Faraday Soc.* **64,** 648 (1968).

16. R. Koningsveld, *Br. Polym. J.* **7,** 435 (1975).

17. K. Solc, L. A. Kleintjein, and R. Koningsveld, *Macromolecules* **17,** 573 (1984).

18. S. Saeki, N. Kuwahara, and M. Kaneko, *Macromolecules* **9,** 101 (1976).

19. G. M. Wilson, *J. Am. Chem. Soc.* **86,** 127 (1964).

20. H. Renon and J. M. Prausnitz, *AIChE J.* **14,** 135 (1968).

21. R. L. Scott, *J. Chem. Phys.* **25,** 193 (1956).

22. J. M. Prausnitz, R. N. Lichtenthaler, and E. G. de Azevedo, *Molecular Thermodynamics of Fluid-Phase Equilibria,* 2nd ed., Prentice-Hall, Englewood Cliffs, N.J., 1986.

23. D. S. Abrams and J. M. Prausnitz, *AIChE J.* **21**(1), 116 (1975).

24. P.-G. de Gennes, *Scaling Concepts in Polymer Physics,* Cornell University Press, Ithaca, N.Y., 1979.

25. K. G. Honnell and C. K. Hall, *J. Chem. Phys.* **90,** 1841 (1989).

26. J.-D. Kim, Y.-C. Kim, and Y.-B. Ban, *Fluid Phase Equilibr.* **53,** 331 (1989); Paper 96, Proceedings of 5th International Conference on Fluid Properties and Phase Equilibria for Chemical Process Design, Banff, Calgary, Canada, April 30–May 5, 1989.

27. G. R. Wheeler and J. C. Anderson, *J. Chem. Phys.* **69,** 3413 (1978).

28. Y.-C. Kim and J.-D. Kim, *Fluid Phase Equilibr.* **41,** 229 (1988).

29. D. Lhuillier and P. J. Jorre, *Macromolecules* **17,** 2652 (1984).

30. A. Bondi, *Physical Properties of Molecular Crystal, Liquids and Glasses,* Wiley, New York, 1968.

31. G. Varhegyi and C. H. Eon, *Ind. Eng. Chem., Fundam.* **16,** 182 (1977).

32. J. M. Prausnitz, T. F. Anderson, E. A. Grens, C. A. Eckert, R. Hsieh, and J. P. O'Connell, *Computer Calculations for Multicomponent Vapor–Liquid and Liquid–Liquid Equilibria,* Prentice-Hall, Englewood Cliffs, N.J., 1980.

33. R. Hook and T. A. Jeeve, *J. Assoc. Comput. Mach.* **8,** 212 (1961).

Index

M.